ON THE ORIGIN OF

STRINGS, DARK ENERGY, TIME, AND THE UNIVERSE

ON THE ORIGIN OF

STRINGS, DARK ENERGY, TIME, AND THE UNIVERSE

PROF RAY R ESHRAGHI

Library of Congress Control Number: 2022903298

Hardcover ISBN: 979-8-9857904-0-5
Paperback ISBN: 979-8-9857904-1-2
eBook ISBN: 979-8-9857904-2-9

I dedicate this book to my children:
Sarah, Jennifer, Tyler, and Hannah

And to my grandchildren:
Max, Porter, Collins, Will, Hudson, Conley, June
Elliot, James, Reece, and Mays

Preface

FIRST AND FOREMOST, THIS BOOK is a celebration of the memory of Mr. Max Planck whose name is a general recurrence in this book. Without his prior contribution to physics, this work would not be possible.

I began this adventure by attempting to publish a paper containing a version of Chapter 1 that I completed in early 2019. Soon I realized, as a scientist and engineer with no prior publications in the physics arena, that this effort may have little chance of success. Perhaps my affiliation with a School of Business was detrimental to this endeavor.

As it turns out, this was a blessing in disguise. As I continued to prepare more material intended for a series of publications in a peer reviewed journal, it became evident that publication of a comprehensive theory extending into every field of physics and published in each corresponding journal could take years, if it were even possible.

As such, I embarked on the quest to gather all my concepts, and the mathematical work supporting it, in one location as a book. This turned out to be a fruitful decision as many of the subjects included in this book revealed themselves to me as I took a deep dive into each topic. Using mathematics as my guide, I utilized the existing knowledge of the physics of the universe as a means of corroborating and validating this theory.

As you will notice throughout this book, I have kept the mathematical concepts simple and their delivery to a minimum. My goal is for a large readership with basic mathematical knowledge to understand and relate this theory to their own work and scientific interests. And for the general reader, to experience the sheer knowledge and joy of understanding of how our universe works.

As I began writing this book, I struggled with how and where to start because I soon realized that no matter where I started, I would need information from future chapters to formulate the mathematics of what I had in hand. As you will see, it is impossible to describe the universe in isolation as all phenomena are interrelated, and thus all subjects and chapters are intertwined throughout the book.

As you read this book, you will notice that I begin with a general approach to the mathematics of the universe. I then use information obtained from this approach to construct the universe in detail, chapter by chapter, covering all subjects that impact the physics of the universe.

Therefore, to understand how and why the species that make up our universe are produced, we must take a step-by-step approach to achieve a detailed understanding of their signature effects leading to formation of their structure.

Perhaps, future editions maybe different. However, I wanted to present this work in the same chronology that it occurred to me. To fully appreciate the contents of this book, one must review the book in its entirety, as many of the early questions cannot be readily answered until the end.

In this book, our three-dimensional universe is described as a network of extra dimension objects that create the fabric of spacetime, encompassing what has been described in classical physics as dark energy, dark matter, and baryonic matter or mass. These extra dimension objects are themselves three-dimensional vacuum species falling into four main categories of space dimensions. You will see that our universe makes four space transformations to create these objects which are significantly smaller than the atomic radius or the radius of fundamental particles such as electrons. A combination of the energy and momentum of these vacuum objects, and the way they interact with one another, creates the fabric of spacetime, matter and fundamental particles, energy, and everything the universe contains. You will find that "time" is intimately connected with the geometric dimension of these vacuum species.

The theory presented in this book covers many of the fundamental phenomena that govern our universe, including gravity, electromagnetism, all fundamental forces, the detailed structure of atoms, dark energy, dark matter, blackholes, new forces not discovered before, and a brief overview of the Big Bang. I have described the detailed structure of atoms, charges, fundamental particles such as electrons, protons, neutrons, photons, Higgs Boson, and magnetic monopoles on the scale of the most fundamental building blocks of the universe, vacuum strings.

Many existing theories in classical physics including Newtonian physics, general relativity, special relativity, fundamentals of electromagnetism, and many laws in classical physics, are simply a natural fall out of the fundamental mathematics of this theory. I have used the proven and accepted physical properties of the universe disclosed in classical physics as a means of corroborating the results of this theory chapter by chapter and where appropriate.

You will find that this theory will address many difficult concepts such as the nature of dark energy and dark matter, quantum entanglement, particle vs wave, blackhole structure, fundamental particles, entropy, and time, just to name a few. I have provided a snapshot of the book content in my closing remarks which may serve as a starting point for certain readers.

I have presented the fundamental mathematics that govern the universe to formulate this theory, however, I consider this book and its theory to be a blueprint for future science and its advancement. It will be impossible to include the detail necessary to address this theory in its entirety with the theoretical and detailed mathematical work it deserves in one book. I believe I have provided sufficient information in each chapter to convincingly point us in a new direction.

It will be up to the scientific community to build on this concept and fill in the many details that will lead to a clearer understanding of the species that make up our universe, their interactions with each other, and their effects on the way the universe operates.

Professor Ray R Eshraghi

Table of Contents

CHAPTER 1

Universe at a glance – setting up the fundamental equations

Introduction

LET US START BY INTRODUCING two three-dimensional species which make up the active ingredients of our three-dimensional universe. A three-dimensional, spherical looped string and a three-dimensional, spherical open string, each with a radius of about 1.61×10^{-35}m known as Planck length. The looped and open strings in three dimensions are both geometrically spherical until they become two-dimensional. Therefore, the notion of a spherical "looped" and spherical "open" string does not make sense until the strings become two-dimensional and take on differentiating shapes as "looped" and "open" strings. In a three-dimensional and spherical shape, open and looped strings do have differentiating properties including the size of the sphere as you will learn later.

Prior to becoming a three-dimensional object, the strings above are two-dimensional looped and open strings of the same dimension with distinct geometric shapes. And prior to becoming a two-dimensional looped and open string with Planck length, the strings are a simple two-dimensional looped string with a dimension that is infinitesimally small, far smaller than Planck length, about 2.3×10^{-51}m. I have referred to this infinitesimally small looped string as the **E-loop** string. The three-dimensional strings, the two-dimensional Planck-size strings, and the E-loop strings are all vacuum species with different energy levels and energy distribution. The dimensional nature, and mechanism of formation of these species will become evident throughout the course of this book.

At the onset of the Big Bang, a series of space transformations converts the tiny two-dimensional E-loops to two-dimensional Planck length, looped and open strings, and then to three-dimensional Planck length looped and open strings. Details of the model describing this space transformation and construction of the three-dimensional looped and open strings is provided in Chapter 8. A conceptual representation of this process is shown in Figs, 1-1, and 1-2.

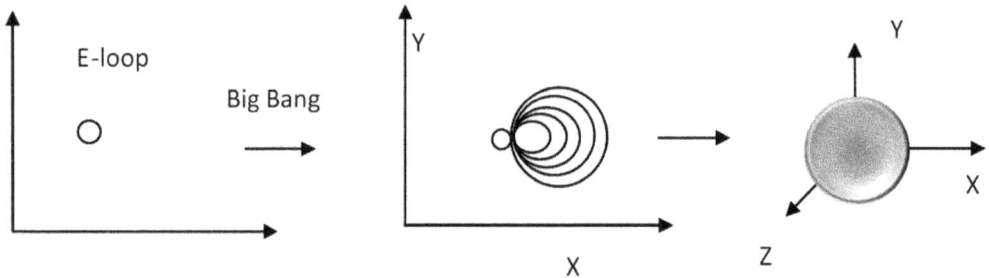

Fig. 1-1 - Conceptual representation of transformation of an E-loop string to two dimensional Planck size looped string and three dimensional (activated) looped string

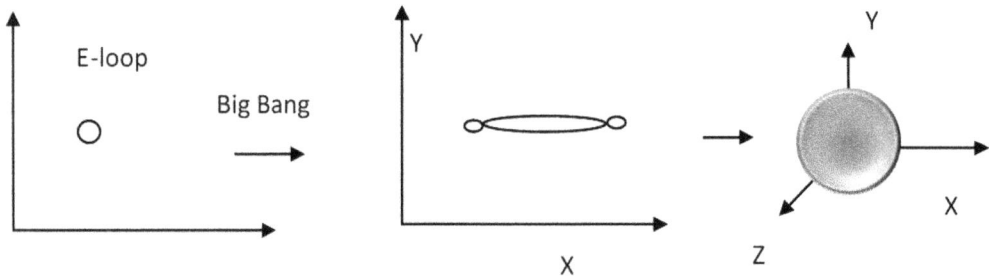

Fig. 1-2- Conceptual representation of transformation of an E-loop string to two dimensional Planck size open string and three dimensional (activated) open string

The looped and open strings described in this theory and throughout the book, are somewhat different than the ones described in classical string theory as shown in Fig. 1-3. In classical representation of strings, an open string appears as a one-dimensional object, and when the two ends of the one-dimensional object are connected, a looped string is formed.

In this theory, the two-dimensional Planck-size looped string is an object with a surface and the open string, although long and narrow, is two-dimensional, which may appear one dimensional, i.e., its width is substantially thinner than its length. The mass of both strings is located at the end of the strings.

Fig. 1- 3- Representation of looped and open strings in classical string theory

Furthermore, where it is referred to two, one, or zero dimensional objects in this book, they are all indeed three-dimensional having substantially smaller thickness (or width) than its other dimensions. For example, a two-dimensional Planck string is indeed a three-dimensional string with substantially smaller thickness (8.9×10^{-160}m) than its other dimension (1.61×10^{-35}m). Detailed calculation of string dimensions will be provided in future chapters. In reality, a true two, one, or zero dimensional string (or vacuum specie) cannot exist in our three-dimensional universe, except for a purely abstract and mathematical purpose.

By the time you finish reading this book, you will see that what appears to be a vast open "space" in our three-dimensional universe with absolute vacuum is indeed "space with a mass" that requires a significant amount of energy to sustain it. You will also learn that what appears to be the background vacuum in the universe, is a random one-dimensional string object which is transformed into two and then three-dimensional vacuum species each with a specific energy content making up our universe. You will see that the vast open space of our universe is in reality a space compacted by quantized three-dimensional vacuum species. We live on the surface of these tiny quantum and sub-quantum scale vacuum species. The process of creating a larger space from infinitesimally small vacuum objects requires a substantial amount of energy. The initial energy for this process arises from a process we have come to know as the Big Bang which I will cover in Chapter 12.

For now, I will remain focused on the three-dimensional species, regardless of its background and how it is formed, and develop the fundamental equations that govern our three-dimensional universe. These equations will become the foundation for the remainder of the book which will allow us to explore the strings and their properties chapter by chapter as we build the universe with matter and energy, seen and unseen.

To do this, I will borrow a few of the important characteristics of the three-dimensional strings from Chapter 8 to get started. The detailed structure and the model of the string formation will be discussed later.

The three-dimensional looped and open strings are formed by simultaneous rotation of a two-dimensional Planck string in two planes perpendicular to each other at the speed of light, c (3x10^8 m/s). As shown in Fig. 1-4, a two-dimensional string orbiting around the origin in the XZ Plane, will also rotate in the XY plane which is perpendicular to the XZ plane creating a three-dimensional object. As you will see in Chapter 8, the rotation in the XZ plane is critical to sustaining the space mass of the two-dimensional string, so I will refer to this as the "mass rotation" and the rotation in the YZ plane as the "spin" of the three-dimensional string.

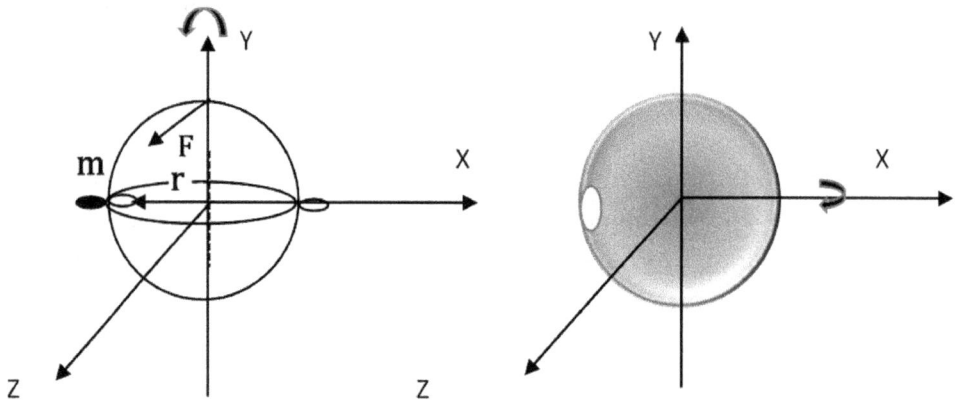

Fig. 1-4 - Conceptual representation of 3D looped or open strings with its axis of orbital rotation and spin. Bold circle represents the mass of the two-dimensional string. On the left, two-dimensional string rotates simultaneously in two planes horizontally and vertically. Vectors, **F** and **r** are parallel. Inner product of the two vectors forms the Hilbert space, scalar object on the right.

The frequency of the orbital rotation of the three-dimensional string is:

$$f = 1/t = c/2\pi l_p \qquad\qquad (1\text{-}1)$$

Using the Planck- Einstein relation:

$$E = hf$$

$$E = h\, c/2\pi l_p \qquad\qquad (1\text{-}2)$$

In which l_p is Planck length, 1.61×10^{-35}m.

$$E = (6.62 \times 10^{-34} kgm^2/s \times 3 \times 10^8 m/s)/2\pi \times 1.61 \times 10^{-35}m$$

$$E = 1.96 \times 10^9 kg\, m^2/s^2 = 1.96 \times 10^9 J$$

And the mass of the three-dimensional strings is:

$$m = h\,/2\pi c l_p = 2.18 \times 10^{-8}\, kg \qquad\qquad (1\text{-}3)$$

This mass is located near the end in the plane of the two-dimensional Planck string in the form of stored E-loop strings. This means that the energy (mass) density of the strings when its radius is at Planck length, 1.61×10^{-35}m (volume$=17.47 \times 10^{-105}$m^3) is about 0.12×10^{97}kg/m^3.

Since the mass of the string is located near the end of the two-dimensional plane of the string at its pole, it creates an angular momentum with a corresponding linear momentum, P_y:

$$E = P_y c = h\, c/2\pi l_p$$

$$E = h\, c/2\pi l_p = 1.96 \times 10^9 J$$

The momentum P_y is perpendicular to the plane of the orbital rotation of the string's core mass (Fig. 1-5).

$$P_y = h / 2\pi l_p \qquad\qquad\qquad (1\text{-}4)$$

Since $f = c/2\pi l_p$ and $\lambda = c/f$,

The wavelength will be, $\lambda = 2\pi l_p$ and substituting in equation (1-4), we will arrive at the famous de-Broglie equation, representing the momentum-wavelength relationship of a particle/wave.

$$P_y = h / \lambda = 6.5 \text{ kg.m/s} \qquad\qquad\qquad (1\text{-}5)$$

While the motion of the two-dimensional string in the XZ plane creates the angular momentum, the motion in the Y direction creates the "spin" of the three-dimensional string.

The two-dimensional surface of the Planck string is sustained by propagation of E-loop strings stored in the two-dimensional string as its mass. Propagation of the E-loop strings creates the surface **gauges** of the two-dimensional Planck string. As you will see in Chapter 8, this propagation is caused by the rotation of the two-dimensional string in the XZ plane as shown in Fig. 1-4. As such, two-dimensional Planck strings can only exist while they possess a constant "mass motion" in the XZ plane at the speed of light, c, with its energy stored as the mass of the string in the form of E-loop strings.

The perpendicular motion of the two-dimensional Planck string creates the three-dimensional structure. This three-dimensional structure entraps a vacuum of about Planck volume. I will refer to these three-dimensional looped and open strings as **"activated strings"** from this point forward so as to distinguish them from their two-dimensional counter parts. In essence, activated strings are activated to entrap and hide the bulk vacuum in our universe in a three-dimensional object that forms the basic constituent of our three-dimensional universe.

The difference between a three-dimensional, spherical looped and open string is that the open string sphere contains two poles which become the site of the attachment of other strings whereas the looped string sphere has only one pole. In addition, as will be shown later, the looped string sphere expands by a factor of 10^{15} to a radius of about 2.3×10^{-20}m while open strings remain at Planck length throughout the universe.

The geometric difference between a looped and open string becomes stark when the strings convert to their two-dimensional state. In that, a two-dimensional open string will collapse into a long and narrow object while the looped string will remain a circular object with a surface. In the first chapter of this book, we will remain focused on the three-dimensional objects in order to generate the fundamental equations.

With the above brief introduction of the three-dimensional open and looped strings, I will continue with the description of our three-dimensional universe.

The "activated" three-dimensional strings form an important composition of the fabric of spacetime in our visible universe, the smallest quanta of energy with a volume of about $17.47 \times 10^{-105} m^3$. The three-dimensional strings, having a large momentum P_y as described above, traverse the universe at the speed of light. The three-dimensional strings have five degrees of freedom, two for the orbital and spin rotation of the strings, and three for its translational movement in x, y, and z direction. I will refer to this state of the "activated strings" as "infinite momentum frame". Not because it has an infinite momentum, but because the strings have the freedom to traverse the universe very large distances in every direction, but more specifically having a large linear momentum in one direction.

Referring to Fig. 1-4, the orbital rotation of the two-dimensional string with a mass "m" creates two vectors, $\mathbf{F}$ and $\mathbf{r}$ which are parallel forming a zero-degree angle. The inner product of these two vectors results in a scalar object:

Inner product $= \mathbf{F} . \mathbf{r} =$ scalar

The magnitude of the scalar will be $= mgr \, \mathrm{Cos} \, 0 = mgr = E$

With the plane of the two-dimensional string rotating horizontally and vertically simultaneously, the three-dimensional string is formed as the inner product of the two vectors which is a scalar object.

In classical physics, this is described as "Hilbert space". Therefore, three-dimensional activated looped and open strings which will become the fundamental building block of our universe and throughout the rest of this book will exist in Hilbert space.

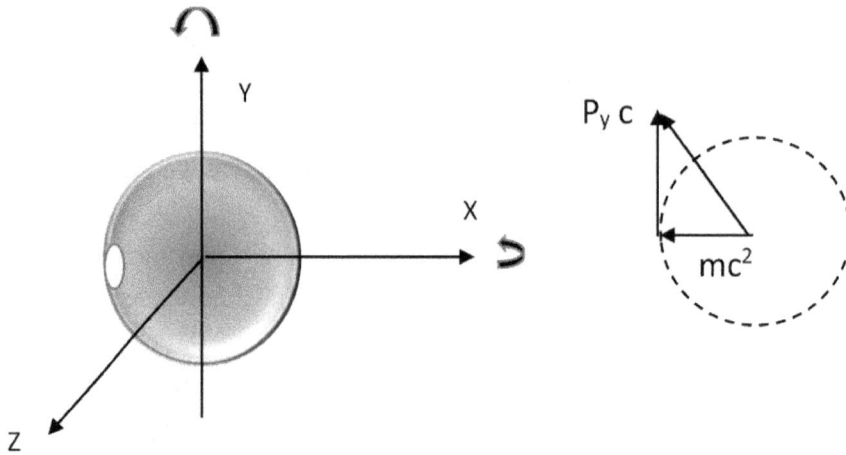

Fig. 1-5- Energy- momentum relationship of activated 3D strings

Referring to Fig. 1-5, if we allow P_y to be the linear momentum of the three-dimensional string in "infinite momentum frame", the total energy of the scalar object will have two energy components, the energy of the mass rotating at the speed of light c; $E_1 = mc^2$ and the momentum energy of the three-dimensional string $E_2 = P_y c$ which are in two planes perpendicular to each other.

The total energy of the string will then be:

$$E_t^2 = E_1^2 + E_2^2$$

$$E_t^2 = (mc^2)^2 + (P_y c)^2 \qquad\qquad (1\text{-}6)$$

Equation (1-6) will then constitute the energy-momentum equation for an activated string in infinite momentum frame. When the radius of the activated strings is at Planck length, the total energy E_t, is 1.96×10^9 J.

This energy is used for two simultaneous actions; one is to sustain the "mass rotation" of the two-dimensional string (mc^2), the other is for maintaining its kinetic energy in infinite momentum frame $(P_y c)$. As you will see later in this book, the energy to sustain the mass is very small, leaving the energy to be predominantly used for its momentum.

The physics of the motion of the three-dimensional string can be described as having an angular momentum because of the rotation of the mass of the string at one end. Since the string is not fixed in one location and is free to move around, its angular momentum around the center of the sphere results in a linear momentum that is large in one direction, P_y, and hence the term" infinite momentum frame". As you will see in the forthcoming chapters, conservation of this angular momentum is key to creation of the torque that leads to momentum of strings in a two-dimensional state.

During the Big Bang, our universe created equal amounts of left- and right-hand spinning three-dimensional looped and open strings. As you will see later, this helps create a balance of forces created by the strings. However, based on the definition of spin in classical physics the strings themselves will be considered as having a "zero spin" value, because of its spherical symmetry.

While creation of equal amounts of three-dimensional looped and open strings with opposite spin rotation is important to balance of the forces in the universe, it also means that under the right conditions they can annihilate each other by the virtue of the friction of its oppositely rotating spin of the two-dimensional plane of XY. In other words, the activated strings decay when two three-dimensional strings with opposing spin come in contact <u>long enough</u> to nullify each other's rotation. Fortunately, as you will see later, the rate of generation and regeneration of activated strings in our universe is high enough to counter the constant loss of the most important ingredient of our universe and sustain its level to give our universe a life of about 120 Billion years.

The half-life of an activated string in infinite momentum frame will vary depending on when and how it encounters another string of the same kind but with opposite rotation. In general, in infinite momentum frame where the strings have 5 degrees of freedom and move at the speed of light c, the contact time between two strings with opposing spin will be too short to cause annihilation. However, under certain circumstances when two strings become immobilized (lose its translational motion) and in contact with each other, the friction of the opposite rotation of the strings annihilate each other converting the three-dimensional string to its two-dimensional state by eliminating one of its rotations, i.e., the rotation in the XY plane. In order for two oppositely rotating three-dimensional strings to annihilate each other's spin they must have at least one rotation (around the Y axis) while in contact with each other.

Let us rewrite Equation (1-2) as:

$E.l_p = h\,c/2\pi$

And substitute for $E = P_y c$;

$$P_y \cdot (l_p) = h\,/2\pi \qquad\qquad\qquad (1\text{-}7)$$

Equation (1-7) implies that the product of P_y and (l_p) which is the angular momentum (L_y) of the three-dimensional quantum specie, the "activated string", must be at least equal to $h/2\pi$ for the three-dimensional string to make one rotation and exist in that state as a three-dimensional object, or:

$$L_y = P_y \cdot (l_p) \geq h/2\pi \qquad\qquad\qquad (1\text{-}8)$$

In essence, equation (1-8) is the Heisenberg uncertainty principle. The state that the three-dimensional strings become immobilized before annihilation is the "restframe" state of the strings.

When a three-dimensional activated looped or open string decays, it loses one of its rotations (spin) in the plane of XY axis while still retaining its orbital rotation in the plane of XZ axis. As such, the string will no longer be three-dimensional, exposing its entrapped vacuum. The decayed string with one rotation reverts to its two-dimensional state and is encased by its own vacuum. We now have a new "state" for the two-dimensional strings. I will refer to this state as "restframe2". The string in restframe2 state is no longer in Hilbert space because it lost its rotation that created the inner product of the two vectors. Let us call this new space as "extra dimension" space. The "extra dimension" space no longer scalar is a vector space.

The energy and angular momentum of the three-dimensional string as shown in equation (1-8) is conserved and transferred to the two-dimensional string in restframe2 state. To our three-dimensional universe, restframe2 is invisible because it is nothing but vacuum and contains two-dimensional strings with a thickness of 8.9×10^{-160}m (see Chapter 8). However, conserved angular momentum and energy transferred to the two-dimensional strings inside this vacuum create three-dimensional objects that while invisible to our universe but as entities with momentum and energy interact with activated strings in Hilbert space. As a result of this interaction, many new objects are created that are a composite of the two states of the strings; three-dimensional strings (in Hilbert space) in equilibrium with two-dimensional strings (in extra dimension space) inside its vacuum. I will refer to these objects as **"extra dimension"** objects from this point forward. As you will see in chapter 3, the extra dimension objects are three-dimensional objects created by the kinetic energy and momentum of the two-dimensional strings that are encased in its own vacuum.

In Chapter 3, I have identified a number of extra dimension objects that play a major role in creation of our universe. However, the number of extra dimension objects that actually exist in our universe are too many to enumerate. This is because of the flexibility of the vacuum of the extra dimension objects and their ability to conform themselves to their environment, reconfiguring themselves in countless shapes like the waves in an ocean. However, the presence of so many possible variations of shapes and configurations of the extra dimension objects does not impact our ability to identify those that are critical to the formation of matter, energy, and the fabric of spacetime in our universe.

Therefore, it is important to understand that in the context of this theory, the extra dimension objects and their vacuum, although invisible to our three-dimensional universe, are in direct interaction and in equilibrium with our universe.

In fact, one of the properties of these extra dimension objects is that its vacuum attracts and adheres itself to the three-dimensional activated strings in our universe. The extent of the vacuum strength and its energy are shown and discussed throughout this book.

When an activated looped or open string attaches to the vacuum of an extra dimension object, it becomes stationary (immobilized) while maintaining its spin. Its angular momentum will be conserved and translated into an orbital rotation. In other words, it loses the three degrees of freedom associated with its translational motion in infinite momentum frame and will have only two degrees of freedom. One for its spin rotation and one its orbital rotation. I will refer to this as "restframe1".

Therefore, "restframe1" is the state of equilibrium between activated three-dimensional strings and two-dimensional strings entrapped by its vacuum. Restframe2, is the state of two-dimensional strings and its vacuum only. The strings in restframe2 will also have two degrees of freedom, one for its translational motion, which is only in one direction, the other for its orbital rotation. By my definition, strings in restframe1 are still in Hilbert space while those in restframe2 are in extra dimension space.

As you will see later, the radius of the field of restframe1 and resframe2 is the same. Therefore, throughout this book, I may refer to either state as simply "restframe", unless I want to specifically refer to the two-dimensional state or the two-dimensional state attached to a three-dimensional string.

To summarize what I have introduced so far, we have three states of the strings; "infinite momentum frame" where activated strings roam freely at the speed of light with five degrees of freedom, "restframe1" where activated strings become immobilized and attached to the vacuum of the extra dimension objects with 2 degrees of freedom, and "restframe2" where the strings exist only as two-dimensional objects encased by its vacuum.

The fabric of spacetime in our universe is a combination of all three states of the strings mentioned above. As you will see in the upcoming chapters, various combinations of the three states mentioned above will result in creation of matter, fundamental particles, and dark matter.

In this book, I will refer to the three-dimensional strings in infinite momentum frame and the corresponding extra dimension objects as what we have come to know in classical physics as "dark energy". Collectively, all the strings including the ones in restfame1 and restframe2 comprise the fabric of spacetime in our universe.

1. Setting up the Fundamental Equations

Background

As will be shown in Chapter 4, matter and the fundamental particles making it are made up of a composite structure that contains strings in restframe2 and activated looped and open strings in restframe1. Activated strings in restframe1 become immobilized losing their translational freedom in xyz direction due to the attachment to the vacuum of the restframe2. This condition significantly increases the residence time of strings in contact with each other and those with opposite rotation will annihilate each other. This is the most prevalent condition within the structure of particles and matter which contain the strings in both states of the restframe. In other words, "matter" and the fundamental particles making it, create the right condition for accelerated decay of the activated three-dimensional strings.

In infinite momentum frame, activated strings with opposing spin may cross each other at high speeds and have a shorter residence time for contact and annihilation. In restframe1 state however, the state comprising all matter and fundamental particles, they lose the mobility, and increase its chances of annihilation. As a result, "matter" in general and the fundamental particles making it become a sink for consumption of activated strings in the universe.

What happens then when an activated string in restframe1 comprising a particle or "matter" decays? First, as described earlier, it becomes a two-dimensional string encased by its own vacuum in resframe2 which becomes essentially invisible to our universe leaving a void. Second, this void is immediately filled by a similar activated string in infinite momentum frame. The energy of an activated string in infinite momentum frame and restframe1 remain the same for the duration of its residence as a restframe1 specie which is in the order of Planck time (0.33×10^{-43}s):

$$E_{imf} = E_{rf}$$

The difference between an activated string in infinite momentum frame and an activated string in restframe is that the momentum energy which is conserved becomes the orbital rotation in restframe. Therefore, activated strings in restframe must have an orbital rotation to sustain the conservation of its momentum.

When an activated string in restframe1 decays, there is a momentary (spontaneous) breaking of this supersymmetry before an activated string of the same kind from infinite momentum frame replaces the decayed and depleted one.

This process of replacement of decayed and depleted activated strings (restframe1) with a new one from the infinite momentum frame preserves the state of "matter" in the universe. In other words, without a continuous supply of activated strings, matter or the fundamental particles making it will decay very rapidly into two-dimensional strings.

The continuous supply of activated looped and open strings from infinite momentum frame to the "matter", establishes a flow of the strings towards fundamental particles forming the "matter" and in general any object in restframe1.

The momentum change created by the flow of activated looped strings towards the fundamental particles forming the matter creates the force we know as "gravity".

The momentum change created by the flow of activated open strings towards the fundamental particles forming the matter creates the force we know as "electromagnetic force".

The above two principles will be used to set up the equations defining conservation of energy and momentum of the activated looped and open strings in infinite momentum frame, restframe1 and restframe2 throughout this book.

Figure (1-6) depicts a conceptual diagram of the process described above, illustrating an energy balance between activated looped and open strings in infinite momentum frame, matter in restframe1, and the two-dimensional strings in restframe2. This implies that if we place an envelope around an arbitrary "matter" with a mass "m", the energies that enter, are converted, and exit the matter within this envelope must balance out. This is a simple concept used in Chemical Engineering called "material and energy balance" of a system.

Matter, for the purpose of this discussion could be any of the stable sub-atomic particles or atoms, or aggregate of atoms and molecules, or large celestial objects such as stars or black holes. Matter could be galaxies distributed across the universe. Therefore, the magnitude of the mass of matter in this energy balance is of no consequence.

As seen in Fig. (1-6), our universe according to this model is then comprised of two regions with the spheres of radius R_a and R.

2D outer surface formed by looped strings

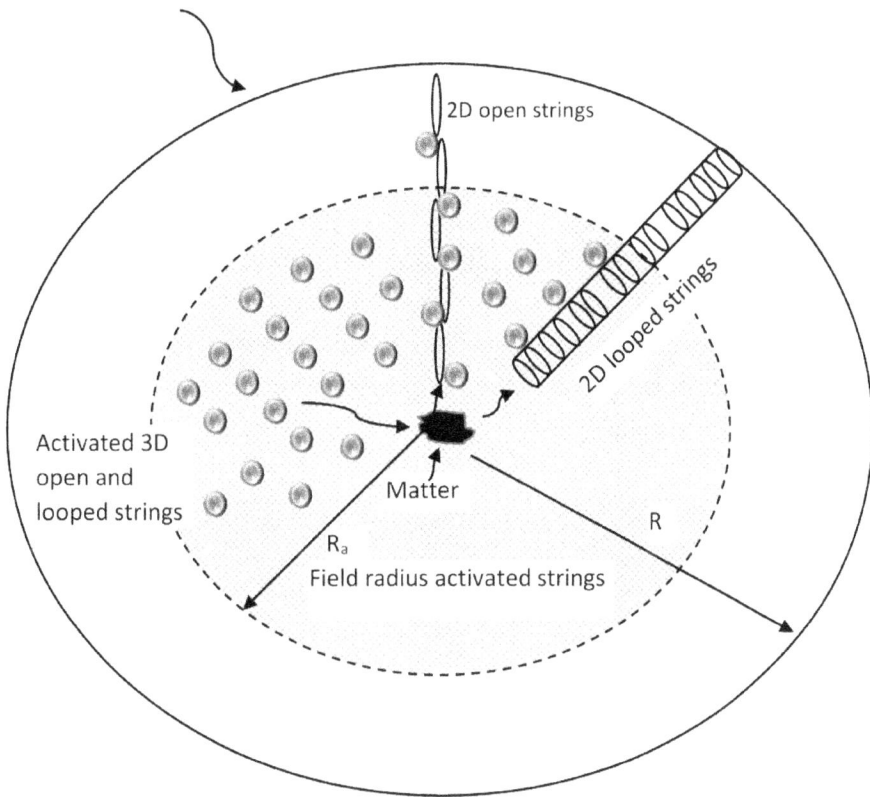

Figure 1-6- Conceptual representation of three and two-dimensional strings in the universe. Small spheres represents three dimensional looped strings. Smaller gray dots represent three dimensional open strings.

R_a is the region of the universe containing the free flowing, three-dimensional, activated open and looped strings in the infinite momentum frame, the radius of the field in which the three-dimensional strings make up the "field quanta". R is the radius of the universe at restframe (1and 2) which contains the region R_a and a region with little or no activated strings in the infinite momentum frame and comprised mainly of two-dimensional strings in restframe2 and restframe1. In essence, R_a is the radius of observable universe.

As demonstrated later in this book, the activated region (R_a) is significantly denser and heavier than the rest of the universe due to the massive amount of energy the activated strings carry in the infinite momentum frame creating the three-dimensional space of our universe.

Activated looped strings upon decay, convert from restfame1 state to restframe2 state as a two-dimensional string. The two-dimensional looped strings exit our universe and form the circumference of the universe as a two-dimensional surface because their tendency to attach to other strings or extra dimension objects is less due to having only one attachment point.

Activated open strings upon decay, also convert from restfame1 state to restframe2 state as a two-dimensional string. The two-dimensional open strings, however, remain in our universe because they attach themselves to other strings or extra dimension objects from both ends. In classical string theory, this is referred to as satisfying the Dirichlet boundary conditions.

The value of R_a and R will always be non-zero, because no matter how small the matter is it must be non-zero for the matter to exist.

To remain consistent with the nomenclature is classical physics, I will refer to the three-dimensional strings (open and looped), and their corresponding extra dimension objects collectively in the regions with the radius R_a, and R as "dark energy".

Looped Strings as Constituent of Dark Energy

Let us first consider the "activated" looped strings as the constituent of dark energy in the infinite momentum frame. Referring to the process described above; a flow of activated strings is established towards a matter of mass "m" to sustain the state of the matter by replacing the decayed activated strings in its structure. A string for string energy balance for this process, i.e., strings in infinite momentum frame, restframe1 (matter), and restframe2 (two-dimensional looped strings) as shown in Fig. (1-6) are as follows:

$$\Sigma cp_i = \Sigma m_i c^2 + \tfrac{1}{2} \Sigma k(\Delta A_i) \qquad\qquad (1\text{-}9)$$

Where,

c = speed of light

p_i = linear momentum of activated looped strings in infinite momentum frame

k = string energy constant, (energy /surface area)

ΔA_i = surface area of the two-dimensional strings

The left side of the equation (1-9) represents the sum of the total energy of all acti-vated looped strings represented by the momentum of the strings flowing towards the matter. On the right side of the equation, the $\sum m_i c^2$ represents the energy of the mass created "or sustained" by the flow of the strings from infinite momentum frame. The last term in equation (1-9) represents the energy of the two-dimensional strings exiting the matter in restframe2. This term is simply the Hook's law energy projected on two-dimensional plane of the strings, i.e., its energy is proportional to (length)^2.

Note that unlike the three-dimensional strings, the two-dimensional strings are no longer a field quanta of the three-dimensional universe. As such, throughout this book it will be treated non-relativistically.

Equation (1-9) describes the PROCESS of conservation of string energy from infinite momentum frame to restframe1 and restframe2, a string for string energy balance.

A derivative of equation (1-9) with respect to time yields:

$$d/dt \sum cp = d/dt\sum mc^2 + d/dt \left[\tfrac{1}{2} \sum k(\Delta A_i) \right] \qquad\qquad (1\text{-}10)$$

Since there is no change in the energy state of "matter" with respect to time (i.e., the flow of strings from infinite momentum frame into the matter maintains the unchanged state of mass);

$d\sum mc^2/dt = 0$

And the term $d[½ \sum k(\Delta A_i)]/dt$ is the change in energy of all infinitesimal areas created by the looped strings having only a two-dimensional state in the XZ plane. The summation and integration of these infinitesimal areas create the two-dimensional surface at the outer edge of the universe with a radius R_a as the two-dimensional strings exit the three-dimensional universe and accumulate at the outer edge of the universe.

In Chapter 3, it will be demonstrated that the energy term $d [½\sum k(\Delta A_i)]/dt$ along with conservation of its momentum described earlier is the basis for formation of extra dimension objects that are invisible to our three-dimensional universe.

The change in momentum of the strings on the left side of equation (1-10) with time is F_i, the force created by the momentum of the flow of the strings in infinite momentum frame into the matter and with the summation of all the string forces having an analog mass (space mass) M_i,

Equation (1-10) can be written as:

$$\sum c(F_i) = (½)k\, d \sum \Delta A_i /dt \qquad\qquad (1\text{-}11)$$

where,

$$F_i = M_i\, dV/dt\,,$$

In which, $V_2\text{-}V_1 = 0 - c = \text{-} c$

The velocity of strings in the infinite momentum frame is c, the speed of light. When the strings become part of the matter (restframe1) as they replace the decayed ones, its velocity becomes zero, hence creating a momentum change and a force towards the matter, i.e., the gravitational force of the matter.

$V=$ speed of strings

$M_i =$ analog mass (space mass) of the strings (as activated looped strings) into the matter

$M_i = vd\rho$

ρ = density of activated looped strings in infinite momentum frame

R_a = radius of the field of the activated looped strings in infinite momentum frame, i.e., our observable universe

$A = 4\pi R_a^2$, surface area of the two-dimensional surface of the universe containing activated strings.

$dA = 8\pi R_a\, dR_a$

$v = 4/3\pi R_a^3$ = volume of the universe containing the activated looped strings in infinite momentum frame

Substituting all the above in equation (1-11) and integrating the summation;

$$\iint_c^0 [cvd\rho]dV = \iint d[(\tfrac{1}{2})k\, dA]$$

$$\int -c^2 vd\rho = \int (\tfrac{1}{2})k\, dA$$

$$\int -4/3c^2\pi R_a^3 d\rho = \int 4k\pi R_a\, dR_a$$

$$\int d\rho = \int -(3k/c^2)dR_a/R_a^2 \qquad\qquad (1\text{-}12)$$

The strings in addition to having momentum, kinetic energy, in infinite momentum frame, and energy described by Hook's law in restframe2, also have an intrinsic energy density that is proportional to the length of the string in restframe (1 and 2), and when integrated over the radius R_i it is proportional to the diameter of the universe. In other words, strings in three and two dimensions in restframe1 and 2 state can stretch with a constant of proportionality that is the string tension, K. As it will be shown in Chapter 8, this increase in length will be proportional to the same quanta of the string as in its corresponding field. This energy comes into play as the strings are stretched in restframe1, and 2 as a component of the 3D universe and extra dimension objects which are constituents of the fabric of spacetime. In essence, mass is energy, mass is length, and length is energy:

$$\rho = KD = 2KR, \tag{1-13}$$

$$d\rho = 2KdR$$

In which K is the string tension constant and R is the radius of the universe (rest frame radius). To maintain conservation of energy, the energy density of the activated strings obtained in equation (1-12) in the infinite momentum frame must equal that of the strings in restframe1 as three-dimensional strings in infinite momentum frame are converted to strings at restframe1:

$$\int 2KdR = \int -(3k/c^2)dR_a / R_a^2 \tag{1-14}$$

As mentioned above, the energy density ρ is a result of the strings stretching in restframe1 and 2, which in turn creates more space expanding the field of infinite momentum frame in the universe. As such, the energy density of the three-dimensional looped (and open) strings at the time of Big Bang when $R = R_a$, is zero ($\rho = 0$), because the restframe state where the strings stretch has not been created yet. The quantity of the strings in infinite momentum frame will begin to increase as the diameter of the universe begins to increase, pulling and stretching the strings in restframe1 and 2. Integrating equations (1-12, and 1-14), and using the boundary condition of $\rho = 0$ at $R_a = R$,

$$2KR = (3k/c^2)/R_a + C$$

$$\rho = (3k/c^2)(1/R_a - 1/R) \qquad R_a \neq 0,\ R \neq 0 \tag{1-15a}$$

$$R = 3k/2Kc^2(1/R_a - 1/R)$$

Choosing natural units ($K=k=c=1$) and rearranging:

$$R_a = 3R/(2R^2 + 3) \tag{1-16}$$

Equation (1-15a) describes the relationship between the energy density of activated looped strings as "quantum species" and in the infinite momentum frame vs the radius of the activated field of looped strings and the radius of the restframe.

Equation (1-16) is the field equation of activated looped strings which describes the relation between the radius of the field of activated looped strings vs. the radius of the universe in restframe. As you will see in Chapter 2, in reality, equation (1-16) represents the energy field of any mass "m" because it is independent of the mass of the object in this energy conservation process.

Fig. (1-7) displays a schematic representation of equation (1-16), the relationship between the radius of the field of activated looped strings in the infinite momentum frame vs the radius of the universe (in restframe).

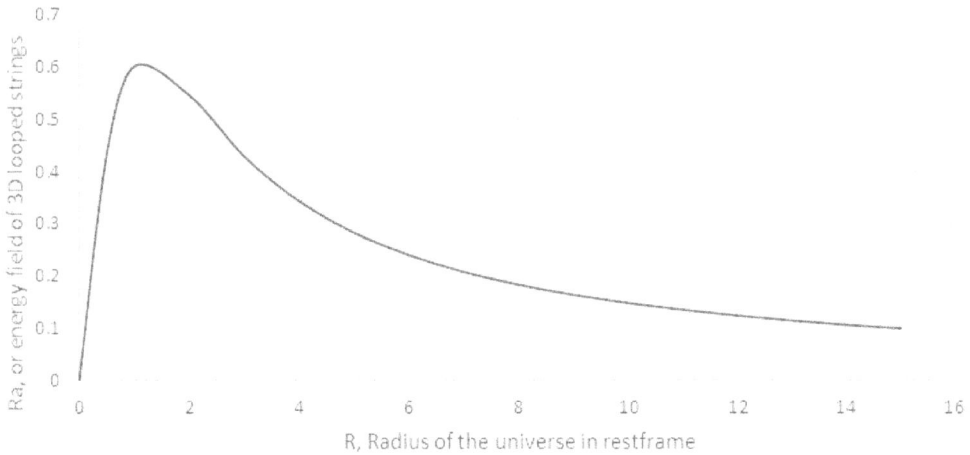

Fig. 1-7 - Radius of the field of three -dimensional looped strings in infinite momentum frame vs Radius of the universe in restframe

If we substitute for R=1.6R$_a$ the slope of the graph, the energy density of activated looped strings in infinite momentum frame (in the linear region) will be:

$$\rho = (1.2k/c^2)(1/R_a) \qquad\qquad (1\text{-}15b)$$

Equation (1-15b) demonstrates that the energy density of the looped strings is inversely proportional to the radius of the field of activated strings (1/R$_a$). Meaning that the energy density of the activated looped strings decreases with an increase in the field radius. You will see in the next chapter that this corresponds to a significant increase in the dimension of the activated looped string which starts off at Planck length.

The graph of R_a vs R is fairly linear between R=0 to R=1, meaning that the strings in the infinite momentum frame follow $\rho = 2KR_a$ in this region, or in other words, a) this portion of strings in the infinite momentum frame are converted to restframe1 (which is a state of matter), and b) new activated strings are generated as the strings in resframe2 are propagated with the expansion of the universe causing the field of the activated strings R_a to increase.

Fig. (1-7) illustrates that the field of activated looped strings increase and peak as more activated strings are generated in infinite momentum frame with expansion of the two-dimensional strings. After peaking, the field of activated strings in infinite momentum frame decreases as the rate of decay of strings exceeds the rate of generation until the activated strings in the universe are largely consumed at about R=12.

The value of k = energy/surface area when a three-dimensional string with Planck radius is converted to a two-dimensional string is k $=1.96x10^9/\pi(1.61x10^{-35})^2 = 0.24x10^{79}$ N/m.

If we substitute for $R_a = 1.61x10^{-35}$m at the time of Big Bang and $R_a =1.15x10^{26}$m the current radius of the observable universe, in equation (1-15b) the energy density of the activated looped string will be respectively $0.12 x10^{97}$ kg/m³ and $0.28 x10^{36}$ kg/m³. This means that the increase in the amount of activated looped strings in infinite momentum frame (the area under the curve) or generation of new energy caused by the expansion comes at the price of decreasing the energy density of the activated looped string as a quantum specie, from a Planck energy density to about $0.26 x10^{36}$ kg/m³. This energy density will remain constant for the remainder of the life of universe until the strings in the infinite momentum frame are totally consumed by the "matter" and converted to restframe2.

The cut-off point for calculation of energy density of the activated looped strings from equation (1-15a) in the infinite momentum frame is the peak point on the curve, or the linear section of the curve. This is because after peaking, the value of R_a will decrease resulting in the energy density of the strings to increase which is not possible.

A seen from equation (1-15a) the radius of the universe in restframe must always be greater than the radius of the field of infinite momentum frame $(1/R_a -1/R)>0$ or the value of the energy density becomes negative which is unacceptable.

Open Strings as Constituent of Dark Energy

Now, let us turn our attention to the energy balance for the activated open strings as the second constituent of dark energy. Again, referring to the process described earlier; a flow of activated open strings is established towards a matter with mass "m" to sustain the state of the matter by replacing the decayed activated strings in its structure. The energy balance for this process, i.e., strings in infinite momentum frame, restframe1 (matter), and restframe2 (two-dimensional open strings) as shown in Fig. (1-6) are as follows:

$$\sum cp_i + \sum cp_i' = \sum m_i c^2 + \tfrac{1}{2} \sum k(\Delta A_i) \qquad (1\text{-}17)$$

As mentioned earlier, unlike two-dimensional looped strings, two-dimensional open strings remain in our three-dimensional universe by attaching to other "activated" strings from both ends upon conversion. This unique property of the open strings become apparent in Chapters 3 and 8. Therefore, a new momentum energy, p_i' is generated when the two-dimensional open strings are carried away by other activated strings as an attachment. In addition, two activated open strings participate in the process, one to replace the depleted one, the other to carry the two-dimensional string as an attachment out of the mass of the "matter" which is then converted to an activated string at restframe1. As such, the momentum energy of the two-dimensional strings exiting the matter will be half of the total energy input into the matter.

$$\sum cp_i' = \tfrac{1}{2} \sum cp_i \qquad (1\text{-}18)$$

The two-dimensional open strings accelerate from a velocity of $V_1 = 0$, at restframe, to $V_2 = c$ in the infinite momentum frame (as an attachment), as compared to the incoming strings that replace the depleted ones having a velocity of $V_1 = c$, and $V_2 = 0$. Substituting equation (1-18) in (1-17), yields:

$$\sum cp_i + \tfrac{1}{2} \sum cp_i = \sum m_i c^2 + \tfrac{1}{2} \sum k(\Delta A_i) \qquad (1\text{-}19)$$

The change in the momentum with time creates the forces entering and exiting the matter on the left side of equation (1-19). The forces entering and exiting the matter created by the momentum change of the activated and two-dimensional open strings are equal in magnitude because the analog mass (space mass) of the strings contributing to the force is the same. Therefore, unlike looped strings which only have an inlet momentum, and an inlet force, the activated open strings have an inlet and outlet momentum (in and out of the matter) with the resultant inlet and outlet forces. Again, differentiating equation (1-19) with respect to time:

$$d/dt \; \Sigma cp + d/dt[\; \tfrac{1}{2} \; \Sigma cp\;] = d/dt\Sigma mc^2 + d/dt \; [\tfrac{1}{2} \; \Sigma k(\Delta A_i)] \qquad (1\text{-}20)$$

Where, $d\Sigma mc^2/dt = 0$

With the exception that $d \; [\tfrac{1}{2} \; \Sigma k(\Delta A_i)] \; /dt$ is rewritten as $d \; [\tfrac{1}{2} \; \Sigma k(\Delta v_i/L_p)] \; /dt$, in which L_p is the Planck length and the summation is the sum of all infinitesimal volumes (Δv_i) created by the two-dimensional open strings attached to activated strings. The summation and integration of these infinitesimal volumes create the additional volume comprising the universe with radius R at rest frame.

In Chapter 3, it will be demonstrated that the change in the energy term $d[\tfrac{1}{2}\Sigma k(\Delta v_i/L_p)] \; /dt$, along with conservation of momentum as described earlier forms the basis for creation of extra dimension objects which are invisible to our three-dimensional universe.

The summation of change in momentum of the strings on the left side of equation (1-20) with time is the F_i, and F_o, the equal inlet and outlet forces created by the momentum of the flow of open strings into and out of the matter. With the analog mass (space mass) of the strings in and out of the matter as M_i,

Equation (1-20) can be written as:

$$\Sigma c(F_i) \; + 1/2\Sigma c(F_o) = (k/2L_p) \; d \; \Sigma \; \Delta v_i/dt \qquad (1\text{-}21)$$

$F_i = M_i \; dV/dt$

In which, $V_2\text{-}V_1 = 0 - c = -c$

And $F_o = M_i \; dV/dt$

In which, $V_2 - V_1 = c - 0 = c$

V = speed of the strings

M_i = total analog mass (space mass) of the activated open strings in infinite momentum frame

$M_i = vd\rho$

ρ = density of activated open strings in infinite momentum frame

R = radius of the universe in restframe

R_a = radius of the field of activated open strings in infinite momentum frame

$v = 4/3\pi R_a^3$ = volume of the activated strings in infinite momentum frame

$dv = 4\pi R_a^2 dR_a$

Substituting all the above in equation (1-21);

$$\iint_c^0 [cvd\rho]dV + (1/2) \iint_c^0 [cvd\rho]dV = \iint d[(k/2Lp)\, dv]$$

$$\int -c^2 vd\rho + 1/2 \int c^2 vd\rho = \int (k/2L_p)dv$$

$$\int -2/3c^2\pi R_a^3 d\rho = \int (2k/L_p)\, \pi R_a^2 dR_a$$

$$\int d\rho = \int -(3k/L_p\, c^2)dR_a/R_a \qquad\qquad (1\text{-}22)$$

Again, to conserve energy, the energy density of the strings in the infinite momentum frame in equation (1-22), must equal the energy density of the strings in the restframe1 as activated strings are converted to strings at restframe.

$$\rho = KD = 2KR, \qquad\qquad (1\text{-}23)$$

$$d\rho = 2KdR$$

In which K is the string tension constant and R is the radius of the universe in rest frame, and ρ representing the energy density of the strings.

$$\int 2KdR = \int -(3k/L_p \, c^2)dR_a/R_a \qquad (1\text{-}24)$$

$$2KR = -(3k/L_p \, c^2)lnR_a + C \qquad (1\text{-}25)$$

Similar to the looped strings, using the boundary condition of $\rho = 0$ at $R_a = R$, in equation (1-22) and (1-25) and rearranging,

$$\rho = -(3k/L_p \, c^2)ln(R_a/R) \qquad R_a \neq 0, R \neq 0 \qquad (1\text{-}26)$$

Where, $k = 0.24 \times 10^{79}$ N/m, $c = 3 \times 10^8$ m/s, $L_p = 1.616 \times 10^{-35}$ m

And

$$R_a = Re^{(-2KL_pc^2/3k)R}$$

Choosing natural units ($K=k=c=L_p=1$),

$$R_a = Re^{-(2/3)R} \qquad (1\text{-}27)$$

Equation (1-26) represents the energy density of the activated open strings in infinite momentum frame as a function of R and R_a. Equation (1-27) represents the relationship between R_a, the radius of the universe containing the activated open strings in infinite momentum frame and the radius of the universe in restframe.

As you will see throughout this book, two equations, (1-16) and (1-27), the energy fields of the three-dimensional looped and open strings, will become the basis of defining everything in the universe from fundamental particles, to blackholes, and all the fundamental forces including the universe beyond our three-dimensional universe.

Similar to looped strings, the field of the activated open strings in infinite momentum frame increase and peak as the strings are stretched along with two-dimensional strings at restframe2, then decrease as the consumption of the strings exceeds energy generation until the field of activated open strings (in the infinite momentum frame) is consumed at R=12 (Fig. 1-8). As reflected earlier in equation (1-19), the consumption of the strings occurs by replacing the decayed strings in the fundamental structure of matter and by conversion of the strings in infinite momentum frame to strings in the restframe1 upon attachment to a two-dimensional open string or the vacuum of a two-dimensional looped string.

As the three-dimensional open strings are converted to two-dimensional strings, they become attached to other three-dimensional activated strings, creating a network of extra dimension objects. The network of the two-and three-dimensional strings form the fabric of spacetime at restframe which then expands as the momentum and kinetic energy of the activated strings (in the infinite momentum frame) are converted to tension energy for stretching the fabric of space time (in restframe). I will cover these extra dimension objects in detail in Chapter 3.

Equation (1-26) shows that the energy density of activated open strings as quantum species and in infinite momentum frame unlike looped strings remains constant throughout the universe since the slope (R_a/R) is constant. Hence, the dimension of the three-dimensional string remains the same as the Planck radius with the expansion of the universe. The magnitude of the energy density of the activated open strings remains the same at about 0.17×10^{97} kg/m^3 from the time of Big Bang throughout the life of universe.

Equation (1-26) representing the energy density of the activated open strings in the infinite momentum frame is constrained to only the linear portion of the graph of the equation or approximately to the point of maxima in Fig. 1-8, from R=0 to about R=1. Therefore, the cut-off point for the energy density of open strings in infinite momentum frame in equation (1-26) is therefore around the peak point which as mentioned above equals Planck energy density.

Fig. 1-8 - Radius of the field of three dimensional open strings in infinte momentum frame vs Radius of the universe in restframe

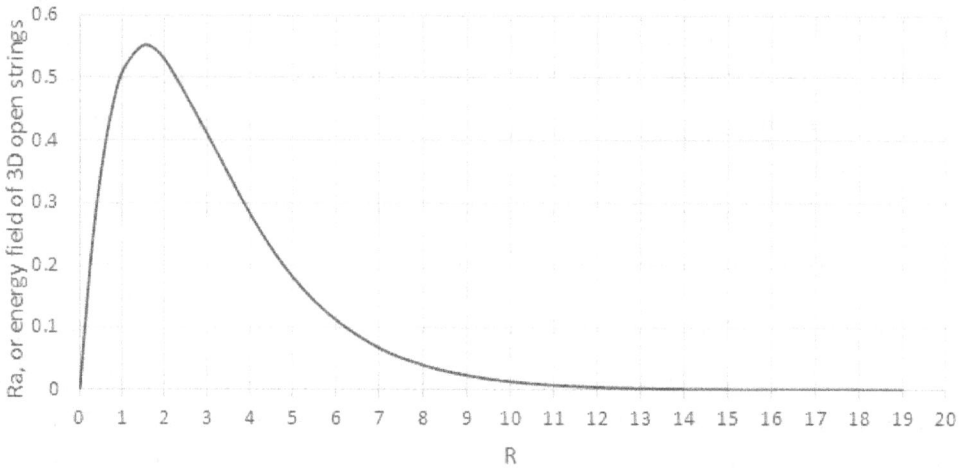

Likewise, as seen from equation (1-26) the radius of the universe in restframe must always be greater than the radius of the field of infinite momentum frame ($R_a/R<1$) or the value of the energy density becomes negative which is unacceptable. This is consistent with the earlier equation obtained for the looped strings in which a condition of equation (1-15) was that the radius of the universe in restframe R was larger than R_a, the radius of the observable universe at all times except at the time of Big Bang when $R=R_a$.

Energy Density of Activated Strings in Restframe1

In the previous section, I defined the energy density of the three-dimensional strings in restframe1 as a function of its tension as:

$\rho = 2KR$

In which K is the string tension constant defined as 2.18×10^{-8}kg/$2 \times 1.61 \times 10^{-35}$m $= 0.675 \times 10^{27}$kg/m. As mentioned earlier, activated strings in restframe1 gain an orbital rotation to conserve its momentum energy in infinite momentum frame. The orbital rotation of the strings with a field radius of R creates a volume of:

$v = 4/3\pi R^3$

The volumetric energy density of the three-dimensional strings in restframe1 will then be:

$\rho = 2KR /v$

in which, R is the radius of the field of restframe.

The volumetric energy density of the strings in restframe1 will then be:

$\rho = 3K /2\pi R^2$ (1-28)

If we use natural units by setting K=1,

$\rho = 3/2\pi R^2$ (1-29)

Fig. 1-9 represents the plot of the energy density of activated looped and open string in restframe1 vs the restframe radius of the universe.

Fig. 1-9- Energy density of activated open and looped strings in restframe

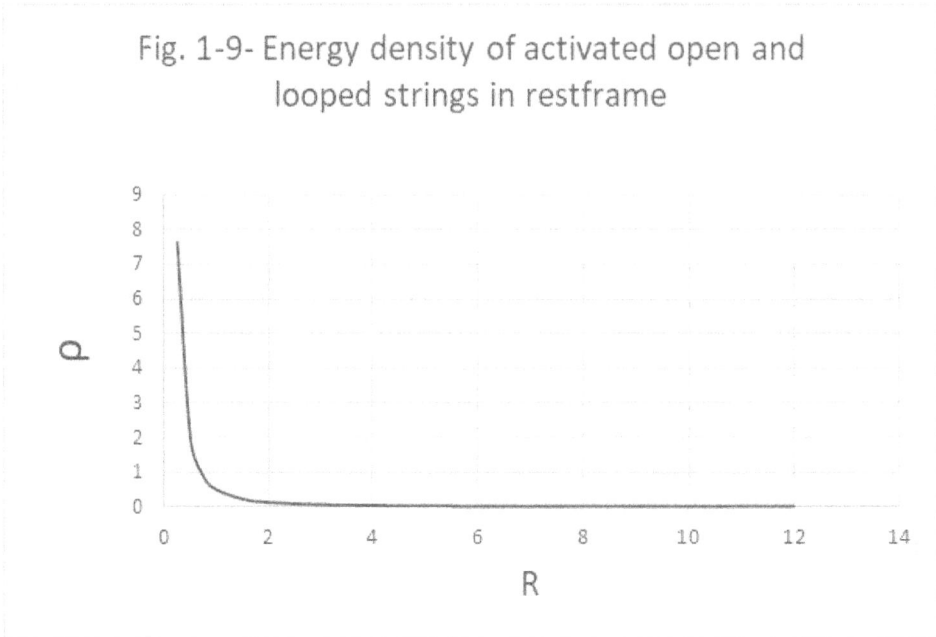

At the time of Big Bang when $R = R_a$, using $R_a = 1.61 \times 10^{-35}$m (Planck length), and substituting the value of R in equation (1-28), the energy density of the activated open and looped strings in restframe1 will be equal, about 0.25×10^{97} kg/m^3, which as expected is about the same energy as the energy density of the activated strings in infinite momentum frame.

As seen from equation (1-28), the energy density in restframe1is inversely proportional to R^2 and decreases rapidly with increasing radius of the universe. Fig. 1-9 demonstrates this sharp drop in the energy density of the activated strings as the universe begins to expand. This sharp drop represents the "inflation" period immediately after the Big Bang. To put this in perspective, the energy density of activated strings at restframe drops from a Planck scale value of 0.12×10^{97} kg/m^3 to about 7×10^9 kg/m^3 within the first second after the Big Bang and to 1×10^{-10} kg/m^3 after only 266 years from the Big Bang. Its current value is about $(5-12) \times 10^{-27}$ kg/m^3. However, after a significant and sharp drop in the early universe, it begins to level off which can be viewed as relatively constant, even though it is still decreasing but not at an extreme rate.

Fig. 1-10 represents the relationship between the radius of the field of activated looped and open strings in infinite momentum frame and the energy density of the activated strings at rest frame as constituents of dark energy, superimposed on the same axis.

Fig. 1-10 - Radius of activated looped and open strings in infinte momentum frame, energy density in restframe vs The radius of the universe in restframe

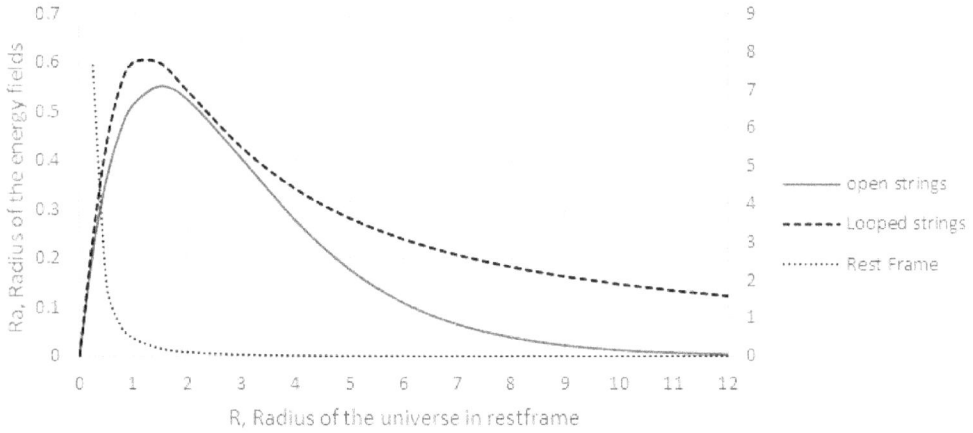

Fig. 1-10 - Radius of activated looped and open strings in infinte momentum frame, energy density in restframe vs The radius of the universe in restframe

As seen above, as the energy density of the activated strings in restframe is rapidly decreasing, the energy density of the strings in infinite momentum frame is rapidly increasing. This is because the creation of more "space" (volume) by the activated strings that constitutes our three-dimensional universe, means that the energy density of the strings at restframe1 (which constitutes matter) decreases at the same rate with increasing volume of the space.

If we set the naturalized energy density of the restframe equal to the naturalized field energy density of the open strings, we will obtain the radius of the field that corresponds to strings with Planck energy density. This is the radius of the field where the energy densities of strings in the infinite momentum frame and restframe are the same and equal to Planck density:

$$\rho = 3 \ /2\pi R^2 = -3ln(R_a/R) \qquad\qquad (1\text{-}30)$$

$R_a/R = 0.6$ (slope of the line)

$R = 0.56$

$R_a = 0.35$

Recall that in the previous section, I described the linear zone between R = 0 and R=1 as the region where activated strings are attached to the resframe2 forming restframe1 which is a state representing matter. The region between R=0 and R=0.56 is where not only it represents resframe1 but that this is the region of the matter where the energy field of the strings forming the matter have the same energy density as open strings or Planck density. As you will see in Chapter 4, this condition only occurs at the nucleus of "matter" comprising baryonic mass of the matter where the energy density of the strings in restframe is equal to that of Planck density. Therefore, the portion of the field between R=0 and R=0.56 and the area under this curve represents the total energy of the universe that becomes the baryonic mass of the universe. The portion of the field between R=0 and R=1 and the area under this curve represents the total energy of the universe that becomes the total mass of the universe (see Chapter 5).

Timescale of Dark Energy

Let us now explore the timeline of activated strings as a constituent of dark energy in the universe. We will develop this timeline for the three-dimensional open strings since the open string's energy density dominates the energy density of the universe at $0.12 \times 10^{97} \text{kg/m}^3$ vs $0.26 \times 10^{36} \text{kg/m}^3$ for activated looped strings.

We can calculate the time scale of the activated strings as a component of dark energy in the infinite momentum frame by tracking its conversion rate to strings at restframe. The energy density of the universe in restframe is trackable and measurable because it represents the mass density of the universe.

The property "mass" in our universe represents a change of energy of the activated strings when it converts from infinite momentum frame to restframe2. As you will see in Chapter 4, this change in energy manifests itself as the mass of fundamental particles (or dark matter) we can measure. The energy associated with the activated strings compromising infinite momentum frame does not change significantly by its mere existence or movement of the strings in infinite momentum frame. As a result, we do not detect nor are able to measure its mass. However, we can track the amount of energy associated with activated strings by tracking its conversion to restframe. I demonstrated above that the energy density of activated strings at restframe was:

$\rho = 3K\,/2\pi R^2$

or

$\rho R^2 = 3K\,/2\pi = constant$ (1-31)

Since the rate of propagation of R in the universe is the speed of light,

$R = ct$

$\rho t^2 = 3K\,/2\pi c^2 = constant$ (1-32)

As you saw above, the region between R=0 and R=0.56 is the amount of energy converted to baryonic mass. The total energy of the activated open strings in the universe (between R=0, to R=12) will be consumed to sustain the baryonic mass in the universe. As you can see from Fig. 1-10 total consumption of the open strings (R=12) coincides with the point where energy density of the strings at restframe levels off at its minimum at R=2.8. Using equation (1-31), we can calculate the energy density of the universe at restframe as a function of R:

$\rho_t = \rho_0 (R_0/R_t)^2$

Substituting for $\rho_0 = 6 \times 10^{-27}$ kg/m³ the current energy density which corresponds to R=0.56, the energy density ρ_t corresponding to dissipation of the open string energy field (R = 2.8) will be:

$\rho_t = 6x10^{-27}\,(0.56/2.8)^2$

$\rho_t = 0.24x10^{-27} kg/m^3$

Using equation (1-32):

$t^2 = 3K\,/(2\pi c^2\,\rho_t) = 0.14x10^{38}$

$t = 3.7\,x10^{18} s = 120B\,years$

The timescale from the time of Big Bang to the point where the energy density or the field energy of the open strings in the infinite momentum frame in the universe disappears is about 120B years. This timescale is one of the "coincidences" of our era since in the cosmological time frame, it is within the scale of the current age of the universe at about 14B years.

Let us now take a brief look at what classical physics tells us.

In classical physics, ρ is derived from Friedmann's equation and also calculated to be about 1×10^{-8} ergs/cm^3 or 11×10^{-27} kg/m^3 which matches the energy density calculated in this model:

$$H^2 \equiv (\dot{\alpha}/\alpha)^2 = (8\pi G/3)\rho - \kappa/\alpha^2 \qquad (1\text{-}33)$$

In which, H is Hubble's parameter, α, the scale factor, κ, spatial curvature.

The energy density of dark energy is also related to the cosmological constant (Λ) as;

The cosmological constant (Λ) is defined as:

$$\Lambda = 8\pi G \rho \qquad (1\text{-}34)$$

In which G is the gravitational constant and ρ is the critical energy in classical physics.

The value of the cosmological constant (Λ) using the zero-zero component of Einstein's gravitational field equation:

$$R_{\mu v} - 1/2g_{\mu v}R + \Lambda g_{\mu v} = (8\pi G/c^4)\, T_{\mu v} \qquad (1\text{-}35)$$

Is calculated to be 2.036×10^{-35} s^{-2}.

In which, $R_{\mu v}$ is the Ricci tensor, $g_{\mu v}$, the metric tensor, R, Ricci scalar, $T_{\mu v}$, stress energy tensor.

If we substitute for ρ the energy density of activated strings in restframe (1-28) in equation (1-34):

Λ = 8πG(3K /2 πR²)

Λ = 12GK/R²   *(1-36)*

In which G= 6.674×10^{-11} N m²/kg² , K = 0.675×10^{27} kg/m, R=2.3×10^{26} m

Interestingly we will obtain a value of 1.02×10^{-35} s⁻² (or 1.1×10^{-52} m⁻²) which matches the value reported in classical physics.

If we assume that the energy density of activated strings in restframe is a close approximation of the critical energy density in classical physics, then equation (1-36) indicates that the cosmological constant is really not constant and decreases with R the radius of the universe at restframe. Although, from Fig. (1-11), one can see that in the timescale of the universe, where the cosmological constant's value is today, the change for the remainder of the life of the universe is very small as compared to its value in the early universe. Therefore, one may argue that compared to its initial values in the early universe, its value now is relatively constant (Fig. 1-11). However, the change in the value of energy density between now and the end of the universe although small, is not trivial. As you saw above, I used this change to calculate the remaining life of the universe as about 100B years.

Looped and Open Strings in the Same Universe

Fig. 1-11 displays the radius of the field of activated looped and open strings as constituents of dark energy in the infinite momentum frame and the energy density at rest frame vs. time.

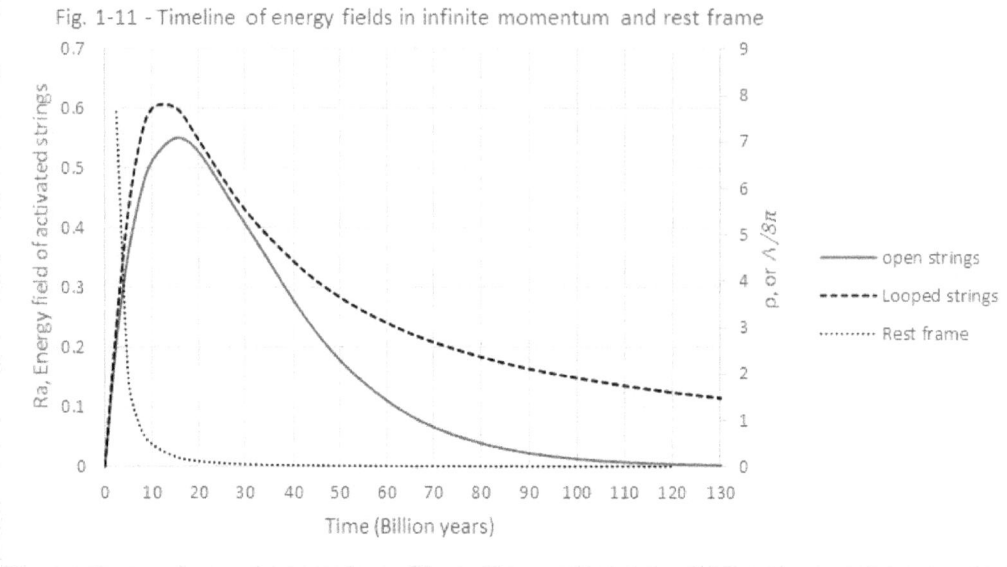

Fig. 1-11 - Timeline of energy fields in infinite momentum and rest frame

Figure 1-11 demonstrates that the field of activated looped and open strings (R_a) in the infinite momentum frame expand concurrently with the expansion of the universe (R) for the first 10-13 billion years after the Big Bang. The expansion of the energy fields implies that new strings in the infinite momentum are being generated, magnifying the amount of energy in the universe. The generation mechanism of strings after the Big Bang will be covered in Chapter 8. At the same time, the energy density of the strings in the rest frame plummets from Planck scale at the time of Big Bang to near its current value at around 10 billion years and then begins to level off but decreases gradually at a much slower rate. The field of the activated open and looped strings peaks at about 13B years which is concurrent with our time and begins to decrease until the energy in the infinite momentum frame is consumed by the baryonic matter and converted to strings at restframe. In the first 10B years, the rate of generation of activated strings due to the expansion exceeds the rate of consumption by the matter and conversion to strings at restframe. After peaking, the rate of consumption and conversion to the restframe exceeds the rate of generation resulting in the decline of activated strings in the universe. This change occurs at about the 13B year mark which is the onset of the current accelerated expansion.

Fig. 1-11 demonstrates that the radius of the energy field of looped strings at the point of maximum is slightly longer than that of open strings. As will be discussed later, this discrepancy maybe important when measuring the rate of the universe's current expansion by the classical methods.

The Magnitude of Energy Densities

Let us now look at the numerical values of energy densities from Fig. 1-10. The rate of expansion of the field of activated open strings in infinite momentum frame is fairly linear between R=0 to R=1 before it peaks. Taking the slope at the mid-point (R= 0.5, Ra= 0.35) and substituting in equation (1-26), the value of the energy density of open strings (on a mass basis) in the infinite momentum frame is calculated to be about 0.18 x 10^{97}kg/m^3 (1.6 x10^{113} J/m^3), which is about the same as Planck density (0.51 x 10^{97} kg/m^3). Since the slope is constant (R_a/R = 0.6 -0.7) throughout the linear region, the value of the energy density of the activated open strings remains the same, meaning that the energy density of open strings is independent of the magnitude of the radius of the universe and remains constant throughout the evolution of the universe, the same as its initial Planck value.

The Hubble time (τ) at zero gravity limit is about 12.16 x 10^9 years which yields a radius of about 1.15 x 10^{26} m. If we take this to be roughly the same as the radius of the observable universe, the radius of the energy field R_a will be 1.15 x 10^{26} m. From Fig. 1-11, using the radius of looped strings as reference for zero gravity limit, the radius of the universe (R) in restframe will be about twice or 2.3 x10^{26} m. Substituting for K (0.675 x 10^{27} kg/m, as string mass/string length) and R in equation (1-28):

$$\rho = 3K/2\pi R^2$$

The energy density or mass density of the activated strings at rest frame, ρ is calculated to be about 6 x 10^{-27} kg/m^3 which is about the same as the observed value of energy density obtained from cosmological distance measurements.

Given that the energy density of activated open strings in restframe is about 6 x10^{-27} kg/m^3, this gives rise to the ratio of the energy density in infinite momentum frame (as open strings) to restframe of about:

$$\rho_{imf}/\rho_{rf} \sim 10^{123}$$

This is the well-known (order of 120) discrepancy between the theoretical and observed value of energy density in the universe as calculated by other methods in classical physics, known as the "cosmological constant problem". As you can now see, this discrepancy is real as the two energy densities represent two different properties of the universe, not one.

A similar value of energy density is obtained for activated looped strings in infinite momentum frame using the slope of $R_a/R = 0.6$ ($R = 1.0$, $R_a = 0.6$) in Fig. 1-10 and substituting for $R_a = 1.15 \times 10^{26}$ m in equation (1-15), the energy density of activated looped strings is calculated to be about 0.28×10^{36} kg/m^3, or 2.52×10^{52} J/m^3. Unlike open strings, the energy density of activated looped strings decreases with increasing magnitude of R_a, and R as it is inversely proportional to the radius ($1/R_a$). Therefore, as the universe expanded, the energy density of the activated looped strings in infinite momentum frame decreased from its initial Planck value to 0.28×10^{36} kg/m^3 in infinite momentum frame which is 10^{61} lower than the energy density of open strings.

The ratio of the energy density of activated looped string in infinite momentum frame to its energy density at restframe is then:

$$\rho_{imf}/\rho_{rf} \sim 10^{62}$$

It is rather remarkable that the field of looped and open strings peaks at about the same time as if the behavior of the two different strings are dependent on one another, and that they both follow the same rules and patterns in the universe even though the equations obtained for these strings are quite independent from each other. Moreover, as seen in Fig. 1-11, the sharp drop in energy density of the strings at rest frame is concurrent with the sharp increase in the radius of the field of infinite momentum frame. The synchronicity of the behavior of the two strings individually and their properties in rest frame and infinite momentum frame may be a testament to the role these strings play in the universe in coordination with each other and as constituents that make up the universe.

Let us now address what appears to be a discrepancy between the energy density of the strings in infinite momentum frame and restframe known as the "cosmological dark energy problem" in classical physics.

The value of 6×10^{-27} kg/m^3 represents the energy density of the activated strings at rest frame. This is essentially the mass density of the universe. As I mentioned earlier and you will see in Chapter 4, this mass density is measurable because it reflects the change of energy when activated looped strings decay in the structure of fundamental particles with a baryonic mass. In other words, baryonic mass is an effect we observe when there is a change of energy as activated looped strings decay. You will also see that this change of energy is very small resulting in small masses for fundamental particles. Given the volume of the universe with a radius of about 2.3×10^{26}m, and a mass of about 1.9×10^{53}kg, this mass density will be very small.

On the other hand, the energy density of 0.12×10^{97}kg/m^3 for activated open strings, and 0.28×10^{36}kg/m^3 for looped strings, is the quanta of energy that makes up the component of the fabric of spacetime in infinite momentum frame. This is the space mass, or energy that creates the three-dimensional space. Although activated strings freely move about in infinite momentum frame at the speed of light, the change in its energy is very small. In fact, the amount of energy stored in each string (1.96×10^9J) is sufficient to provide the kinetic energy needed in the order of 100B years. Without a significant change in energy, the energy that makes up the mass of space, it will not be measurable or observable. Therefore, for now we will have to rely on theoretical calculation to obtain its value.

We can measure the signature effects of the quantum energy of activated strings when there is a significant change in its energy such as gravity for activated looped strings, and electromagnetic force for activated open strings. These will be covered in Chapter 6, and 7. But for now, we do not have an experimental method to measure it directly.

As you will see later (Chapter 8) "space" comes at a premium. The energy of an activated string is the energy that is required to create a three-dimensional space equivalent to the volume of the activated string. What appears to be empty space in our universe in reality is a space created by three-dimensional strings which require a significant amount of energy to uphold. This energy is the energy of the infinite momentum frame.

Curvature of the Universe

The variation of the energy field of activated looped and open strings in infinite momentum frame verses the energy field of the strings in restframe as shown in equations

(1-16), and (1-27) is a representation of the curvature of spacetime. As you will see in Chapter 11, "time" is a geometric dimension of space which is imbedded in the geometric dimension of the three-dimensional strings which form the Hilbert space. The energy fields of activated looped and open strings represent the constituents that make up the three-dimensional universe. The radius of the field of infinite momentum frame designates a three- dimensional space structure where the strings are fluid and free to move with five degrees of freedom. The radius of the field of restframe represents a space where the strings are immobile with two-degrees of freedom and belong to the extra dimension space. One might envision them as the invisible structures that make up the fabric of spacetime.

As you will see in the next chapter, the dimension of activated looped strings which is a dominant constituent of space changes with R_a. Its dimension will remain the same for a constant R_a or in a direction perpendicular to R_a which is R. Variations in the dimension of activated looped strings which is the space itself with respect to R where the dimensions remain constant represents itself as the curvature of spacetime. In essence, the radius of restframe becomes the reference where time can be measured as a baseline with no variations. Where [If] there are no variations in the dimension of the activated looped strings with R, (R_a is parallel to R) the spacetime is flat.

The concept of spacetime will become more evident as we go through this book chapter by chapter, and in Chapter 11 where "time" is defined as a geometric dimension associated with the strings. From this point forward, the graphs of R_a, vs. R as presented in this book also represent the curvature of spacetime. Similar to Minkowski space as represented in general relativity, in this theory, a constant interval separates the events in space. This constant is related to the dimension of the activated looped strings in infinite momentum frame. As such, "time" becomes an integral component of the dimension of space and imbedded in it. Therefore, geometric demonstration of the curvature of R_a vs. R will also represent the spacetime itself.

To observe the curvature of spacetime in three dimensions, we need to plot the above equations (1-16, 1-27) not only for positive values of the X and Y axis but also in the negative values where it is applicable. In essence, the universe has an inherent symmetry due to the presence of equal amounts of left and right handed spinning activated looped and open strings. If we apply equations, 1-16, and 1-27 to the strings of one spin, we must apply it to the strings of the opposite spin to describe the system as a whole. If we take the restframe axis, as the axis of reference, then we must plot the equations for negative values of R_a to account for the same energy field for strings of opposite rotation. If we plot the graph of the above equations on the positive and negative axis for R_a, the field of strings in infinite momentum frame vs only the positive values of R, the radius of the restframe, we will obtain the overall curvature of spacetime for the universe as it pertains to the field of open and looped strings, Fig. 1-12.

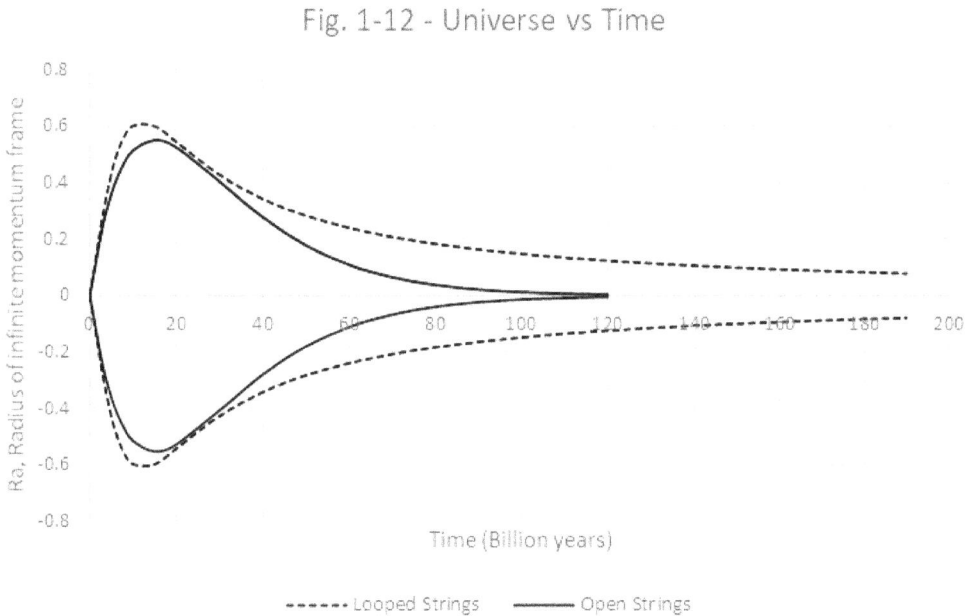

Fig. 1-12 - Universe vs Time

Event Horizons

As shown in Fig. 1-11, there is a concurrent field of three-dimensional looped and open strings with three event horizons in the universe. The first is designated by the field radius of activated open strings, the second by the field radius of the activated looped strings which is about 9% longer than the activated open strings at its peak and the third is the radius of the outer edge of the universe where the two-dimensional surface of the sphere of the universe lies at its restframe radius. The event horizons of the activated open and looped strings were closer in the first 5B years after the Big Bang but begin to diverge and are currently at maximum divergence. This means that concurrent with our time, the field of activated looped strings has expanded 9% more than the field of open strings. Although this appears small on a cosmological scale, it could present inconsistencies when the rate of universe's expansion is measured by Hubble method, due to the energy density differences of the two regions.

As seen in Fig. 1-11, this discrepancy did not exist in the early universe (first 5B years) and will disappear again when the age of the universe is between 20-30B years. The ratio of the radius of the universe to the field of infinite momentum frame which is now about 2 (using the field of looped strings as reference) will continue to get larger for the next 100B years until the field of strings in the infinite momentum frame disappears. Another way to interpret Fig. 1-11 is that in the earlier universe, the radius of the observable universe, and the radius of the universe (at restframe) both expanded simultaneously, but after peaking, the radius of the observable universe will begin to decline as the universe as a whole continues to expand. This means that as the radius of the observable universe continues to shrink, future generations may be able to see fewer galaxies and stars as compared to today's skies.

An important impact that this will have on cosmological measurements is that photons only travel through a medium that is comprised of activated open strings. As you will see in Chapter 4, the structure of photon is comprised of activated opens strings. This means that distances that are larger than the field of activated open strings will represent a different light transmission speed if any than within its field radius.

CHAPTER 2
Matter and Spacetime

Energy field of a mass

IN CHAPTER 1, I INTRODUCED the field equations representing the variation of the radius of the field of activated strings in infinite momentum frame vs the radius of the field of the strings in restframe for both, looped and open strings using natural units:

$R_a = 3R/(2R^2 + 3)$ Looped strings   (1-16)

$R_a = Re^{-(2/3)R}$ Open strings   (1-27)

These equations were obtained by creating an envelope around an arbitrary matter with a mass "m", further developing a string for string energy balance around the mass of this matter. It was also demonstrated that the peak of the looped and open strings in infinite momentum frame coincided with the current era with a radius of about 1.15×10^{26} m.

As I mentioned in Chapter 1, the variation of the field of activated open and looped strings in infinite momentum frame verses the field of the strings in restframe as shown in equations (1-16), and (1-27) is a representation of the curvature of spacetime created by the forces resulting from the flow of the strings and its momentum change.

If we now plot the graph of these equations with respect to positive and negative values of R, the restframe energy field but increasing values of R_a, we will obtain a three-dimensional view of the spacetime curvature created by the mass "m" for which we set up the energy balance equations. Again, as explained in Chapter 1, the universe has an inherent symmetry due to the presence of equal amounts of left and right handed spinning activated looped and open strings. If we apply equations, 1-16, and 1-27 to the strings of one spin, we must apply it to the strings of the opposite spin to describe the system as a whole. If we take the R_a axis, as the axis of reference, then we must plot the equations for negative values of R to account for the same energy field for strings of opposite rotation in restframe, since the mass resides in the restframe.

Figure 2-1 represents the curvature of spacetime for the energy field of any matter of mass "m" because the equations that created it are independent of the mass of the matter. In Chapter 6, you will find out that this curvature is caused by the gravitational force generated by the flow of the activated looped strings into mass "m" for which equations (1-16) and (1-27) were developed.

The graph of the spacetime curvature of mass "m" was obtained from naturalized field equations for activated looped and open strings in which all Planck related constants were set equal to 1. Therefore, the graph represents not only any mass, but also can be scaled accordingly to represent the scale of the corresponding mass. This means that the value of R_a, at the peak point and the point representing the origin (the maximum and the minimum) can be scaled to the dimension representing the mass.

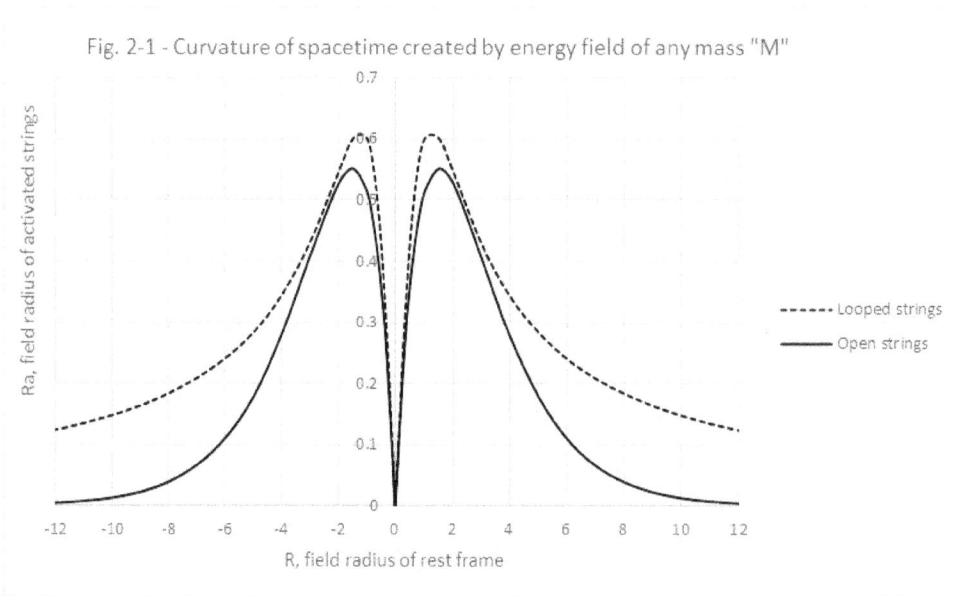

Fig. 2-1 - Curvature of spacetime created by energy field of any mass "M"

If we set the peak value at $R_a = 1.15\times10^{26}$m, the curvature of space time will represent the energy field of the mass of the universe. For example, if we set the peak value to be $R_a = 4.6\times10^6$m, the curvature of spacetime will represent the mass of the Earth and its quantum field. Fig. 2-2 is a representation of the curvature of spacetime created by a mass greater than the atomic scale, or macro-masses. Again $R_a = 0$ is a mathematical reference point of the value of R_a which means a small non-zero number, because for matter to exist, R_a cannot be zero. Let us designate this shape as the "spacetime well" for larger masses greater than the atomic scale. In Chapter 6, I will present the precise coordinates of any mass "m" with respect to its position within the spacetime well.

As you can see, the energy field of a smaller mass will be the same as a larger one but with its own corresponding coordinates which will fall within the energy field of the larger mass. For example, the energy field of Earth will fall inside that of the Sun, and the Sun inside the Milky way (as one mass), and the Milky way inside the universe with a radius of 1.15×10^{26}m.

In Chapter 6, you will learn about the gravitational field of any macro-mass "m" with respect to the curvature of spacetime of its energy field shown here. In this chapter, I will keep our focus on the energy field of subatomic particles at quantum scale.

Fig. 2-2 Curvature of Spacetime of macro-mass "M"

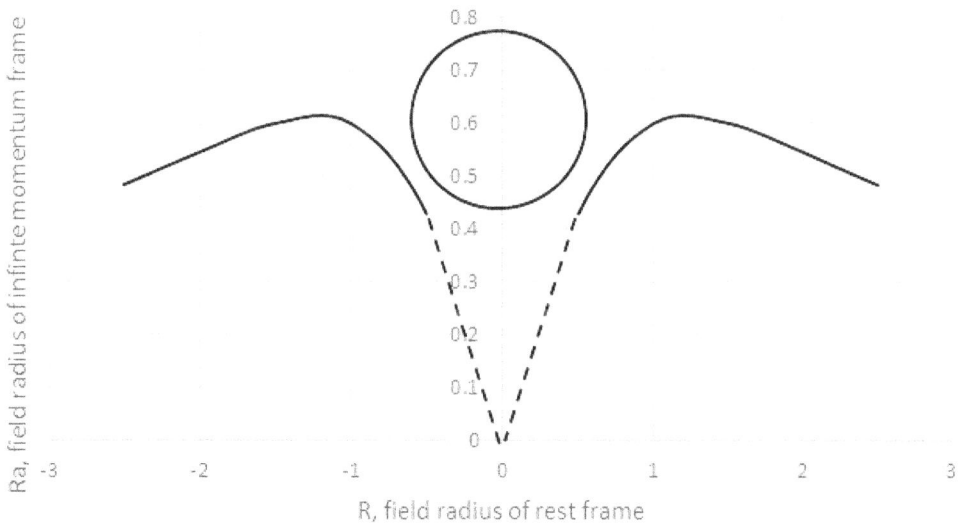

Spacetime cavity

As mentioned above, the graph in Fig. 2-1 represents the energy field of any mass "m", including subatomic particles at quantum scale. To obtain the value of R_a at the peak point for the subatomic particles, we must now scale the graph corresponding to the geometric dimension of mass "m" at quantum scale.

To rescale the graph, we will set the origin to be $R_a = 1.61 \times 10^{-35}$m on the energy field of the graph in Fig. 2-1. This means that the smallest radius of matter "m" will be equal to the Planck length, the radius of an activated string as described in Chapter 1.

To obtain the corresponding peak value for the new origin, we will multiply the lowest range $R_a = 1.61 \times 10^{-35}$m by 1.15×10^{26} to maintain the same proportionality of space as we had for spacetime curvature of macro-masses in the universe. The new maximum range (peak) will now become 1.85×10^{-9}m.

$1.61 \times 10^{-35} \times 1.15 \times 10^{26} = 1.85 \times 10^{-9}$ m.

Fig. 2-3 represents the energy field and the curvature of spacetime created by sub-atomic "masses" covering a range of 1.61×10^{-35}m to 1.85×10^{-9}m where $R_a = 0$ becomes the nucleus of the matter having a dimension of 1.61×10^{-35}m. All other dimensions inside the cavity are represented on the logarithmic scale calculated proportional to the naturalized radius of R_a, and R encompassing a cumulative range of 1.15×10^{26}.

Fig. 2-3 Curvature of spacetime created by subatomic mass "M"

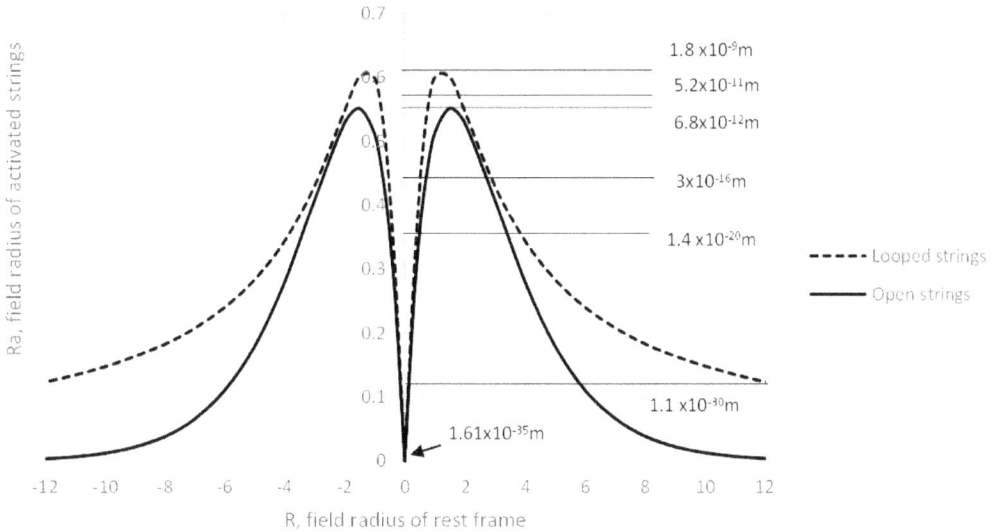

Let us designate the above spacetime curvature as the "spacetime cavity" since it represents matter at sub-atomic and quantum levels.

In essence, if we start from a dimension of 1.61×10^{-35}m, the space makes a transformation at 1.8×10^{-9}m to infinite momentum frame which then extends to 1.15×10^{26}m by transformation of the energy field of larger masses. More importantly that the spacetime curvature of the energy field of large macro-masses beyond 1.8×10^{-9}m will include two energy fields, one for the mass itself and one for its constituents at subatomic and quantum scale. This will be further demonstrated in Chapter 6.

We will see in Chapter 4 that the spacetime cavity created at quantum scale below 1.8×10^{-9}m is in reality the "cradle of matter" for fundamental particles with a mass and atoms.

In classical physics, a similar field of energy is known as "Higgs field" commonly referred to as the "Mexican hat". The Standard Model of particle physics is generally limited to dimensions in the order of about 10^{-20}m. The spacetime cavity shown above, and in this theory, extends the field of energy to 1.6×10^{-35}m and as you will see in Chapters 8 and 10 to 2.3×10^{-51}m, and 8.5×10^{-160}m. A three-dimensional graph of the curvature of spacetime cavity (R_a Vs R) is shown in Fig. 2-4A and B.

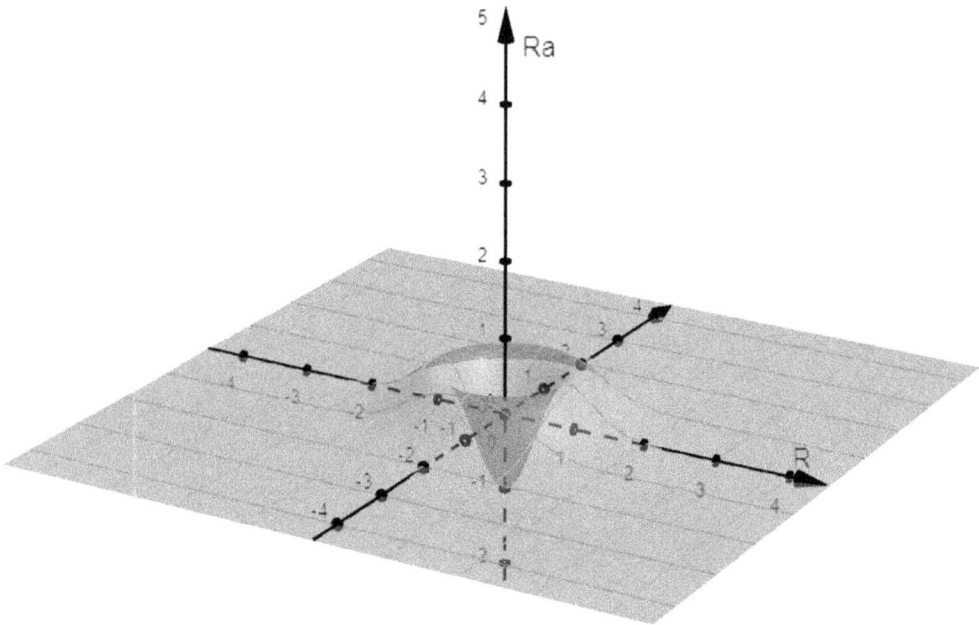

Fig. 2-4A – Three-dimensional view of spacetime cavity of sub-atomic mass "M".

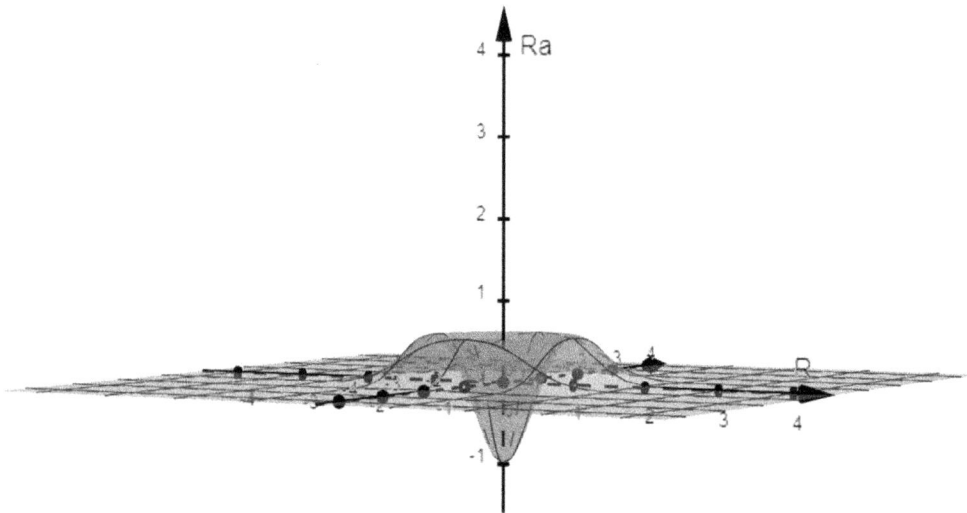

Fig. 2-4B – Three-dimensional view of spacetime cavity of sub-atomic mass "M".

The curvature of spacetime created by large celestial and quantum scale objects as shown in "spacetime well" and "spacetime cavity", are caused by the gravitational effect of activated looped strings which in turn is caused by the momentum change of strings replacing the decaying strings.

This may be in contrast to the classical understanding of gravity which is presumed to be created by the curvature of spacetime. In other words, the curvature of space-time does not create gravity, rather the gravitational effect of activated looped strings creates the spacetime curvature. As you will see in Chapter 7, activated open strings while they do have an acceleration creating the electromagnetic forces in the universe, the net effect of its force on the restframe is zero. Therefore, activated open strings by themselves do not contribute to creation of curvature of spacetime but follow the curvature created by the looped strings.

As you will see later, the gravitational field equations for both activated looped and open strings developed in this theory are applicable to both large and quantum scale matter with applicable adjustments for spacetime transformation as described above. In essence, using the same fundamental equations developed at string level, one can bridge the quantum scale gravitation to macro-scale gravitation by incorporating the space transformation at about 1.8×10^{-9}m (see Chapter 6).

The spacetime cavity as shown in Fig. 2-3, 2-4A and B is the fundamental building block of spacetime encasing any matter with a baryonic mass, including atoms, protons, electrons, Higgs Boson, and all other subatomic particles (with a baryonic mass) as described in the Standard Model of particle physics. We will take a deep dive into this "cavity" when the fundamental particles that make up "matter" are described in Chapter 4.

Beyond the opening of the "spacetime cavity" which is the cradle of "baryonic mass", lies the fabric of spacetime forming the universe which, as mentioned in Chapter 1, is composed of activated looped and open strings in infinite momentum frame and restframe with its own space mass. You will see that the change in the energy of the strings is significant and measurable inside the spacetime cavity revealing the mass of the particles. Whereas as mentioned before, the change in energy of the strings form-ing the fabric of spacetime in infinite momentum frame is too small to be measurable or observable.

Let us now describe some of the properties of this "spacetime cavity" which is the cradle of all matter with a baryonic mass.

Using equation (1-28) in Chapter I, we can see that the energy density of the strings in restframe continues to increase with the depth of the "cavity" and peaks at 0.25×10^{97} kg/m^3 at the nucleus of the cavity when it reaches the string length at 1.61×10^{-35}m:

$\rho = 3K / 2\pi R^2$

This means that the conditions inside the cavity as you move into the cavity are exactly the reverse of the energy densities immediately after the Big Bang. In Chapter 1, it was demonstrated that the energy density of the strings in the restframe plummeted from the Planck scale down to its current value of 6×10^{-27} kg/m^3 whereas inside the "cavity" the energy density increases from its current value of 6×10^{-27} kg/m^3 outside the cavity back to Planck scale at the nucleus of the cavity.

A graph of the energy density of the strings in restframe inside the "cavity" is given in Fig. 2-5. In this graph, only the "spacetime cavity" created by the looped strings is shown.

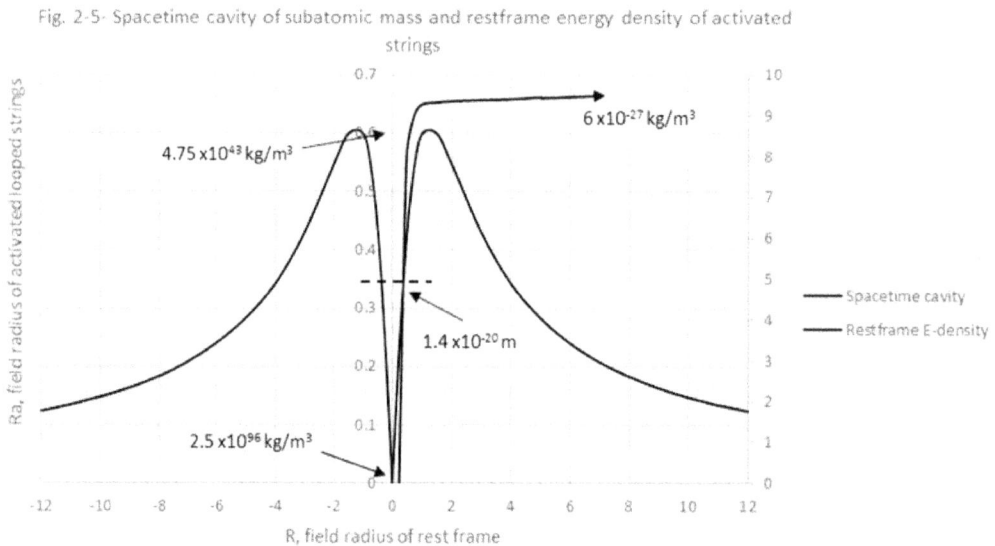

Fig. 2-5- Spacetime cavity of subatomic mass and restframe energy density of activated strings

In Fig. 2-5, the energy density of the strings at restframe outside the "spacetime cavity" at a large distance away from the cavity is 6×10^{-27} kg/m^3. However, this energy density begins to change (equation 1-28) as the strings flow closer to the vicinity and at the opening of the cavity upon exposure to the gravitational field of the cavity. This field is created by the flowing strings replacing decaying ones but has a more complex dynamic due to the orbital rotation of the strings at restframe and presence of various extra dimension objects, and the flow of two different types of strings each having right- and left-handed spin. I will cover these complex dynamics gradually in the up-coming chapters, but to keep the matter simple to understand I will stay focused on the fundamentals as it relates to the subject in each chapter.

If we substitute for $R_a = 1.85 \times 10^{-9}$m, the dimension of the field radius at the opening of the spacetime cavity, the restframe radius will be about $2R_a$, or $R = 3.7 \times 10^{-9}$m. Substituting R in equation (1-28), the energy density of the strings in the restframe will be:

$$\rho = 3K\,/2\pi R^2 = 3 \times 1.347 \times 10^{27}/2\pi \times (1.85 \times 10^{-9})^2$$

$$\rho = 4.75 \times 10^{43}\ kg/m^3$$

This means that the energy density of the strings at restframe which is 6×10^{-27} kg/m^3 in infinite momentum frame far away from the cavity will increase to 4.75×10^{43}kg/m^3 upon exposure to the field of the spacetime cavity. The energy density will then increase to 0.2×10^{97} kg/m^3 at the nucleus of the cavity (R=1.61×10^{-35}m).

Recall that in Chapter 1, the intersection of the curve of the energy density at restframe and the radius of the field of the activated open and looped strings in infinite momentum frame ($R_a = 0.35$, R =0.56) signified the portion of the field that strings existed in Planck density at restframe. We obtained this point by setting the field energy density of the strings in restframe to be equal to activated open strings. Similarly, the intersection of the restframe energy density and the field radius of activated looped strings represents the portion of the field that the activated looped strings are at restframe. This point occurs at a depth of about 1.4×10^{-20}m inside the spacetime cavity corresponding to $R_a = 0.35$ as shown in Fig. (2-5). This means that activated looped strings in the field between $R_a=0$ and $R_a=0.35$ exist predominantly in restframe1 state. In other words, the field of infinite momentum frame ends at this point. This point inside the spacetime cavity has several significant implications which I will discuss at length in the forthcoming sections and chapters.

Looped and Open Strings vs Field Radius of a Mass

Let us now address the energy density of the looped and open strings in infinite momentum frame inside the energy field of mass "m" or inside the spacetime cavity for subatomic masses.

In Chapter 1, it was demonstrated that the energy density of activated looped strings in infinite momentum frame dropped from its Planck scale value of 0.12×10^{97} kg/m³ to 0.26×10^{36} kg/m³ at $R_a = 1.15 \times 10^{26}$m with the expansion of the universe.

The energy density (on a mass basis) of the looped strings was derived to be (1-15):

$$\rho = (3k/c^2)(1/R_a - 1/R)$$

and on an energy basis:

$$\rho = (3k)(1/R_a - 1/R)$$

Inside the energy field of a mass, the slope of the graph of looped strings is $R_a/R = 0.6$; we can eliminate R by substituting in the above equation and obtain the energy density of the strings in infinite momentum frame:

$$\rho = (1.2k/R_a) \qquad\qquad (2\text{-}1)$$

Substituting for $R_a = 1.85 \times 10^{-9}$m, the energy density of looped strings in infinite momentum frame at the entrance of the cavity will be 0.53×10^{71} kg/m³ (on a mass basis) or 0.48×10^{88} J/m³. This means that the energy density of the activated looped strings increases from 0.26×10^{36}kg/m³ beyond the opening of the cavity to 0.53×10^{71} kg/m³ just inside the cavity due to the strong gravitational field of the "spacetime cavity". Again, as it will be demonstrated later this field is created by the gravitational force of the mass inside the cavity.

In Chapter 1, I described the energy of an activated three-dimensional string (looped or open) as a quantum specie as:

$$E = h\,c/2\pi r_a$$

In which $r_a = l_p$ at the time of big bang. The energy density of an activated string will then be:

$\rho = E/V$

where $V = 4/3\pi r_a^3$

$$\rho = 3hc/8\pi^2 \, r_a^4 \qquad\qquad\qquad (2\text{-}2)$$

If we set equation (2-1) equal to (2-2), we can calculate the change in the radius of the activated looped string as a function of the radius of the field of infinite momentum frame.

$$3hc/8\pi^2 r_a^4 = (1.2k/R_a)$$

$$r_a^4 = 2.62x10^{-105}R_a$$

$$r_a = 0.71x10^{-26} (R_a)^{1/4} \qquad\qquad\qquad (2\text{-}3)$$

The mass of the activated looped string as a function of R_a using Planck equation and substituting for r_a will be:

$$m = 0.5x10^{-16}/R_a^{1/4} \qquad\qquad\qquad (2\text{-}4)$$

The Energy of the activated looped string as a function of R_a will be:

$$E = h \, c/[2\pi \, (0.71x10^{-26} (R_a)^{1/4}]$$

$$E = 4.45/R_a^{1/4} \qquad\qquad\qquad (2\text{-}5)$$

Equation (2-3) demonstrates the relation between the radius of the activated looped string r_a, as a quantum specie and the field radius of the looped strings of a mass, R_a. This means that unlike activated open strings, the radius of the three-dimensional looped string changes with the energy field radius of the mass in infinite momentum frame. In other words, the dimensions of the looped string inside the spacetime cavity will be smaller than the dimensions of the string outside the cavity. Substituting for the current radius of the universe, $R_a = 1.15 \times 10^{26}$m, r_a, the radius of the activated looped string will be about 2.3×10^{-20}m. The radius of the three-dimensional quantum species of activated looped strings expands from the Planck length of 1.61×10^{-35}m at the nucleus of the spacetime cavity to 2.3×10^{-20}m, a factor of 10^{15}. Equation (2-3) represents the relationship of the radius of activated looped strings vs energy field of a mass anywhere in the universe from the nucleus of spacetime cavity to the edge of the universe. Note that the spacetime cavity essentially simulates the energy densities of the activated strings after the Big Bang but in reverse. The dimension of activated looped strings are recompressed as it is pulled into the nucleus of the spacetime cavity.

Fig. 2-6 is a schematic representation of the recompression of the activated looped strings as a function of the depth of the "spacetime cavity". The dimensions of the activated looped strings are shown on the left and the dimensions of the depth of the cavity on the right.

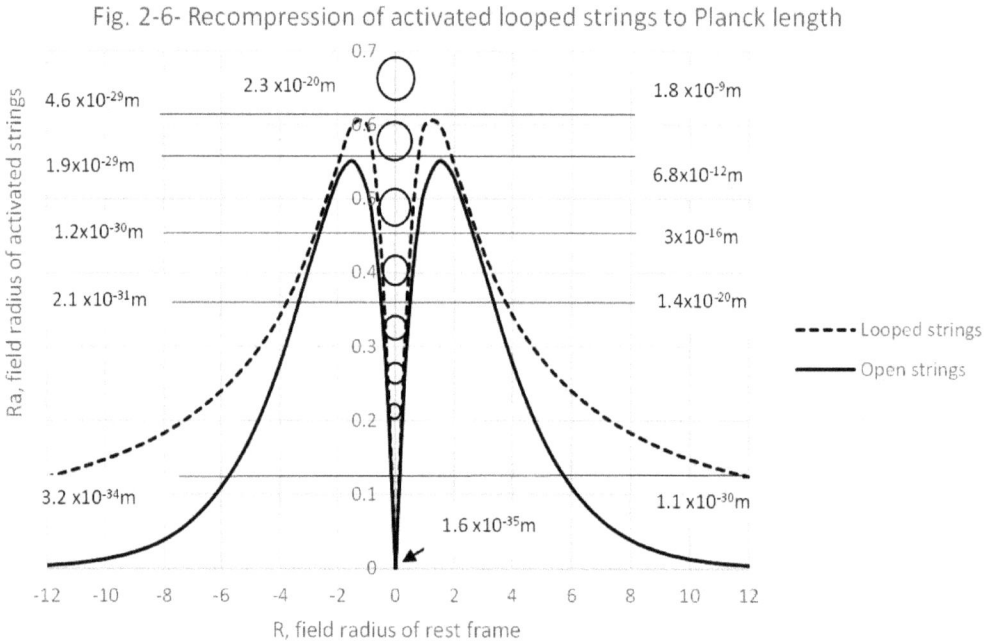

Fig. 2-6- Recompression of activated looped strings to Planck length

The recompression of the dimension of activated looped strings from 2.3×10^{-20}m to its Planck dimension, 1.61×10^{-35}m, inside the spacetime cavity means that the space itself shrinks by the same proportion inside the spacetime cavity. This is because the looped strings make up a significant volume of the space. The shrinkage of space inside the spacetime cavity means that "time" for looped strings increases with a reduction in length. This is in contrast with the "time"/"dimension" relationship outside the space-time cavity, where "time" decreases as we move away from the energy field of a mass. This is where quantum gravity deviates from field gravity of large-scale masses. I will cover this subject extensively in Chapter 11, under "Strings and Time".

We will see this same exact effect inside the event horizon of a blackhole where "time" is dilated as a result of the space shrinkage.

As shown in Chapter 1, since the energy density of activated open strings remains the same, at Planck scale throughout the life of the universe and anywhere in the universe, the dimension of the activated open strings remains the same as at the time of Big Bang or 1.61×10^{-35}m. This means that the dimensions of activated open strings do not change once exposed to the gravitational field of the spacetime cavity or large macro-masses.

Activated three-dimensional strings with opposing spin annihilate each other and decay, converting to two-dimensional strings and releasing their vacuum. This decay is more pronounced inside the spacetime cavity because of the relative increase of the vacuum volume released during this decay which acts to bind the activated strings and immobilize them in restframe1 state. Without the mobility they possess in infinite momentum frame, strings with opposing spin decay faster as the frictional residence time between the strings increase. In fact, the rate of decay and production of released vacuum significantly increases below 1.4×10^{-20}m creating a condition that is restframe1 state.

As I discussed above, the gravitational field of the spacetime becomes stronger as the strings are pulled into the spacetime cavity and towards the nucleus of the matter where the energy density of the strings continues to increase to its maximum which is Planck energy density. From a distance of about 1.4×10^{-20} (the point of intersection of the curve of energy density at restframe and the field of infinite momentum frame) down to the nucleus of the matter, the activated strings are no longer continuously in infinite momentum frame due to the increased decay from a three-dimensional state to a two-dimensional state. From this point to the nucleus of the matter, the strings exist only in restframe state or fluctuate between restframe and infinite momentum frame. The reason for this fluctuation is creation of an extra dimension object I have termed as a "pulsating extra dimension cylinder" which pulsates between a Planck length and 2.3×10^{-51}m, dramatically changing the volume inside the cavity below 1.4×10^{-20}m. I will discuss this extra dimension object in the next chapter in detail. However, the strings between $R_a = 1.61 \times 10^{-35}$m and $R_a = 1.4 \times 10^{-20}$m will all be in restframe1 one moment and in infinite momentum frame the next moment due to the volumetric fluctuation. As such all particles that are formed in this region will have a very short half-life because its state is constantly changing.

The spacetime cavity, which is essentially the housing for the atomic structure and fundamental particles, is also a place for accelerated decay of activated strings. As such, matter becomes a sink for consumption of available activated strings in the universe. Activated strings also decay in infinite momentum frame but this decay is significantly less pronounced because the ratio of vacuum/space volume is significantly lower than inside the spacetime cavity.

CHAPTER 3

Extra Dimension Objects and Their Role in the Universe

Background

BEFORE I BEGIN THIS CHAPTER, I must emphasize that a better understanding and appreciation of the subjects covered in this chapter is realized after completion of Chapter 8 in this book. However, since we must verify the validity of the theory with proven physics in Chapter 4, a brief overview of extra dimension objects becomes necessary to achieve this goal.

In Chapter 1, I began by introducing two three-dimensional strings (activated looped and open strings) and set up the fundamental equations (1-10) representing energy conservation of the three-dimensional strings to a two-dimensional string:

$$d/dt \sum cp_i = d/dt \sum m_i c^2 + d/dt \left[\frac{1}{2} \sum k(\Delta A_j) \right]$$

In this chapter, we will explore the many different forms of two-dimensional strings and the objects they create. The energy term on the far right side of equation (1-10) is the energy expression for these objects which we will pursue in detail.

I described a few of the important characteristics of the three-dimensional strings having a core mass comprised of a two-dimensional string with its mass at the end, simultaneously rotating in two Planes XZ, and XY at the speed of light, c (3×10^8m/s) as shown in Fig. 3-1.

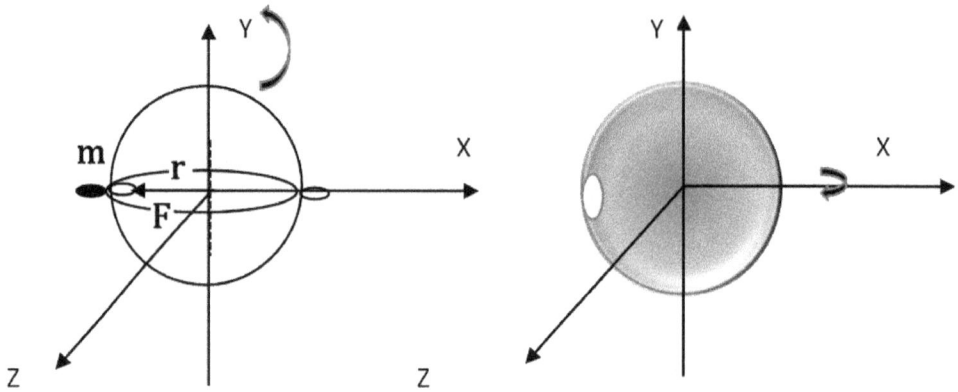

Fig. 3-1 - Conceptual representation of 3D looped or open strings with its axis of orbital rotation and spin. Bold circle represents the mass of the two-dimensional string. On the left, two-dimensional string rotates simultaneously in two planes horizontally and vertically. Vectors, **F** and **r** are on the same line. Inner product of the two vectors forms the Hilbert space, scalar object on the right.

I also described the total energy of three-dimensional strings as the scalar product of the two vectors, the force (mg) and the radius (R) perpendicular to the plane of the rotations (XZ and XY) of the two-dimensional string:

$$E_t^2 = (mc^2)^2 + (P_y c)^2$$

This energy momentum equation will become our guide in constructing the extra dimension objects in this chapter because the energy and momentum of the three-dimensional strings are conserved in its two-dimensional state.

A detailed model for the construction of the three-dimensional strings is presented in Chapter 8. The three-dimensional strings contain a vacuum with Planck volume entrapped by the two rotations of the two-dimensional strings.

The activated looped strings will possess one pole which is an E-loop string with a dimension of 2.3×10^{-51}m. The activated open string will have two poles opposing each other, Fig. 3-2. The linear momentum of the three-dimension string is always perpendicular to the plane of the rotation of the two-dimensional string comprising its core mass.

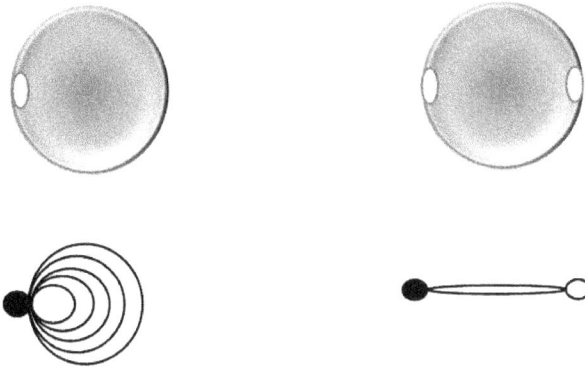

Fig. 3-2- Three and two-dimensional looped string (left) with one pole (E-loop string), open string with two poles (right). Bold circle showing the location of its mass.

The three-dimensional strings will decay when two strings with opposing spins annihilate each other by the friction of the two-dimensional string rotating in the XY plane. When the three-dimensional strings decay, it will be left with a vacuum with the two-dimensional Planck-size string comprising its core inside the vacuum rotating in the XZ plane.

The energy and momentum of the string described above will be conserved and transferred to the two-dimensional string encased by the vacuum. As will be shown later, the vacuum of the string which is about $17.47 \times 10^{-105} \text{m}^3$ will have no specific shape without the conservation of the momentum and energy of the three-dimensional string. Since the momentum and energy are conserved, the vacuum of the decayed three-dimensional string maintains its three-dimensional state by virtue of the movement and kinetic energy of the two-dimensional string inside it creating a "space volume". The two-dimensional string will have two energy components, one is the rotation of the string in the XZ plane which remains unchanged after the string decays, the other is the momentum energy component which is perpendicular to the plane of XZ rotation. The difference in the "space volume" created by the two-dimensional strings and three-dimensional strings is that it contains strings with two degrees of freedom (one rotation, and one lateral movement) reducing the flexibility of its movement as an object. This state of a two-dimensional string inside the vacuum space was referred to as "restframe2" state.

The two-dimensional Planck-size string and its vacuum with Planck volume (restframe2) create many different forms of objects when attached to the three-dimensional strings to create the restframe1 state of the strings. I referred to these objects as **extra dimension objects**. The extra dimension objects are essentially invisible to our universe because the thickness of the strings is far too small (at about 8.9×10^{-160}m, see Chapter 8) as compared to Planck dimensions. The momentum and kinetic energy of the two-dimensional strings make up the "mass" of the "space" it occupies which appears as "vacuum space". These objects are an essential part of our three-dimensional universe when they integrate with the three-dimensional strings, but not visible and measurable by any standard, at least for now. We can prove their existence by their signature effects reliably but as you will see their dimensions are far too small to be measurable for the foreseeable future.

When two Planck-size activated strings of opposite spin rotation decay by the virtue of annihilation, two different extra dimension objects are created. One two-dimensional string creates an extra dimension object in which the mass of the string always remains in one Planck volume of vacuum as it propagates in one direction. Cumulative addition of each Planck volume of vacuum due to the continuous decay of strings create a much larger and longer extra dimension object or space. The energy of the three-dimensional string is converted to propagation of the strings in the new "extra dimension space". I will refer to this one as "high vacuum" or "high-vac" extra dimension. The entire mass of the three-dimensional string is conserved as a two-dimensional string which is then propagated as the string traverses the universe. The second decayed string creates a "low-vac" extra dimension object. In this case, the Planck volume of the vacuum along with the two-dimensional string spreads out across the allowable distance, creating a low-vac extra dimension object. In other words, the Planck volume of the vacuum spreads with the two-dimensional string as it propagates, creating a low vacuum to string ratio. In this case, the predominant portion of the energy of the string is converted to the kinetic energy for displacement of the high-vac object and the rest for propagation of the string itself. However, the two-dimensional strings created by the decay of the three-dimensional strings become entangled, each having a spin opposite the other similar to their parent string. A proposed mechanism for this quantum entanglement process is described later in Chapter 8.

The decay of activated looped strings in infinite momentum frame and outside the spacetime cavity is a special situation which I will cover later because the activated looped strings are significantly larger by about $\times 10^{15}$ than the Planck scale activated strings described above. However, two-dimensional looped strings, regardless of the dimension of its parent three-dimensional strings, will always be in Planck dimensions. As you will see later in this chapter, the force created by the strings in infinite momentum frame will necessitate the two-dimensional strings to only exist in Planck dimension. The three-dimensional open strings have a Planck dimension to begin with and its two-dimensional string begins with Planck dimension and propagates by Planck length.

In classical terms, the strings in a three-dimensional state are in Hilbert space, a vector space with an inner product that is scalar. When converted to two-dimensional state by the virtue of annihilation, it becomes a vector space with a cross product of the same two vectors orthogonal (perpendicular) to the plane of its two vectors, a sort of duality of the strings in the Hilbert space.

The two-dimensional strings and their vacuum housing are in equilibrium with three-dimensional strings in infinite momentum frame. As such, the forces created by the strings inside the vacuum must equal the forces generated by the activated strings in infinite momentum frame, otherwise one state will cause the other to collapse under its dominating force. I will use these "force balancing" criteria throughout this book in obtaining information about these extra dimension objects when it is needed. The energy density, acceleration, and spin orientation of the two-dimensional strings determine the type of extra dimension object that is created each having a significant role in creation of fundamental particles, atoms, and the properties they exhibit.

An important distinction between the restframe2 state, and infinite momentum frame is that the speed of the strings in infinite momentum frame is limited to the speed of light c, while in restframe2, the linear speed of the strings can significantly exceed the speed of light in some extra dimension objects. The linear speed of the activated strings in restframe1 is zero.

As you will shortly see, velocities above the speed of light become a necessity in one restframe2 state where a constant acceleration "g" is required to maintain a "force balance" between different states of the strings which will dictate speeds above the speed of light. From a physics standpoint this is acceptable because some two-dimensional strings exist inside its vacuum with no restrictions and unhindered by the speed of the activated strings which is limited to c. You will see that other extra dimension objects that become attached to activated strings will become restricted to the speed of light c, in our three-dimensional universe.

An important concept to discuss here, before I proceed with describing the key extra dimension objects in our universe, is the conservation of energy momentum which applies to all extra dimension objects. When a three-dimensional string decays and converts to a two-dimensional string, the energy momentum equation described above must also be satisfied in the two-dimensional state:

$$E_t^2 = (mc^2)^2 + (P_y c)^2$$

This means that the term mc^2 will become the mass of the two-dimensional string that rotates at the speed of light c and the term $P_y c$ will be the kinetic energy of the string with an angular momentum which is always perpendicular to the plane of the orbital rotation of the plane of the mass.

When a three-dimensional string converts to a two-dimensional string, its angular momentum is conserved because the two-dimensional string retains its rotation in the plain of its mass (XZ plane), i.e., the plane of the two-dimensional string will continue to orbit around its center. The vectors of the force of the mass and the radius which were parallel with a zero-degree angle and an inner product which created a scalar object in a 3D state will become perpendicular, Fig. 3-3. The cross product of the two vectors then produces a vector (torque) that is perpendicular to the plane of the mass and radius (Chapter 8). This means that the angular momentum of the two-dimensional strings will be perpendicular to its plane of mass.

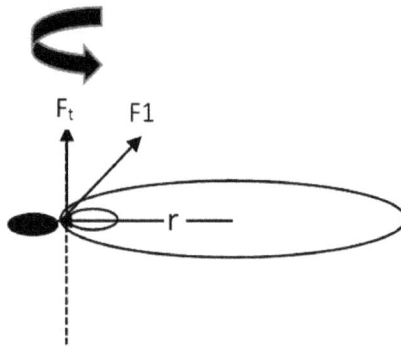

Fig. 3-3- Vector F_1 is perpendicular to **r**. Vector F_t (torque) is perpendicular to the plane of F_1 and **r**.

The magnitude of the torque on a string level will be:

F_t = mgrSin90 =mgr

Where "m" is the mass of the string, "g" the acceleration and "r" the radius of the string's orbital rotation.

As you will soon see, since the strings propagate across a field with a restframe radius of "R", this torque expands to the entire field and becomes:

F_t = m**gR**

As you can see above, the units of torque and energy or "work" are the same except that torque is a vector and has a direction, whereas energy or work is scalar. Since the angular momentum of the two-dimensional strings is perpendicular to the plane of its mass in the same direction as the torque, the magnitude of the torque becomes "work" for displacement of the string the distance of "R".

E= mgR

This energy will become the kinetic energy of the two-dimensional strings in the rest-frame. This means that all two-dimensional objects (and other extra dimension objects) in the universe enjoy the displacement energy of the torque which becomes the energy of its angular momentum "L" which is changing as the mass of the string changes:

F_t = d**L**/dt

Availability of this kinetic energy allows the strings, in some cases, to use its energy (or mass) for propagation of its mass.

As you will see in Chapter 8, the rotation of the plane of the two-dimensional string at the speed of light c, is a necessity for the two-dimensional string to maintain its Planck length. Two-dimensional strings without this rotation will collapse into an E-loop string. Therefore, one common property of all extra dimension objects is that their two-dimensional string must have an orbital rotation at the speed of light as shown above in Fig. 3-3.

As shown above, since the angular momentum and the torque are perpendicular to the strings plane of rotation, a two-dimensional looped string placed inside its vacuum housing will create a cylindrical structure by the virtue of its movement. This will be the fundamental geometric structure of a Planck-size two-dimensional looped string in restframe2 state.

A common property of the vacuum housing of the two-dimensional strings in restframe2 state is that it acts to attract and attach to the activated strings in infinite momentum frame. This attraction is simply driven by Fick's law of diffusion because the concentration of the activated strings as species, inside the vacuum is zero, creating an enormous driving force towards the vacuum. In essence, the stickiness of the vacuum of restframe2 on the one hand helps create the extra dimension objects that are critical to the formation of our universe, and on the other hand immobilizes the activated strings which then decay faster as a result of this stickiness which in turn creates more vacuum.

A unique attribute of the vacuum state of restframe2 is its flexibility, and one might say elasticity, to conform to many different configurations in a medium that is a homogenous mixture of activated three-dimensional looped and open strings. This is particularly more so with two-dimensional open strings because the strings themselves are more adaptable to conforming to the shape of its vacuum. As a result, the number of different extra dimension objects that can be created are enormously large. However, the existence of this huge variety of extra dimension objects is not an impediment to understanding those that play a critical role in our universe. One may liken this to a ship navigating its way through an ocean with countless number of waves. The existence of so many waves is of no consequence if there are a few islands to guide the ship.

Looped string extra dimension objects

I must briefly borrow a few important characteristics of the two-dimensional strings from Chapter 8 to allow me to discuss the properties of these strings from hereon.

As mentioned earlier, two-dimensional looped strings are of Planck length. It contains a surface created by E-loop gauges. This is unlike depictions of looped strings in classical string theory where a line-like open string is looped on itself to create a looped string without a surface. An open string of proper length can create a loop, but it will not be a looped string because it has no surface and does not create "space". It is simply one of many varieties of shapes that an open string with flexible length can create.

The two-dimensional looped string contains a mass which is located near the end (periphery of the loop) and uses it to propagate into many quantized strings of Planck length, Fig. 3-2. The correct mechanism of the storage of the mass of the two-dimensional string will be revealed in Chapter 8.

Two-dimensional looped and open strings described in this chapter are created as a result of the transformation of space from a significantly smaller (E-loop Space) to "Planck Space" which then becomes the foundation of our three-dimensional space.

With this introduction, I will now cover the extra dimension objects that play an important role in construction of our universe.

a. Pulsating extra dimension cylinders

This is the most fundamental type of extra dimension object created when activated three-dimensional looped strings decay into two-dimensional strings inside the space-time cavity.

We will use the energy momentum equation as our guide to construct this extra dimension object:

$$E_t^2 = (mc^2)^2 + (P_z c)^2$$

The first term implies that the mass of the two-dimensional string must be orbiting at the speed of light c, to maintain the two-dimensional string at Planck length. The second term implies that the two-dimensional string must have a momentum energy moving in a direction which is perpendicular to the plane of its mass, creating a cylindrical structure we will call "an extra dimension cylinder". This extra dimension cylinder in the absence of any rotation in a media that is comprised of activated open and looped strings will only see the forces created by the much larger activated looped strings. In other words, the larger activated looped strings which will attach to the vacuum of extra dimension cylinders will shield the extra dimension object from the force of the much smaller (Planck-size) activated open strings. The force of the activated looped string in infinite momentum frame will cause the two-dimensional looped string with a Planck radius to collapse into an E-loop string. We will shortly calculate the dimensions of this tiny, looped string. When the two-dimensional looped string collapses into an E-loop the energy of the mass of the two-dimensional string is conserved and transferred to the E-loop. For example, if an activated looped string is at the nucleus of a spacetime cavity where it has a mass of 2.18×10^{-8}kg and energy of 1.96×10^9J, this energy is transferred to the E-loop which becomes rotational energy of the E-loop. The E-loop in turn attaches to two two-dimensional open strings that are present as a result of the decay of activated open strings which then stretch the E-loop back to its initial Planck-size looped string. As a result, a two-dimensional looped string inside an extra dimension cylinder that has no orbital rotation becomes a pulsating string collapsing from a Planck dimension to its smallest dimension and is stretched back to its Planck length as shown in (Fig. 3-4). This action occurs by a momentary attachment of a two-dimensional open string to an E-loop string which then is released after the string has been stretched back to the Planck length. When the Planck string collapses into an E-loop, a portion of its kinetic energy becomes the energy of a new activated open string which is generated in this process. The pulsation mechanism of the two-dimensional looped string and generation of an activated open string will become evident to the reader after Chapter 8 has been studied.

Fig. 3-4- Pulsating sequence of a two-dimensional looped string from Planck length to an E-loop string and back to Planck length via stretching by two-dimensional open strings.

In essence, in the absence of an orbital rotation of the cylindrical extra dimension object, the two-dimensional Planck-sized looped string collapses to its smallest stable dimension which the universe will allow, the E-loop string. As you will see in the next section, a cylindrical extra dimension object with an orbital rotation will remain intact as a non-pulsating extra dimension object.

Conceptual representation of a pulsating extra dimension cylinder is shown in Fig. 3-5.

Fig. 3-5- Conceptual representation of a pulsating extra dimension cylinder

The extra dimension cylinder created by the pulsating string, is a three-dimensional object with a radius of about 1.61×10^{-35}m moving forward as a pulsating plug in a virtual cylinder. The momentum energy of the torque described above propels the pulsating two-dimensional string moving at the speed of light, c. This extra dimension object momentarily attaches to a two-dimensional open string that spins-off a new three-dimensional open string. As such, its movement is restricted by the speed of the three-dimensional strings at c, 3×10^{8}m/s.

The distance R that the plug moves forward with every pulse will be the same as the radius of the two-dimensional string because the speed at which the string collapses and moves forward is the same, c. Meaning that by the time the string dimension reaches its collapsed state (E-loop), it has moved forward by the same distance:

$R = 1.61 \times 10^{-35}$m

As described earlier, when two Plank-sized activated looped strings decay two pulsating extra dimension objects are created, one is a high-vac, the other low-vac, both entangled. The high-vac pulsating string carries at least one Planck volume of vacuum, 17.47×10^{-105}m which then in combination with a portion of the energy of the pulsating string creates the conditions required to spin-off an activated open string (Chapter 8). The low-vac pulsating string with a spin opposite that of high-vac string propagates in the medium of infinite momentum frame with random availability of bulk vacuum. As such, it will likely not regenerate activated open strings as it pulsates.

Let us first look at the properties of the high-vac pulsating extra dimension cylinder. In the case of the high-vac plug, the mass of the string is conserved and remains inside its own quantized vacuum and is consumed as the source of energy for propagation of the pulsating action and rotation of the E-loop string as discussed above.

The pulsating string moving as a plug creates a pulsating cylindrical extra dimension object with the restframe radius of R. If we consider decay of an activated looped string at the nucleus of the spacetime cavity, the energy of the mass of the string will be 1.96×10^9 J and the energy density of the pulsating extra dimension cylinder will be:

$$1.96 \times 10^9 = \rho_{rf} (\pi l_p^2) R$$

Substituting for R = 1.6Ra, the slope of the activated looped string graph (Fig. 1-10) and rearranging,

$$\rho_{rf} = (0.15 \times 10^{79}/R_a) \qquad\qquad (3\text{-}1)$$

Which is the same as the relationship we obtained for the energy density of an activated three-dimensional string in infinite momentum frame, equation (1-15):

$$\rho = (3k/c^2)(1/R_a - 1/R)$$

Substituting for R = 1.6R_a, and multiplying by c^2, to obtain energy density in Joules.

$$\rho_{imf} = (1.2k/R_a) = 0.29 \times 10^{79}/R_a$$

The contraction and expansion of the looped string in a pulsating extra dimension cylinder creates a g-force. In reality, the looped strings contract to a very small dimension at the speed of light until its contraction velocity becomes zero and then back to its Planck radius at the speed of light. This change in velocity creates the diagonal acceleration of the strings in restframe2. As I mentioned in Chapter 1, since the two-dimensional string is no longer a quanta of the field of infinite momentum frame, we will treat it as non-relativistic. Its Energy will be:

$$E = 1/2mV^2 = F \, xR = mg_{rf}R$$

$$g_{rf} = V^2/2R$$

In this case, since its velocity is restricted by the speed of light c;

$$g_{rf} = c^2/2R \qquad\qquad (3\text{-}2)$$

The pulsating plug moving forward as an extra dimension cylinder has two g-forces, one in a radial direction as explained above to create the pulsating effect (g_1), the other in a lateral direction (g_2, moving forward) creating the cylindrical object as shown in Fig. 3-5.

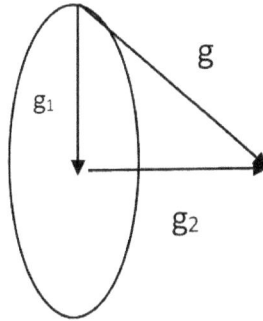

Fig. 3-6 - Vectors of acceleration in a pulsating string

The net g-force produced by the two motions creates the spacetime inside the cylinder:

$$g^2 = g_1^2 + g_2^2 \qquad\qquad (3-3)$$

$$g_1 = c^2/2r$$

g_1, the radial acceleration remains constant since r is the radius of the looped string which remains constant as 1.61×10^{-35}m.

$$g_2 = c^2/2R$$

The lateral acceleration which varies with R the radius of the field of restframe.

$$g^2 = (c^2/2r)^2 + (c^2/R)^2$$

$$g^2 = (c^4/4r^2)(1 + r^2/R^2)$$

$$g = (c^2/2r)(1 + r^2/R^2)^{1/2} \qquad\qquad (3-4)$$

Since the radius of the field of the restframe R is significantly larger than r, the radius of the looped string, the term r^2/R^2 will be a very small number.

We know from Taylor series, the binomial approximation of a small number €, is:

$$(1 + €)^{1/2} = 1 + €/2 \qquad\qquad (3-5)$$

Therefore,

$$g_{rf} = (c^2/2r)\ (1 + r^2/2R^2) \qquad\qquad (3-6)$$

Equation (3-6) is the net gravitational acceleration of the pulsating plug.

We will now consider the "force balance" between an extra dimension cylinder in the restframe2 state that spans the length of the universe and an activated looped string in infinite momentum frame when they collide. The force created by the activated open strings do not impact the extra dimension cylinder in this case, because it is shielded by the much larger activated looped strings. As mentioned earlier, the force of the activated looped string will cause the extra dimension cylinder to collapse to its smallest possible dimension. The two forces in equilibrium must then balance out:

$$F_{imf} = F_{rf}$$

$$V_{imf}\ \rho_{imf} g_{imf} = V_{rf}\ \rho_{rf} g_{rf} \qquad\qquad (3-7)$$

in which,

ρ_{imf} = mass density of the string in infinite momentum frame

g_{imf} = gravitational acceleration of strings in infinite momentum frame (see Chapter 6)

ρ_{rf} = mass density of strings in restframe

g_{rf} = gravitational acceleration of strings in the restframe

V_{imf} = volume of the activated string in infinite momentum frame = $4/3\pi(2.3\times10^{-20})^3 = 51\times10^{-60}\text{m}^3$

V_{rf} = volume of the extra dimension cylinder in restframe = $\pi(1.61\times10^{-35})^2 \times 2.3\times10^{26} \times 2 = 37.44\times10^{-44}\text{m}^3$

As you will see in Chapter 6, $\rho_{imf}g_{imf} = 0.96k/R_a^2$

Substituting all the above in equation (3-7) including the gravitational acceleration of the pulsating plug:

$51 \times 10^{-60} \times 0.96k/R_a^2 = (c^2/2r)(1 + r^2/2R^2)(0.29 \times 10^{79}/c^2 R_a) \times 37.44 \times 10^{-44}$ (3-8)

$k = 0.24 \times 10^{79}$

$c = 3 \times 10^8 m/s$

$r = 1.61 \times 10^{-35} m$

After rearranging:

$R_a = 2.16 \times 10^{-16} r[2R^2/(2R^2 + r^2)]$ (3-9)

Equation (3-9) reveals an important dimension regarding the pulsating string.

If we substitute for $R = r = 1.61 \times 10^{-35} m$, the distance the pulsating plug moves forward, the value of **$R_a = 2.3 \times 10^{-51} m$**. This is the new radius of the Planck-sized two-dimensional looped string when it collapses down to an **E-loop string**. This very small dimension is just one of the many dimensions that comprise the extra dimensions I will cover later in this book.

The **E-loop string** with a radius of $2.3 \times 10^{-51} m$ is the smallest two-dimensional looped string that can exist in the universe as a stand-alone object having no sweeping orbital rotation as in a Planck-size string. The E-loop is the origin of all strings including the activated looped and open strings (Chapter 8). As such it is the most important object in the universe, the source of creation of all matter and energy. I have termed this the "E-loop" string because it is the most fundamental ingredient for "Existence" in the universe.

As I showed earlier, the radius of an activated looped string varies with the radius of the field of activated strings in the universe according to equation (2-3):

$r_a = 0.71x10^{-26} (Ra)^{1/4}$

The frequency of the activated looped strings as a function of the radius of the infinite momentum frame will then be:

$f = c/2\pi r_a$

$f = 3x10^8/2\pi x0.71x10^{-26} (R_a)^{1/4}$

$$f = 0.67\,x10^{34}(1/R_a)^{1/4} \tag{3-10}$$

We can now obtain the relationship between the radius of the field of looped strings in two-dimensional state at restframe and the radius of infinite momentum frame by a simple string for string energy balance.

Energy of an activated string in a three-dimensional state = the energy of the looped string in two-dimensional state after decay:

$$E = hf = (\pi l_p^{\,2})R_s \rho_{rf} \tag{3-11}$$

The left side of the equation represents the energy of a three-dimensional activated string, the right side, is the volume of the extra dimension cylinder multiplied by the energy density of the pulsating two-dimensional string. Where R_s is the field radius of the two-dimensional string at restframe, i.e., the distance it uses the energy to pulsate.

$$6.62x10^{-34} x\, 0.67\,x10^{34}(1/R_a)^{1/4} = (\pi l_p^{\,2})R_s \rho_{rf} \tag{3-12}$$

Substituting for ρ:

$\rho_{rf} = (0.29x10^{79}/R_a)$

$$R_s = 1.88\,x10^{-9}(R_a)^{3/4} \tag{3-13}$$

Equation (3-13) describes the relationship between the radius of the field of the two-dimensional looped strings in the restframe and the radius of the field of infinite momentum frame R_a. If we substitute for $R_a = 1.15 \times 10^{26}$m, the radius of the field of activated looped string in the universe:

$R_s = 0.66 \times 10^{11}$m

This will be the restframe radius of the high-vac pulsating extra dimension cylinder. Recall, earlier in this Chapter I described the quantized nature of the three-dimensional activated strings forming our three-dimensional universe. The two-dimensional strings propagate in restframe only as a quantized species, meaning that as the strings stretch in the restframe it will always be multiples of a single looped string or open string having a radius of 1.61×10^{-35}m. Therefore, the radius of the field of two-dimensional strings is:

$R_s = n \times 2 \times 1.61 \times 10^{-35}$m

If we divide $R_s = 1.2 \times 10^{11}$m obtained above by $2 \times 1.61 \times 10^{-35}$m, we will find the quantized number of pulsating looped strings that comprise the radius of the two-dimensional strings in the restframe:

$n = 0.66 \times 10^{11} / 2 \times 1.61 \times 10^{-35} = 2 \times 10^{45}$

In other words, as the high-vac pulsating string traverses our universe, it propagates to 2×10^{45} two-dimensional strings. This will be the number of high-vac pulsating cylindrical extra dimension strings in the first grouping of such extra dimension objects in the universe. Since we know the field radius of the restframe in the universe is about 2.3×10^{26}m, the number of groupings of pulsating strings beyond the 0.66×10^{11}m radius is:

$2.5 \times 10^{26} / 0.66 \times 10^{11} = 3.8 \times 10^{15}$

Therefore, the total number of quantized strings traversing the universe is then:

$3.8 \times 10^{15} \times 2 \times 10^{45} \approx 8 \times 10^{60}$

In essence, if 8×10^{60} two-dimensional looped strings with a Planck radius are lined up side by side it will have a radius of about 2.3×10^{26}m which is the restframe radius of the universe.

If we begin with a string decay at the nucleus of the spacetime cavity, with a mass of 2.18×10^{-8}kg, the mass of each string at R= 2.3×10^{26}m will be:

$2.18 \times 10^{-8}/8 \times 10^{60} = 2.7 \times 10^{-69}$kg

This means that the final mass of a fully depleted string is about 2.7×10^{-69}kg. However, since the string uses a portion of its mass energy to spin-off new activated open strings in this process, the distance traveled by the string will shorten to 0.66×10^{11}m.

Since two activated strings with opposing spin decay in the process of creation of the pulsating strings, similarly, there will be 8×10^{60} low-vac pulsating extra dimension cylinders in the universe, Fig, 3-7A.

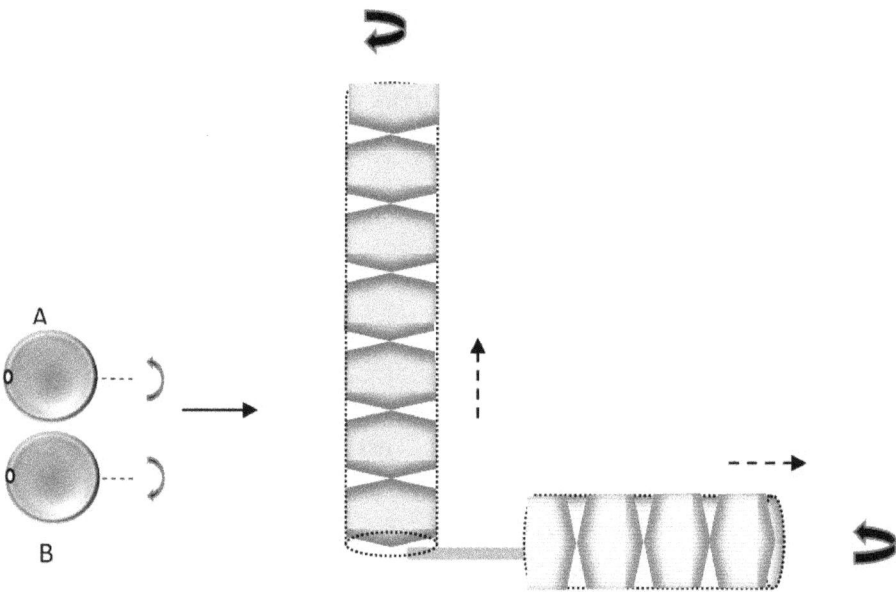

Fig. 3-7A – Conceptual representation of annihilation and propagation of two three-dimensional looped strings into High and low-vac pulsating extra dimension cylinders. The high and low-vac strings remain connected thru E-portals (Chapter 8).

One major distinction between the high-vac and low-vac pulsating extra dimension cylinder is that availability of excess bulk vacuum allows creation of the cylindrical extra dimension path which conserves the energy of the string for propagation and regeneration of activated strings. The low-vac pulsating string on the other hand propagates in a media with random availability of bulk vacuum as such a portion of its energy is consumed due to the hindrance created by the interference of other activated strings.

The high-vac pulsating extra dimension cylinder traverses the space as a wave. The energy for the disposition of this wave is supplied by the low-vac extra dimension cylinder. Therefore, the second important distinction between a high and low-vac extra dimension object throughout this book is that the energy of the low-vac string is substantially used for displacement of the wave created by the high vac extra dimension object. In classical terms, this energy will become the Hamiltonian of the wave function in Hilbert space. I will discuss this in more detail in Chapter 11 and provide some examples for calculation of the Hamiltonian of the wave function.

The high and low energy-vac pulsating strings will have a spin in the same direction as its parent three-dimensional string. Therefore, the high and low-vac strings will have opposite spin rotations. This is an important point to remember because as you will see later, one spin rotation will be responsible for the angular momentum of what will create matter, the other for expansion of the universe which are opposite each other. The process of quantum entanglement of the decayed strings is a critical process for sustaining the universe (see Chapter 8).

Theoretically, a low-vac string should be able to propagate until its mass is depleted to 2.7×10^{-69} kg. If we rewrite the equation for "force balancing" between the activated looped strings and the pulsating strings and substitute for a string mass of 2.7×10^{-69} kg in equation (3-8):

$$51 \times 10^{-60} \times 0.96 k / R_a^2 = (c^2/2r)(1 + r^2/2R^2)\, 2.7 \times 10^{-69} \qquad (3\text{-}14)$$

$$k = 0.24 \times 10^{79}$$

$$c = 3 \times 10^8 \text{m/s}$$

$$r = 1.61 \times 10^{-35} \text{m}$$

$$0.78 \times 10^{37} / R_a^2 = 1 + r^2/2R^2,$$

For the maximum value of the term $1+r^2/2R^2$, R=r;

$R_a = 3.9 \times 10^{18}$m

This is the length of the pulsating, low-vac extra dimension objects in the universe. Which means we have about "n" groups of these to make up the length of the universe at restframe:

$n = 2.3 \times 10^{26}/3.9 \times 10^{18} = 0.63 \times 10^{8}$

Again, if we divide the length of the restframe by Planck length,

$n = 3.9 \times 10^{18}/2 \times 1.61 \times 10^{-35} = 1.22 \times 10^{53}$

The total number of two-dimensional strings covering the restframe radius of the universe will also be:

$1.22 \times 10^{53} \times 0.63 \times 10^{8} \approx 8 \times 10^{60}$

One may contribute the longer distance a low-vac pulsating string travels to the fact that it does not use its energy to regenerate activated open strings as do high-vac pulsating strings.

Pulsating extra dimension cylinders have a few unique attributes. The energy of the collapsed two-dimensional Planck-size string is transferred to the E-loop which then attaches to other two-dimensional open strings and becomes the axis of rotation and one of the energy sources for the orbital rotation of all extra dimension objects in the universe. I will briefly point to the frequency and energy of the E-loops here to make the point and leave the rest of this important discussion to Chapter 8.

The frequency generated by the E-loop when the Planck-size two-dimensional string collapses into an E-loop string is:

$$f_2 = 3 \times 10^{8}/2\pi r = 0.207 \times 10^{59} \qquad\qquad (3\text{-}15)$$

in which $r = 2.3 \times 10^{-51}$m

and its energy will be:

$E = hf = 6.626x10^{-34}x0.2x10^{59} = 1.37 \, x10^{25}J$ (3-16)

As you can see the energy generated by an E-loop is $0.7 \, x10^{16}$ times higher than Planck energy ($1.96x10^{9}$J). This energy will be used for orbital rotation of other extra dimension objects in the universe.

Pulsating extra dimension cylinders are the vehicles for transformation of angular momentum of a three-dimensional string to the field of two and three-dimensional strings when the three-dimensional strings decay. This is because the two-dimensional open strings attach to the "E-loop string" of a collapsed string and use a portion of its rotational energy to create angular momentum for all other two-dimensional objects. This action which creates the orbital rotation of all objects in the universe in restframe is due to a momentary interaction of a pulsating string and a two-dimensional open string.

Again, all of this will be covered in future chapters, but it is important to understand moving forward that the pulsating extra dimension cylinders become the axis of rotation of other extra dimension object that will be discussed from this point forward.

The second important feature of the pulsating extra dimension cylinder serving as the axis of rotation is that the direction of its rotation is based on the spin of the decayed and entangled strings which are opposite of one another. This is critical to the existence of our universe. While one rotation creates an inward angular momentum to create and hold the matter together, the opposite rotation of the E-loop creates the orbital rotation and angular momentum that causes our universe expand.

b. Orbiting extra dimension cylinders

Let us now discuss a second type of extra dimension object, the non-pulsating and **orbiting** cylinder. Similar to pulsating strings, the non-pulsating extra dimension cylinders are comprised of vacuum released by the three-dimensional looped strings containing two-dimensional Planck-size looped strings but do not pulsate as the name implies. The non-pulsating extra dimension cylinders orbit at the speed of light c, using the orbital energy of the pulsating extra dimension cylinders, Fig. 3-7B. In this case, the extra dimension cylinders become exposed to the force created by the activated open strings which is significantly higher than the looped strings. However, this force and the force of the activated looped strings are negated by the String Resistance Forces (Chapter 10) created by the activated looped and open strings. As such, the structure of an orbiting extra dimension cylinder remains intact, i.e., the two-dimensional looped string will not collapse to create the pulsating effect.

Similar to the pulsating extra dimension cylinders, decay of two activated looped strings with opposite rotation creates a high and low-vac orbiting extra dimension cylinder. The high-vac orbiting extra dimension cylinder traverses the space as a wave. The energy for the disposition of this wave is supplied by the low-vac extra dimension cylinder. In classical terms, this energy will become the Hamiltonian of the wave function in Hilbert space.

Before I describe the formation of orbiting extra dimension cylinders, I must jump ahead briefly and describe two-dimensional open strings. Concurrently, activated three-dimensional open strings also decay, and create a two-dimensional open string encased by the vacuum it releases. I will discuss open strings in the next sections in detail, but a brief mention is needed here to understand the nature of the orbiting extra dimension cylinders. As I discussed in Chapter 1, unlike two-dimensional looped strings, the open strings attach themselves to other activated strings which are in turn attached to the vacuum of the extra dimension cylinders.

Three-dimensional open strings upon decay, convert from a spherical geometry to a long and narrow two-dimensional object with two vacuum poles on each end (E-loop strings). The vacuum sites (poles) at each end act to attach the two-dimensional open strings to other extra dimension objects.

The two-dimensional open strings in the restframe then attach to other activated strings (looped or open) which in turn are attached to the vacuum of other two-dimensional looped strings. The two-dimensional open strings which can extend following the $\rho = 2KR$ rule, are attached to the vacuum of the two-dimensional looped string on one end, and to the rotating E-loop of a pulsating string on the other. As a result, the vacuum of a two-dimensional looped string which would otherwise pulsate, will orbit at the speed of light, c, around the axis created by the pulsating extra dimension cylinder, Fig. 3-7B.

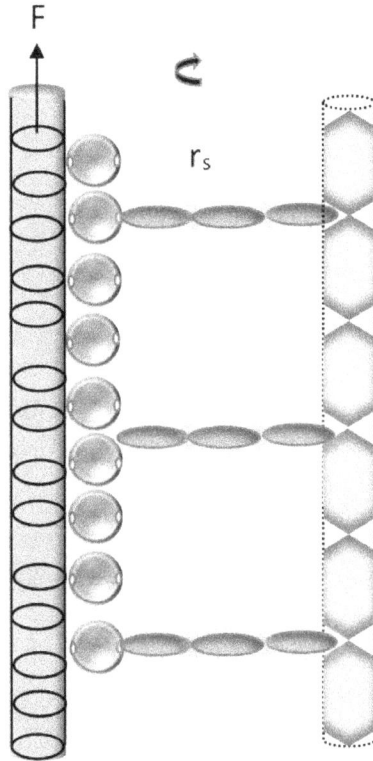

Fig. 3-7B- Conceptual representation of a pulsating extra dimension cylinder as the axis of orbital rotation for an orbiting extra dimension cylinder

The orbital rotation of the vacuum (with two-dimensional looped string inside) changes the balance of force acting on the extra dimension cylinder as described above. As such it will not collapse into an E-loop as did the pulsating string. As you will see shortly, the angular momentum of the rotation provides the kinetic energy of the Planck-size two-dimensional looped strings inside an orbiting vacuum, whose movement creates a three-dimensional cylindrical extra dimension that is non-pulsating and orbiting, Fig. 3-7B. Since each high-vac string carries its own Planck volume of vacuum as it decays, a long, uniform, and cylindrical object is created. In essence, the orbiting extra dimension cylinder becomes a smooth pathway for the exit of the two-dimensional looped strings that are created as a result of the decay of the three-dimensional looped string with no hindrance from other objects in Hilbert space.

Extra dimension cylinders are vector fields where the vectors of the force of the mass of the two-dimensional strings and the radius are perpendicular to each other. Unlike Hilbert space, the cross-product of these vectors produces a torque that is perpendicular to the plane of the two vectors (see Chapter 8).

Again, to create a cylindrical extra dimension object, the two-dimensional string must satisfy the energy momentum equation:

$$E_t^2 = (mc^2)^2 + (P_y c)^2 \qquad\qquad (3\text{-}17)$$

The first term will be the energy of the two-dimensional Planck-size string for sustaining its mass by the virtue of the orbital rotation at the speed of light c, and the second term the momentum energy of the two-dimensional looped string which moves in a direction perpendicular to the plane of the mass. This means that the two-dimensional strings moving inside its vacuum create a three-dimensional cylindrical extra dimension object which will be invisible to our universe. In other words, "the vacuum" will not be able to sustain its "space" which is a three-dimensional cylinder without the energy components described above. If we remove the orbital rotation of the cylinder it will collapse into an E-loop and becomes a pulsating extra dimension object.

The two-dimensional string uses its mass to propagate inside the extra dimension cylinder. The direction of the movement and the energy for propelling the strings inside the extra dimension cylinder is determined by the torque of the orbital rotation the strings. As you will see in the remainder of this section, most extra dimension objects in the universe capitalize on this orbital energy for propulsion and momentum energy, allowing the energy of the string to be used to sustain its Planck dimension throughout the length of the universe.

Let us now look at some of the properties of the orbiting extra dimension cylinders.

The orbiting extra dimension cylinder by the virtue of its vacuum, attracts other activated strings (looped and open) as an attachment, creating the restframe1 state. The activated strings when attached to the extra dimension cylinders, lose their mobility in infinite momentum frame but continue to spin around the extra dimension cylinders until annihilated. In fact, as I mentioned earlier, the rate of decay of strings significantly increases, because the residence time for the contact between two adjacent strings increases as compared to infinite momentum frame. Given that there are equal numbers of activated strings in the universe with opposing spins, the odds of two adjacent strings being of opposite rotation is about 50/50.

In Chapter 6, you will see that the effect of gravitation on the curvature of spacetime occurs on the restframe2 of which orbiting extra dimension cylinders are one of the objects comprising it. The velocity of the activated string in infinite momentum frame, which is normally c, becomes zero on the restframe2, creating the change in momentum, and the gravitational force. This, in addition to the orbital rotation of the activated string which creates the orbital gravity, will comprise the total gravitational effect of the activated strings in restframe which will cause it to curve.

Activated strings attached to the orbiting extra dimension cylinders have two orbital rotations. One is around the cylinder itself created by the spin of the three-dimensional string attached to the vacuum of the cylinder. The other is the orbital rotation of the entire string/cylinder assembly around the axis of the pulsating string as described above. I will cover both of these rotations as we proceed throughout this book. However, when two activated strings of opposite spin rotation attach to the vacuum of an extra dimension cylinder, it must rotate at least once before it decays into a two-dimensional state. In other words, its angular momentum must be equal to or greater than $\hbar/2$ before it decays.

$$L_z = P_z \cdot l_p \geq \hbar/2 \qquad\qquad (3\text{-}18)$$

Similar to pulsating strings, when activated strings decay as a result of annihilation of two strings with opposing spins, two forms of orbiting extra dimension cylinders are formed. One is high-vac, the other is low-vac, Fig. 3-8.

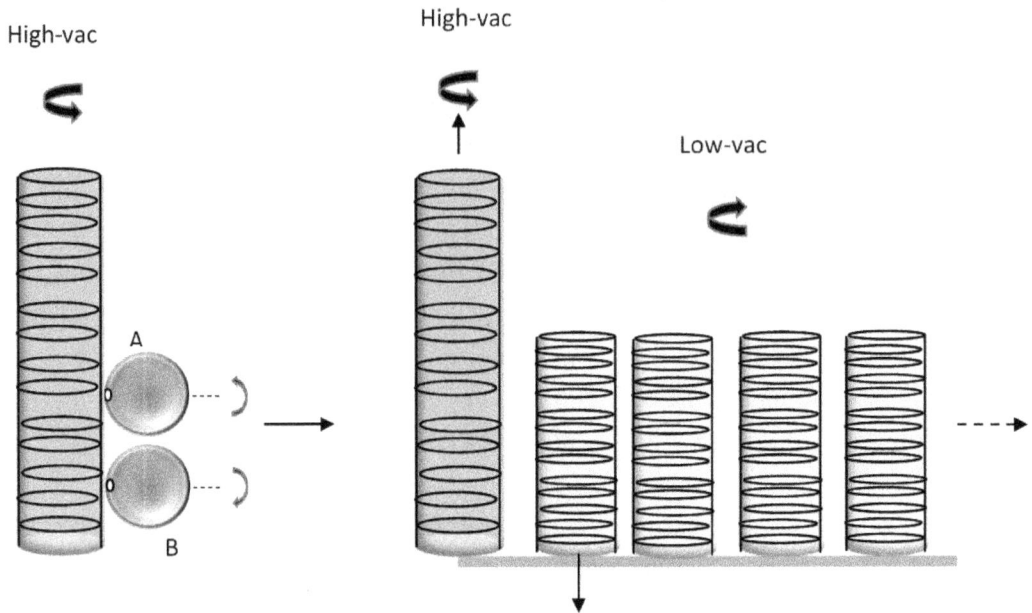

Fig. 3-8 – Conceptual representation of annihilation and propagation of two three dimensional looped strings into High and low-vac orbiting extra dimension cylinders. The high and low-vac strings remain connected thru E-portals (Chapter 8).

The two extra dimension cylinders are entangled (Chapter 8) and will maintain their opposite rotation as two-dimensional strings. The two-dimensional string inside the high-vac extra dimension cylinder will propagate as it moves through the vacuum of the cylinder in one direction and unhindered, while the low-vac cylinder propagates as quanta of multiple strings which then propagate similar to a high-vac cylinder in one direction. The propagation mechanism of both high and low vac orbiting extra dimension cylinders are dominated by E-portals which will be covered in more detail in Chapter 8.

Let us now describe the properties of the extra dimension cylinders. Recall that in Chapter 1, I introduced the energy density of the activated strings in restframe to be:

$\rho = 3K/2\pi R^2$

for both open and looped strings. Now, if we set this energy density to be equal to the energy density of a string as a quantum specie in the restframe, we will obtain the radius of the strings in restframe for both open and looped strings as a function of R_a, the radius of the energy field of a mass in infinite momentum frame:

$$\rho = 3hc/8\pi^2 \, r_s^4 = 3K/2\pi R^2$$

$$8\pi^2 \, r_a^4 = (h/Kc)(2\pi R^2)$$

$$r_s = (h/4\pi Kc)^{1/4}(R)^{1/2}$$

$$r_s = 3.37 \times 10^{-18}(R)^{1/2}$$

Substituting *for R=1.6R$_a$*

$$r_s = 4.26 \times 10^{-18}(R_a)^{1/2} \qquad\qquad (3\text{-}19)$$

Equation (3-19) represents the radius of the strings in restframe for both open and looped strings as a function of R_a, the radius of the energy field of a mass in infinite momentum frame. This radius (r_s) represents the non-quantized value for the orbit of extra dimension cylinders as shown in Fig. 3-9. As can be seen, the orbit of the extra dimension cylinders (r_s) increases with increasing value of R_a. Fig. 3-10 represents the dimensions of the looped or open strings in restframe on the left, and depth of the spacetime cavity on the right.

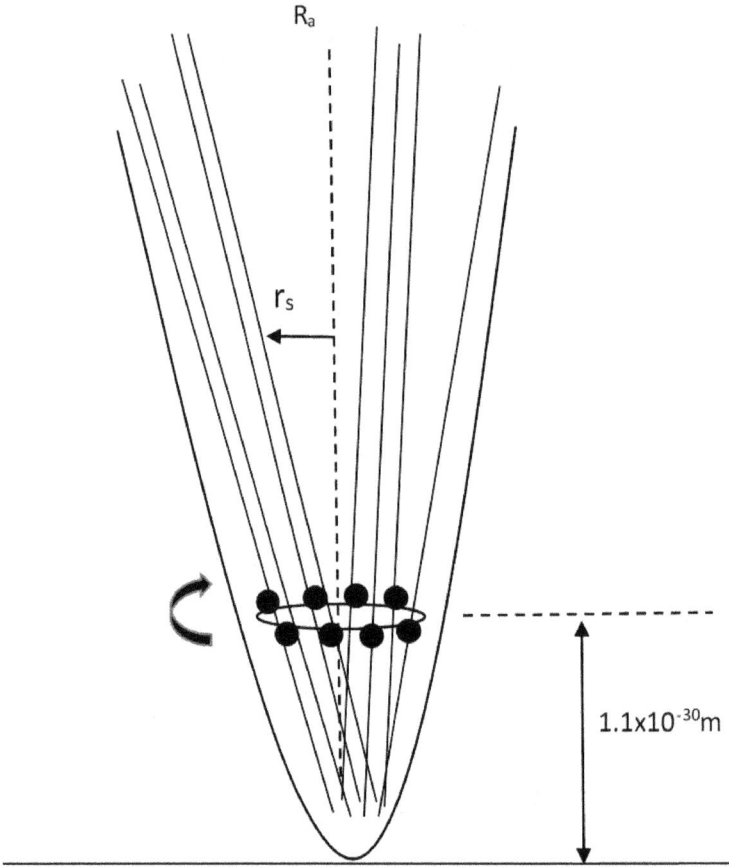

Fig. 3-9- Conceptual representation of orbiting extra dimension cylinders inside the spacetime cavity

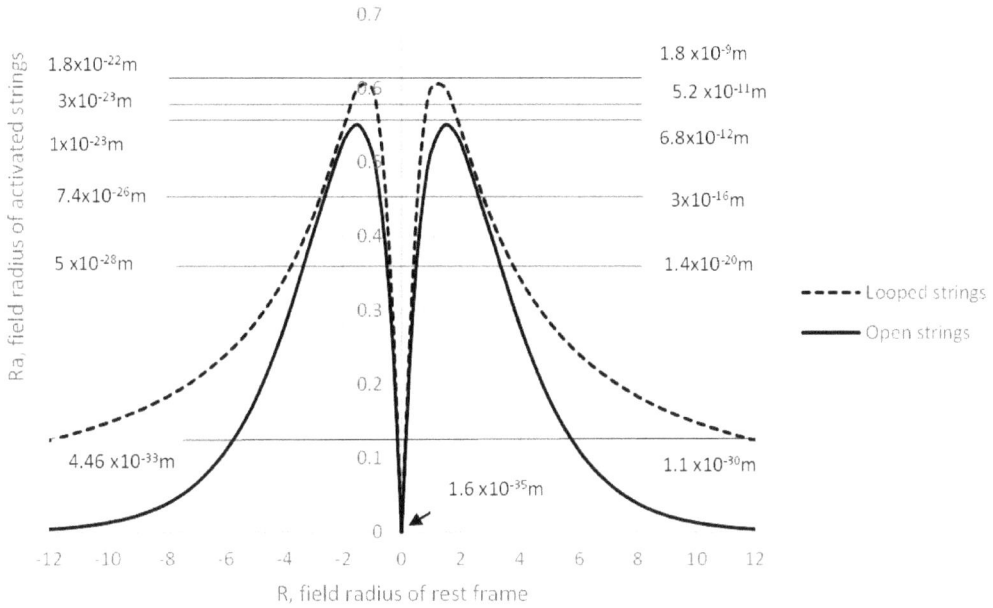

Fig. 3-10- Radius of looped or open strings in restframe2 vs cavity depth

Fig. 3-11 represents a two-dimensional view of the spacetime cavity and orbiting extra dimension cylinders.

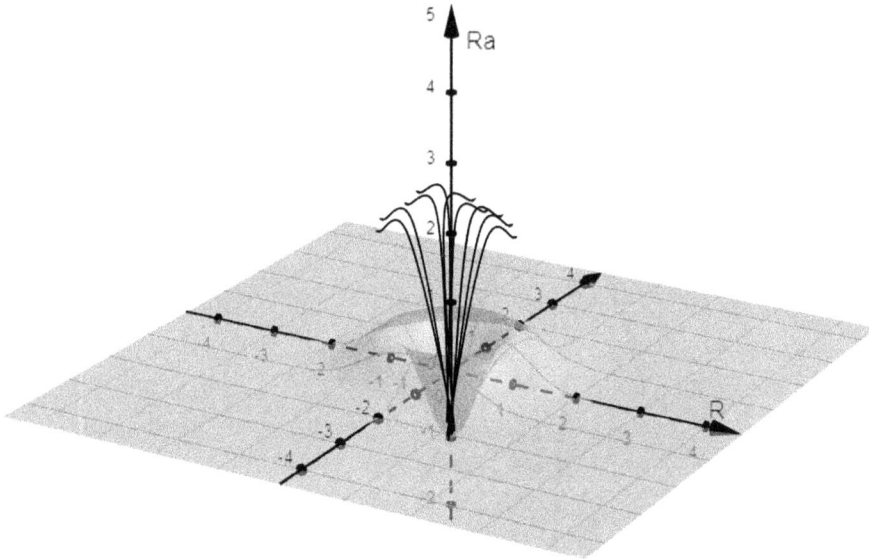

Fig. 3-11 – Conceptual representation of extra dimension cylinders extending out of spacetime cavity.

Table 3-1 displays the radius of strings in restframe for both looped and open strings for a few R_a depths of interest. Note that since in restframe all extra dimension objects orbit, the restframe radius r_s is the same as the orbital radius of the extra dimension cylinders.

Table 3-1

R_a (depth of the cavity), m	r_s, Radius of the string in restframe, m
1.1×10^{-30}	4.46×10^{-33}
1.4×10^{-20}	5×10^{-28}
3×10^{-16}	7.4×10^{-26}
6.8×10^{-12}	1×10^{-23}
5.2×10^{-11}	3×10^{-23}
1.85×10^{-9}	1.8×10^{-22}
1.15×10^{26}	4.6×10^{-5}

As seen from Table 3-1, the radius of the strings in restframe continues to increase as the radius of the field in infinite momentum frame increases. However, since the radius at restframe, r_s, is quantized, we will divide by the diameter of the string ($2 \times 1.61 \times 10^{-35}$) to obtain the number of two-dimensional strings created.

From Table 3-1, if we divide the radius of the looped strings at restframe (4.6×10^{-5}m) when $R_a = 1.15 \times 10^{26}$m, the radius of the field of infinite momentum frame in the universe, by the diameter of the quantized looped string in the restframe, we will obtain the number of non-pulsating extra dimension cylinders:

$$n = 4.6 \times 10^{-5} / 2 \times 1.61 \times 10^{-35} = 1.43 \times 10^{30}$$

This is the number of high-vac orbiting extra dimension cylinders that will reach the edge of the universe with a radius of 1.15×10^{26}m. Since the total radius of the restframe in the universe is $R = 2.3 \times 10^{26}$m, the total number of orbits at $R_a = 1.15 \times 10^{26}$m is:

$$2.3 \times 10^{26} / 4.6 \times 10^{-5} = 5 \times 10^{30}$$

Therefore, the total number of orbiting extra dimension cylinders in the universe will be about

$$5 \times 10^{30} \times 1.43 \times 10^{30} \approx 8 \times 10^{60}$$

Which is the same as the total number of pulsating extra dimension cylinders in the universe.

The total number of high-vac orbiting extra dimension cylinders in the universe as calculated above has an important significance. These are the extra dimension objects with the highest amount of vacuum energy that will extend the entire length of the universe without interruption. As you will see in Chapter 5, these extra dimension cylinders determine the mass of the universe and are associated with "dark matter". The source of these extra dimension cylinders is likely to be the singularity of blackholes.

A high-vac extra dimension cylinder starting at the nucleus of the spacetime cavity will have an energy density of:

$\rho = 2.18x10^{-8}/V$

$V = \pi l p^2 R$

$\rho_{rf} = 0.27x10^{62}/R$

The field of activated open strings which are responsible for the electromagnetic forces in the universe is the dominating force of the universe (see Chapter 7). Let us now consider the "force balance" between the field of activated open strings in infinite momentum frame and the high-vac orbiting extra dimension cylinder forming the restframe2. Since the volume of activated open strings are the same those in the restframe2 state, for any given equal volume of the strings in both states:

$$V\rho_{imf}g_{imf} = V\rho_{rf}g_{rf} \tag{3-20}$$

in which,

ρ_{imf} = energy density of the activated looped string in infinite momentum frame

g_{imf} = gravitational acceleration of strings in infinite momentum frame

ρ_{rf} = energy density of strings in restframe

g_{rf} = gravitational acceleration of two-dimensional strings in the restframe (inside the orbiting cylinder)

As you will see in Chapter 7, $\rho_{imf}g_{imf} = 0.25x10^{114}/ R_a$

Substituting for the values in equation (3-20):

$$0.25x10^{114}/ R_a = (0.27x10^{62}/R) g_{rf} \tag{3-21}$$

$R_a = 0.6R$

$g_{rf} = 1.48x10^{52} m/s^2$

This is the average field gravitational acceleration of the two-dimensional strings in restframe2.

Extra dimension cylinders with their attachments, i.e., activated looped and open strings, behave like a wave. In this book, for the sake of simplicity, I have shown them as straight cylinders. As a wave, it has a frequency and an amplitude which will be briefly discussed in Chapter 11 for a few fundamental particles. The energy for this wave is supplied by the decay of a string which becomes the low-vac extra dimension object (for both looped and open strings). When two oppositely spinning strings decay, the mass (energy) of one remains intact and is used for propagation inside the high vac-extra dimension cylinder. The mass of the other string propagates in a media with many hindrances, however in this process it loses the predominant portion of its energy as the displacement energy of the wave. A smaller portion of its energy will propagate which is calculated from Planck-Einstein relation:

$$m = h/2\pi r_s c \qquad\qquad (3\text{-}22)$$

In which r_s is the radius of the orbital rotation of the extra dimension cylinder to which the decaying string is attached. The energy of the torque created by two-dimensional strings with the above mass orbiting with an extra dimension cylinder will spread across the field of the restframe with the radius R:

$$E = mg_{or}R$$

In which g_{or} is the orbital acceleration of the two-dimensional string in the restframe:

$$g_{ob} = c^2/2r_s$$

Substituting for all the above, the energy of the orbital rotation applied to the extra dimension cylinder is:

$$R = 1.6R_a$$

$$E = (h/2\pi r_s c)(c^2/2r_s)1.6R_a$$

$$E = (hc/4\pi r_s^2)R_a$$

Substituting for r_s from equation (3-19),

$r_s = 4.26 \times 10^{-18}(R_a)^{1/2}$

$E = hcR_a/2\pi(4.26 \times 10^{-18})^2 R_a$

$E = 1.44 \times 10^9 J$

Which means that the extra dimension cylinder to which the activated strings were attached will receive a constant energy of $1.44 \times 10^9 J$ as a result of the orbital rotation of the extra dimension cylinder. Since activated open strings with a constant radius of Planck length attach to the extra dimension cylinders, the cylinders will have yet another orbital rotation around the activated open string which gives them precisely an energy of $1.96 \times 10^9 J$ ($E = hc/2\pi r_a$). This is the largest energy of the two orbital rotations. This energy is the magnitude of the torque (work) I discussed earlier in this chapter and applies to all extra dimension objects in the universe equally including two-dimensional open strings discussed in the next section. It is essentially the conservation of the angular momentum of the three-dimensional string when it is converted to a two-dimensional string.

The above energy is transferred to the two-dimensional strings as kinetic energy of the strings moving inside the extra dimension cylinder in the direction of R_a:

$E = mg_{rf}R_a = 1/2mV^2$

We will obtain:

$g_{rf} = V^2/R_a$

$V = (2g_{rf}R_a)^{1/2}$ 　　　　　　　　　　　　　　　　　　(3-23)

Using the average field acceleration of $g_{rf} = 1.48 \times 10^{52}$ m/s calculated above (for example at $R_a = 1.61 \times 10^{-35}$ m which is at the nuclease of the spacetime cavity), $V = 5.6 \times 10^8$ m which is in the same range as the speed of light, c. At $R_a = 1.4 \times 10^{-20}$ m, $V = 1.4 \times 10^{16}$ m/s, and by the time the two-dimensional strings inside the orbiting cylinder reach the mouth of the spacetime cavity ($R_a = 1.8 \times 10^{-9}$ m), its speed reaches $V = 7 \times 10^{21}$ m/s and at R_a =1.15x10^{26}m (radius of the observable universe) V= 1.4x10^{39}m/s.

As you can see above, the velocity of the two-dimensional looped strings inside the vacuum of the high-vac extra dimension cylinder quickly exceeds the speed of light. This is in contrast to the speed of the pulsating strings which was limited to the speed of light c. The two-dimensional strings in a high-vac orbiting extra dimension cylinder are not hindered from high speed propagation due to the long and organized vacuum path and thus not restricted by the speed of three-dimensional strings in Hilbert space. As you will see later in this book, the high velocity of the two-dimensional strings is critical in maintaining the balance of "mass" and energy inside the spacetime cavity, and in the absence of these orbiting extra dimension cylinders, a spacetime cavity will turn into a blackhole.

The mass of the two-dimensional looped string as it travels through the high-vac orbiting extra dimension cylinder changes as a function of R_a. If we start at the nucleus of a spacetime cavity, the total energy of a two-dimensional string will be the sum of 1.96x10^{9}J, the energy of the string itself, and the orbital energy calculated above (1.96x10^9 J).

$E = 2x1.96x10^9 = mg_{rf}R_a$

Substituting for g_{rf} = 1.48x10^{52}m/s^2,

$m = 2.65x10^{-43}/R_a$ (3-24)

If we substitute for R_a = 1.15x10^{26}m in equation 3-24,

$m= 2.3x10^{-69}$ kg

This will be the mass of the two-dimensional high-vac looped string at the edge of the universe.

The mass of the two-dimensional strings propagating as a low-vac extra dimension cylinder before propagation will be:

$m = h/2\pi r_s c$

In which "m" is the total mass of the two-dimensional string. The quantized number of strings (looped or open) at any given r_s is:

$$n = r_s/2l_p$$

in which l_p is the quanta length of a string or 1.61×10^{-35}m. Since two strings of equal mass decay, the quantized mass of a single string will be:

$$m_s = \tfrac{1}{2}\, m/n = (h/2\pi r_s c)/(r_s/2l_p) = h\, l_p/2\pi r_s^2 c \qquad (3\text{-}25)$$

Substituting for r_s from equation (3-19) in equation (3-22) and (3-25),

$$r_s = 4.26 \times 10^{-18}(R_a)^{1/2}$$

The mass of an un-propagated and propagated single string in a low-vac extra dimension cylinder can be calculated as a function of R_a:

$$m = 0.82 \times 10^{-25}/R_a^{1/2} \qquad\qquad (3\text{-}26\text{A})$$

$$m_s = 3.1 \times 10^{-43}/R_a \qquad\qquad (3\text{-}26\text{B})$$

If we substitute for $R_a = 1.15 \times 10^{26}$m, the mass of a single string propagated as a low-vac extra dimension cylinder at the edge of the universe will also be 2.7×10^{-69}kg. The mass obtained from equation 3-26B is more accurate as it does not rely on the average field acceleration of the strings as in equation (3-24). This means that the mass of a single string regardless of whether it is a high-vac or low-vac extra dimension cylinder is about the same at $R_a = 1.15 \times 10^{26}$m.

For a high-vac extra dimension cylinder, the mass of the two-dimensional string which started at 2.18×10^{-8}kg at the nucleus of the spacetime cavity becomes 2.7×10^{-69}kg by the time it reaches the edge of the universe at $R_a = 1.15 \times 10^{26}$m, which is on the average equivalent to 8×10^{60} single two-dimensional looped string of 2.7×10^{-69}kg. In other words, on the average, a single two-dimensional string of mass 2.7×10^{-69}kg is consumed for every Planck length it travels.

Recall that the mass of a pulsating looped string was also calculated to be 2.7×10^{-69}kg and propagated into 8×10^{60} single two-dimensional strings at the edge of the universe

$(R = 2.3 \times 10^{26} \text{m})$.

The energy density of the low-vac orbiting extra dimension cylinder will then be:

$\rho = m/V$

$V = \pi l p^2 R$

$\rho = m/V = (3.1 \times 10^{-43}/R_a)/ \pi(1.61 \times 10^{-35})^2 R$

$R = 1.6 R_a$

$\rho = 1.9 \times 10^{27}/ \pi R^2$ \hfill (3-27)

Not surprisingly, this is about the same as the restframe energy density of the three-dimensional strings as shown in Chapter 1:

$\rho = 3K/2\pi R^2$

Which also yields,

$\rho = 2 \times 10^{27}/\pi R^2$

Which means that the energy densities of the three-dimensional strings in restframe1 and the two-dimensional strings in restframe2 are equal as they should be since the two states are in equilibrium with each other.

The "force balance" between the field of activated open strings in infinite momentum frame and orbiting low-vac extra dimension cylinders forming the restframe2:

$\rho_{imf} g_{imf} = \rho_{rf} g_{rf}$

in which,

ρ_{imf} = energy density of the activated looped string in infinite momentum frame

g_{imf} = gravitational acceleration of strings in infinite momentum frame

ρ_{rf} = energy density of strings in restframe

g_{rf} = gravitational acceleration of two-dimensional strings in the restframe (inside the orbiting cylinder)

As you will see in Chapter 7, $\rho_{imf}g_{imf} = 0.25 \times 10^{114}/ R_a$

Substituting for the values in equation (3-20):

$$0.25 \times 10^{114}/ R_a = (0.63 \times 10^{27}/R^2)\, g_{rf} \qquad\qquad (3\text{-}28)$$

$R_a = 0.6R$

$g_{rf} = 0.6 \times 10^{87} R$

Since the low-vac strings are instantaneously quantized to Planck length when prop-agated, $R = 1.61 \times 10^{-35}$m,

$g_{rf} = 0.96 \times 10^{52}\, m/s^2$

$V = (2g_{rf}R_a)^{1/2}$

$V = 3.3 \times 10^{8}\, m/s$

Which means that unlike high-vac strings, the velocity of the low-vac strings will be the same as the speed of light in restframe.

As you have seen in this segment, one of the properties of orbiting high-vac extra di-mension cylinders is high speed transport of energy (as a two-dimensional string) in our universe. The speed of the strings inside the high-vac cylinders starts at around the speed of light and accelerates to about 1.8×10^{39}m/s by the time it reaches the edge of the universe. In addition, the low-vac extra dimension cylinder uses a substantial portion of its energy for the orbital rotation of the extra dimension cylinder and kinetic energy of the wave created by the high-vac string extra dimension cylinder (Chapter 10). Only a small portion of its energy is consumed for propagation of two-dimensional strings. The high-vac extra dimension cylinder on the other hand, uses its energy predominately for propagation of two-dimensional strings.

In the above section, I described the two main types of extra dimension cylinders and their energy and momentum characteristics. The extra dimension cylinders come in a variety of shapes and sizes, too many to enumerate. However, there are a few that are of special interest that play an important role in our universe.

The extra dimension cylinders can break up in smaller lengths, rejoin and reconnect to make a variety of shapes and sizes. Fig. 3-12 shows a sample of a few possible configurations.

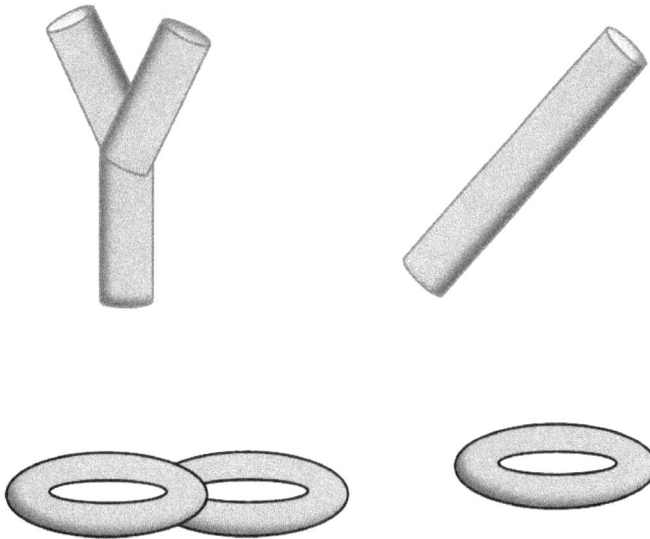

Fig. 3-12 – Examples of the many variety of orbiting extra dimension cylinders that may exist.

c. An important feature of orbiting extra dimension cylinders

In this chapter, we learned that a photon and a charge have length of about 16×10^{-12}m. I demonstrated the structure of these extra dimension cylinders in relation to three-dimensional open and looped strings that attach to its vacuum. As you will see in Chapter 11, these extra dimension cylinders and their attachments exist as waves which is the fundamental property that defines its motion in the universe.

Spacetime cavity of fundamental particles such as proton, neutron, and atoms produce many of these extra dimension cylinders of about the same length which are emitted into the open space of infinite momentum frame. These extra dimension cylinders with their structured vacuum also become an important ingredient of "Distributed Dark Matter" which I will discuss in Chapter 5. In fact, they act like a glue forming the restframe backbone of "Distributed Dark Matter" in the universe. We will call these orbiting extra dimension cylinders "vacuum threads" because they essentially weave the fabric of spacetime. These threads have a diameter of 3.2×10^{-35}m, and a length of about 16×10^{-12}m, but its length may vary depending on circumstances it may have encounter.

Given the importance of these vacuum threads in construction of fundamental particles, dark matter, and that they essentially thread the restframe structure of the universe, it is important to discuss some of its key properties. For example, what is the life cycle of a vacuum thread? This is important because the life span of a photon and distributed dark matter species depend on this object.

The first and most important property of the vacuum thread is that the mass of activated looped or open strings that attaches to its vacuum will decrease by a factor of 4.3×10^{16}. The reason for this substantial change is that the activated strings must rotate at least once around the cylinder based on its energy and momentum. An activated three-dimensional string which has 5 degrees of freedom will become immobilized and have two degrees of freedom, one is its spin, the other is its orbital rotation. The energy consumed to make one rotation when attached to the vacuum of the cylinder is equivalent to the reduction of its mass (its energy reservoir). The velocity of the activated string free flowing in infinite momentum frame is c, and when attached to the vacuum thread, it becomes zero. The mass of an activated string in infinite momentum frame is m_1 and after making one rotation around the extra dimension cylinder it becomes m_2;

$m_2 gr = \Delta E = m_1(c^2 - 0)$

$m_2 \times 3.9 \times 10^{67} \times 2 \, \pi \times 1.61 \times 10^{-35} = m_1 \times 9 \times 10^{16} - 0$

$m_1 / m_2 = 4.3 \times 10^{16}$

In the above equation, $g = 3.9 \times 10^{67}$ m/s² is the acceleration of the mass of the activated string which is stored in its E-portal (see Chapter 8).

This momentum of this string is transferred to the single two-dimensional looped string inside the vacuum thread which has a propagation speed of 3.5×10^{16} m/s (Chapter 8), increasing its velocity. Activated open and looped strings (when attached to the vacuum of extra dimension cylinder has a constant momentum because the radius of its rotation is constant at Planck length.

$P = h/\lambda = h/2\pi r = \text{constant}$

$P = m_1 V_1 = m_2 V_2$

$V_2 = V_1 (m_1/m_2)$

$3.5 \times 10^{16} \times 4.3 \times 10^{16} = 15.2 \times 10^{32}$ m/s

As seen above, by the virtue of attachment of a three-dimensional open or looped string to an orbiting extra dimension cylinder or vacuum thread, the mass of the string is reduced by 4.3×10^{16}, and the velocity of the two dimensional string inside the extra dimension cylinder is increased by 4.3×10^{16}.

The increase in the velocity of the two-dimensional looped string inside the vacuum thread is significant because it provides the necessary speed for a Planck size two-dimensional string to sweep across the length of the vacuum thread (16×10^{-12}m) within the timeframe a two-dimensional Planck string will dissipate. The dissipation time of a two-dimensional string equals one full rotation:

$t = 2\pi r/c = 2\pi \times 1.6 \times 10^{-35}/3 \times 10^8 = 3.37 \times 10^{-43}$s

As you will see in Chapter 11, time is about 5 times faster in the two-dimensional stage than in the three-dimensional universe. The velocity required for one string to travel the length of the extra dimension cylinder within the dissipation time of a two-dimensional string is:

$$V = 16 \times 10^{-12}/(3.37 \times 10^{-43})(1/5) = 2.3 \times 10^{32} m/s$$

As you can see from above we obtained a velocity of $15 \times 10^{32} m/s$, just about the right velocity for one two-dimensional string to sweep across the length of a vacuum thread and sustain its cylindrical structure.

So, why is this so important? This means that a vacuum thread can conserve its energy in the most efficient way by consuming only one two-dimensional looped string per Planck time cycle to sustain its cylindrical vacuum shape. When a vacuum thread is detached from an atom, proton, or a neutron, it will have a mass of $2.2 \times 10^{-8} kg$ in its two-dimensional string and is capable of spinning 8×10^{60} Planck scale two-dimensional strings (Chapter 8). This means the life cycle of a vacuum thread can be as long as 87.5 B years:

$$3.37 \times 10^{-43}s \times 8 \times 10^{60} = 2.7 \times 10^{18} s = 87.5B \text{ years}$$

Open string extra dimensions

A detailed model of the two-dimensional looped and open string is provided in Chapter 8. However, I must briefly point out a few important characteristics of the two-dimensional open strings that I will cover from hereon.

Unlike looped strings, when a three-dimensional open string decays, its structure converts into a narrow and long string, as shown in Fig. 3-13A. The string will retain its Planck length and has a width of about $1 \times 10^{-66} m$ which is 10^{31} times smaller than its length. Similar to looped strings, it will have a thickness of about $8.9 \times 10^{-160} m$, hence it is considered two-dimensional. The two-dimensional open string does not have a surface or surface gauges. Using its stored mass which is also located at the end of the string, it can stretch end over end to long distances, quantized by Planck length.

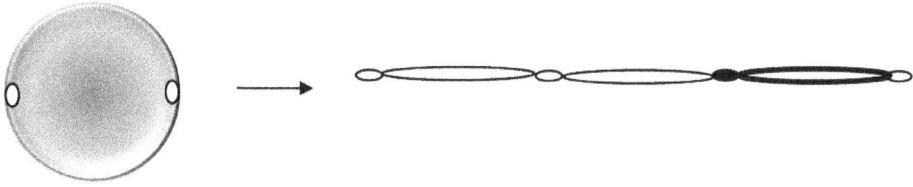

Fig. 3-13A - Conceptual representation of three dimensional open string converting to two-dimensional open string(s). Bold string represents the mass of 2D strings.

The E-poles of the three-dimensional string will remain as the E-loop strings at the end of each string as a point of vacuum attachment to other objects and surfaces.

Two-dimensional open strings remain in the Hilbert space by the virtue of attachment from both ends to activated strings.

Since the mass of the string is at the end of the chain of orbiting strings attached to pulsating extra dimension objects or other activated open strings, the vector of its force **F** will be on the same line and direction of its radius vector of orbital rotation, $\mathbf{r}_s$ when the string is propagating. Therefore, the inner product of the two vectors is scalar:

$$\mathbf{F}.\,\mathbf{r}_s = \text{scalar}$$

As such, the two-dimensional open string, similar to its parent three-dimensional open string, remains in Hilbert space, Fig. 3-13B. This means that the propagation of two-dimensional strings will be limited to the speed of light, c.

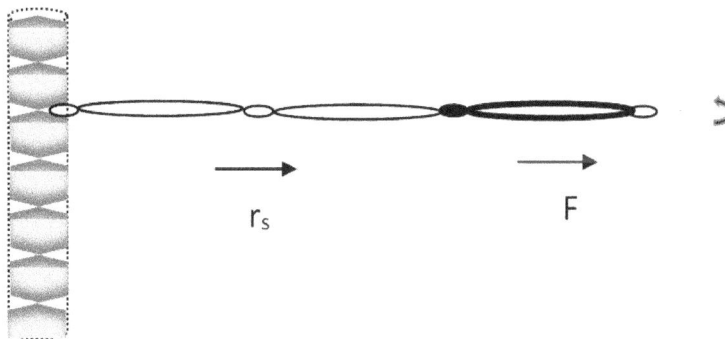

Fig. 3-13B - Conceptual representation of two-dimensional open string(s), attachment to a pulsating string, and orbital rotation of its mass.

The two-dimensional Planck-size open string is an object, unlike a one-dimensional line presented in many other descriptions of classical string theory. It is a narrow object with a fixed surface area, although it does not have a surface, or surface gauges like a two-dimensional looped string. Open strings with multiple of quantized Planck length can loop, but it will not be a looped string.

A line-like, one-dimensional open string as depicted in some descriptions of classical string theory cannot physically exist in the universe because it does not create "space".

The two-dimensional looped and open strings described in this book (Chapter 8) are created as a result of the transformation of space from a significantly smaller space (E-loop Space) to "Planck Space" which then becomes the foundation of our three-dimensional space.

As shown in Fig. 2-3, the field of open strings ends at R =12. If we draw a vertical line from this point where the field of open strings at restframe ends, we will find the coordinate of this point on the axis of infinite momentum frame to be $R_a = 1.1 \times 10^{-30}$m, Fig. 2-3. This means that below this coordinate, inside the spacetime cavity, open strings do not exist in restframe2 as a stable specie (because the coordinate marks the end of the field of restframe). In other words, as the three-dimensional activated open strings convert to two-dimensional open strings, they recombine with their vacuum to recreate the three-dimensional string in infinite momentum frame. This phenomenon occurs with the aid of high energy density pulsating extra dimension cylinders described above, harnessing the rotational energy of the E-loops and spinning the two-dimensional strings back into the three-dimensional state (See Chapter 8). This means that stable species of two-dimensional strings begin to exist above the coordinate of 1.1×10^{-30}m inside the spacetime cavity.

Fig. 2-3 Curvature of spacetime created by subatomic mass "M"

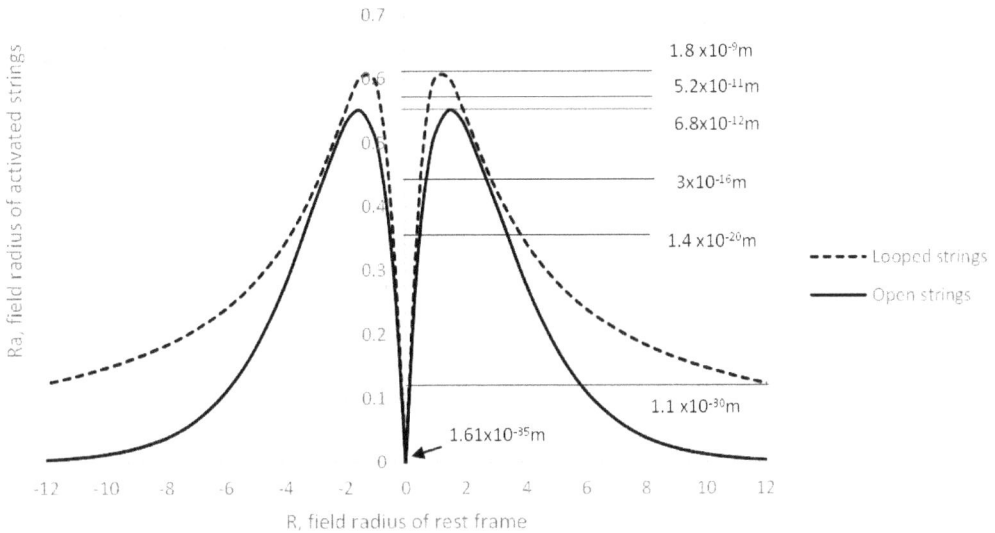

With this introduction, I will now cover the open string extra dimension objects that play an important role in construction of our universe.

Similar to activated looped strings, activated three-dimensional open strings decay into a two-dimensional string by the same mechanisms as described earlier. This includes the friction of two oppositely spinning activated strings causing annihilation of one another. When an activated open string is converted to a two-dimensional string, the vacuum inside the three-dimensional sphere of the string is exposed which becomes the housing for the two-dimensional open string. The two-dimensional string with conserved energy and momentum from the three-dimensional state will create the backbone for restframe2 state and the foundation of the extra dimension objects created by the open strings. These extra dimension objects like the pulsating and non-pulsating extra dimension cylinders created by the looped strings, are themselves three-dimensional structures that are in equilibrium with the energy densities of the three-dimensional activated strings in infinite momentum frame and restframe1.

Similar to the looped strings, the number of configurations that the two-dimensional open strings create in the universe is enormous, but a few of these extra dimensions are important to the creation of our three-dimensional universe. The equation describing the relationship between the radius of two-dimensional open strings in restframe2 vs

R_a the radius of the energy field of a mass in infinite momentum frame equally applies to open strings since when deriving this equation, I did not specify the type of string (3-19):

$r_s = 4.26 \times 10^{-18}(R_a)^{1/2}$

The table below shows the radius of the strings in restframe for a few different values of R_a:

R_a (depth of the cavity), m	r_s, Radius of the string in restframe, m
1.1×10^{-30}	4.46×10^{-33}
1.4×10^{-20}	5×10^{-28}
5.1×10^{-15}	3×10^{-25}
6.8×10^{-12}	1×10^{-23}
5.2×10^{-11}	3×10^{-23}
1.85×10^{-9}	1.8×10^{-22}
1.15×10^{26}	4.6×10^{-5}

A single activated open string when converted to a two-dimensional string can stretch the entire length of the universe quantized by Planck length:

$8 \times 10^{60} \times 2 \times 1.61 \times 10^{-35} = 2.5 \times 10^{26}$m

Similar to looped strings, annihilation of two activated open strings with opposing spin rotation will create two sets of two-dimensional open string extra dimension objects, which are entangled.

However, the two-dimensional strings do not create a high and low-vac extra dimension object similar to looped strings. Both decayed strings propagate in opposite direction while attaching to other three-dimensional activated open, looped, and pulsating strings.

The most basic extra dimension objects created by an open string are shown in Fig. 3-14. The strings begin with a Planck mass, inside a Planck volume of vacuum, and continue to stretch the Planck volume of the vacuum as the mass of the string is propagated.

The three-dimensional extra dimension objects created by two-dimensional open strings are relatively flat and narrow. Unlike two-dimensional looped strings which have a surface and can create a three-dimensional cylindrical structure, open strings do not create a distinguishable object except for two inverted cone-like structures which I will discuss shortly.

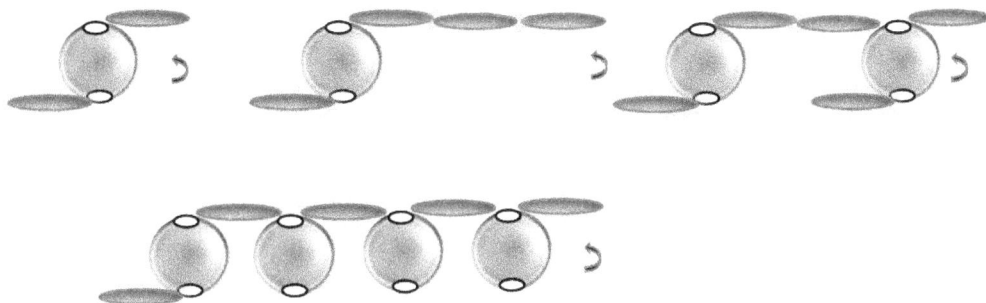

Fig. 3-14 – Examples of different modes of attachment and propagation of 2D open strings to activated open strings.

Let us now look at some of the general characteristics of this extra dimension. The open strings when converted to a two-dimensional string attach to the vacuum of the E-pole of the three-dimensional activated open or looped strings. As I mentioned earlier, the E-pole is the smallest looped string in the universe with a radius of 2.3×10^{-51}m. Since the dimension of an activated open string does not change, the dimension of the E-pole remains the same. The dimension of the activated looped strings increases from a radius of Planck length 1.61×10^{-35}m to about 2.3×10^{-20}m, a factor of 1.42×10^{15}. This means that the radius of the E-pole on the activated looped string outside the spacetime cavity increases to about:

$$1/2 \times 2.3 \times 10^{-51} \times 1.42 \times 10^{15} = 1.6 \times 10^{-36}m$$

Which is about one tenth of Planck length.

Two-dimensional open strings are attracted to the vacuum port of the E-pole where it attaches to the surface of the three-dimensional open or looped strings. But in the case of activated open strings, this attachment is transient, and the string can move from one activated string to the next by attaching and releasing from one E-pole to the next. Therefore, while the open strings rotate with the spin of the string to create a cylindrical structure as it propagates, it also moves or jumps from string to string to create its lateral movement. The driving force for this lateral movement is the torque created by the orbital rotation of the strings.

In the case of activated looped strings, the attachment becomes more permanent because the size of the E-pole has increased significantly (1.6×10^{-36}m) creating more surface area and vacuum for attachment of two-dimensional open strings to its surface. As you will see later in this section, this attachment creates a heavier mass than the string attachment to the activated open string.

The main difference between the attachment mechanism of an activated open and looped string is that an open string has two E-poles on opposite sides vs looped strings which have only one E-pole. Attachment to the E-poles of an activated open string means that there will always be unequal attachment tensions on an activated open string because the attachment on either side is in turn attached to other moving and extra dimension objects with unequal pull. This unequal tension along with the small dimension of the E-pole vacuum will cause the attachment to break off easily. In an activated looped string, there is only one E-pole, therefore there will be less tension on the attachment along with a significantly larger surface area for attachment.

Unlike activated looped strings, two-dimensional open strings cannot pulsate because two-dimensional open strings immediately attach themselves to other spinning activated open or looped strings which will create the orbital rotation regardless of whether they are attached to an orbiting extra dimension cylinder. Therefore, as a general rule, two-dimensional open strings are always orbiting either by the virtue of attachment to the E-loop of a pulsating extra dimension cylinder, or an orbiting extra dimension cylinder, or an activated string.

In general, two-dimensional open strings propagate according to equation (3-19), regardless of whether it is attached to an orbiting extra dimension when it decays or away from an extra dimension cylinder:

$r_s = 4.26 \times 10^{-18}(R_o)^{1/2}$

The energy and momentum of the three-dimensional string is conserved and converted to the kinetic energy of the two-dimensional string inside its vacuum bubble. The general movement of the string must satisfy the energy momentum equation.

$$E_t^2 = (mc^2)^2 + (P_y c)^2$$

The two-dimensional open string has two energy components that are conserved. One component is its kinetic energy, the other, its mass. Similar to the looped strings, the two-dimensional open strings use the angular momentum of the orbital rotation, which is constant, as its source of kinetic energy, $(P_y c)$. The direction of this momentum will have to be perpendicular to the plane of rotation comprising the mass of the two-dimensional string, Fig. 3-15. This leaves the energy of the string to be used for the propagation of its mass. The mass of the two-dimensional open string will propagate as it moves through infinite momentum frame creating the orbital radius for the orbiting extra dimension cylinder and other extra dimension objects. Its mass will vary by:

$$m = h/2\pi r_s c$$

where, r_s, the radius of the restframe is also the radius of its orbital rotation. Similar to the looped string, the quantized mass of a single string as a function of R_a will be:

$$m_s = 3.1x10^{-43}/R_a$$

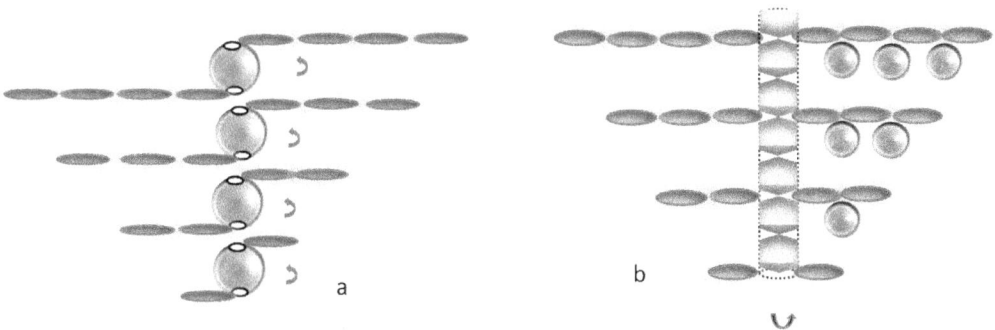

Fig. 3-15 – Conceptual representation of propagation of 2D open strings.
a) Attachment of 2D strings to activated open string, b) attachment of 2D strings to pulsating extra dimension cylinder.

Fig. 3-16 represents propagation of two decayed activated open strings in opposite direction, creating two inverted cone like structures.

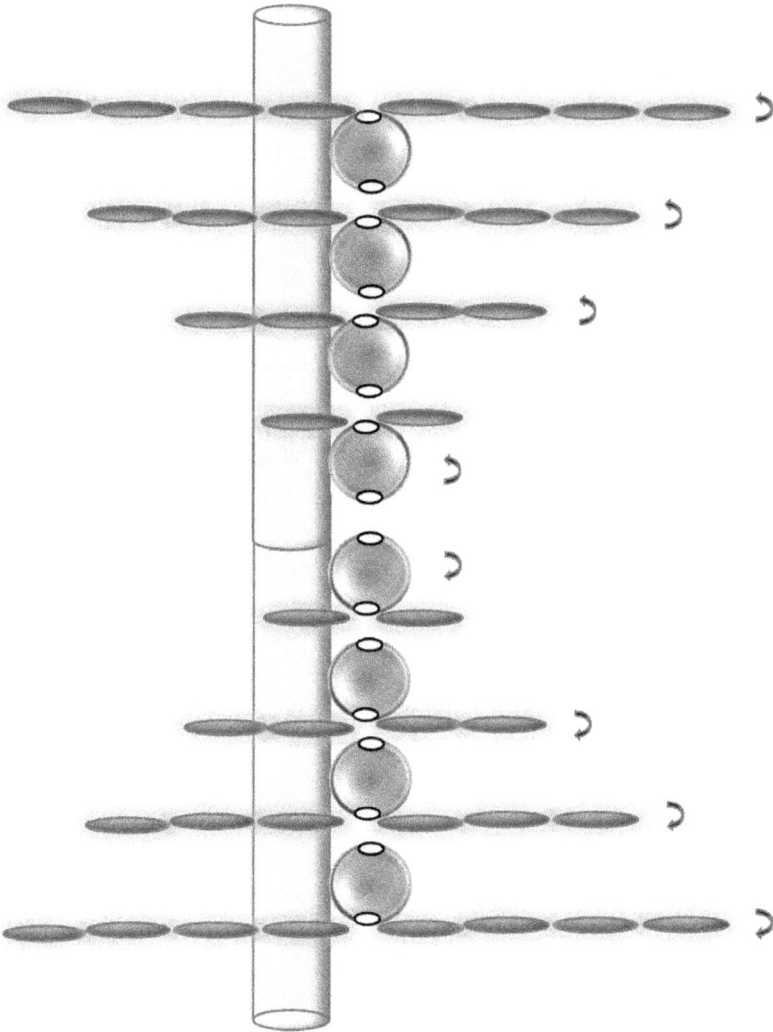

Fig. 3-16 – Conceptual representation of decay of two open strings with opposite rotations and propagation of 2D open strings in a conical shape.

The volume of the cone created by each decayed string is:

$$r_s = 4.26 \times 10^{-18}(R_a)^{1/2}$$

$$r_s = 3.37 \times 10^{-18}(R)^{1/2}$$

$V = (1/3)\pi r_s^2 R$

$V = 11.88x10^{-36}R^2$

Since the mass of the activated open string is constant regardless of its location along the length of the extra dimension cylinder, the energy density of the cone will be:

$\rho = 2.18x10^{-8} / 11.87x10^{-36}R^2$

$\rho = 1.7x10^{27}/R^2$ \hfill (3-31)

This is the field energy density of two-dimensional open strings which is the same as the field energy density of low-vac orbiting extra dimension cylinders.

a. Magnetic Monopole

Attachment of a two-dimensional string to activated looped and open strings at each end

A special case of the two-dimensional open strings occurs inside the spacetime cavity of an atom or fundamental particle. As mentioned earlier, open strings begin to form at a length of about $4.4x10^{-33}$m and at a depth of about $1.1x10^{-30}$m inside the spacetime cavity. Availability of a preponderance of two-dimensional strings with a length of $4.4x10^{-33}$m results in the formation of three important extra dimension objects, photons, charge, and magnetic monopoles as I will discuss below in detail.

The open string extra dimension object described above, having a length of $4.4x10^{-33}$m with Planck mass and volume will attach to two activated strings from each end. If one end is attached to an activated open string and the other to an activated looped string, a new extra dimension object is created. This extra dimension object is a "**magnetic monopole**", Fig. 3-17. This extra dimension object is very important in many ways and exists in both infinite momentum and restframe as a specie of its own. However, likely, monopoles will attach to one or two extra dimension cylinders as they move about in infinite momentum frame.

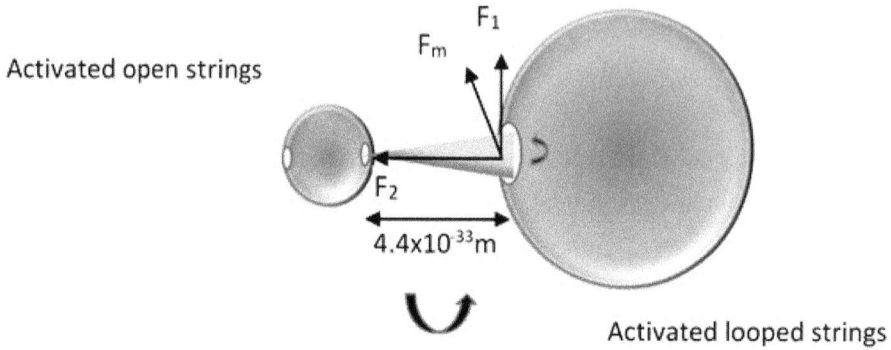

Fig. 3-17 – Conceptual representation of a magnetic monopole.
F_m is the magnetic force a geometric average of F_1 and F_2.

As I mentioned earlier, the E-pole of the three-dimensional activated looped strings grows with the expansion of the string. Recall that the activated looped string expanded from its Planck dimension to 2.3×10^{-20}m outside the spacetime cavity. Inside the cavity, at around the depth of 1.61×10^{-35}m, the size of the pole is the same as the pole on the activated open string or 2.3×10^{-51}m. The size of the E-pole grows to about 1.6×10^{-36} m outside of the spacetime cavity with the expansion of the three-dimensional looped string. The mass of the activated looped string is reduced as a result of this expansion from 2.18×10^{-8}kg to 1.3×10^{-23}kg.

The two-dimensional open strings attached to the pole of an activated looped string have a stronger attachment partly because of the significantly larger surface area for attachment than the activated open string. As shown in Fig. 3-17, two forces are created as a result of orbital rotation of the magnetic monopole. One is as a result of the mass of the two-dimensional open string attached to the E-pole of the activated looped string orbiting the E-pole of the looped string, F_m, and the other is as a result of the orbital rotation of the activated open string around the looped string with an orbital radius of R, F_2.

The force F_m is the magnetic force of the monopole. Its direction can be up or down depending on the spin rotation of the two-dimensional string that attaches the two activated strings. Using the right-hand rule, if the two-dimensional open string orbits counterclockwise with respect to the looped string as the center, it creates a force that is up or vice versa.

Similar to the two-dimensional string attached to the activated open string as described in the previous section, the length of the string is 4.4×10^{-33}m. This means that we have 273 single strings making up the length of the string attaching the activated open and looped strings:

$4.4 \times 10^{-33} / 1.61 \times 10^{-35} = 273$

The mass of the single string attached to the activated looped string will then be:

$2.18 \times 10^{-8} / 273 = 0.8 \times 10^{-10}$kg

In a monopole, when the mass of the two-dimensional string is added, 0.8×10^{-10}kg is significantly heavier than the mass of the looped string it attaches to (1.3×10^{-23}kg) by a factor of about 10^{12}. Therefore, we observe this energy change as a massive particle. As you will see in the next segment, the same string mass is attached to an activated open string forming the structural component of a photon. Why is the photon massless but a monopole significantly heavier? The answer is that the mass of the two-dimensional string, 0.8×10^{-10}kg is smaller than the mass of an activated open string, 2.18×10^{-8}kg. When added, this change in energy is not noticeable.

As you will see in the next chapter, the baryonic mass of fundamental particles we observe is merely the change in its energy not the cumulative masses of the components comprising it.

The acceleration of this mass attached to the pole of the activated looped string will then be:

$g_1 = c^2/2r = 9 \times 10^{16} / 2 \times 1.6 \times 10^{-36} = 2.8 \times 10^{52}$m/s^2

In which r is the radius of the E-pole of the looped string. The force generated by the monopole will be:

$F_1 = 2.8 \times 10^{52} \times 0.8 \times 10^{-10} = 2.2 \times 10^{42}$ N

As the dimension of the activated looped strings become smaller around and inside the spacetime cavity, the diameter of its E-pole becomes smaller, which will cause an increase in the above g_1, and the force created by the monopole. As you will see later, this force is the fundamental force behind the electromotive force of a magnetic field.

One can calculate the magnetic force by substituting for r, the radius of the E-pole of the activated looped string based on its dimension. For example, in the upper region of the spacetime cavity where the dimension of an activated looped string is about 1.9×10^{-29}m, the radius of the E-pole, the value of "g" and F_1 will be:

$$r = 2.3 \times 10^{-51}(1.9 \times 10^{-29}/1.61 \times 10^{-35}) = 2.7 \times 10^{-45} m$$

$$g_1 = c^2/2r = 9 \times 10^{16}/2 \times 2.7 \times 10^{-45} = 1.66 \times 10^{61} m/s^2$$

$$F_1 = 1.3 \times 10^{51} N$$

Referring to Fig. 3-17, the radius of the activated open string orbital rotation, R, in general, is constant at 4.4×10^{-33}m. The acceleration of the orbital rotation of the activated open string is then:

$$g_2 = c^2/2r = 9 \times 10^{16}/2 \times 4.4 \times 10^{-33} = 1 \times 10^{49} m/s^2$$

$$F_2 = 1 \times 10^{49} \times 2.18 \times 10^{-8} = 2.18 \times 10^{41} N$$

The magnitude of the force F_m, and g_m outside the spacetime cavity is the geometric average of the two vectors:

$$F_m = (2.18 \times 10^{41} \times 2.2 \times 10^{42})^{1/2} = 6.9 \times 10^{41} N$$

And

$$g_m = (g_1 \times g_2)^{1/2} = (2.8 \times 10^{52} \times 1 \times 10^{49})^{1/2} = 5.29 \times 10^{50} m/s^2$$

The magnitude of the force F_1 is about 10 times higher than F_2. You will see shortly that this creates a charge 10 times more than F_2. Therefore, attachment of a two-dimensional open string to an activated looped string creates a heavy carrier of the force which will be the magnetic force in the universe.

The energy density of the cone of the two-dimensional string attached to the open string on one end and to the looped string on the other is:

$$\rho = 2.18 \times 10^{-8} / 17.7 \times 10^{-105} = 0.12 \times 10^{97} \text{kg/m}^3$$

The mass of the activated string will convert to the mass of the two-dimensional string and the volume of the cone will be equal to the vacuum of Planck volume.

b. Tubular Monopole

When the "magnetic monopoles" extra dimension described above are attracted to two orbiting (non-pulsating) extra dimension cylinders, a new extra dimension object is created. The activated open string on one end of the monopole and the activated looped string on the other attach to the vacuum of two extra dimension cylinders as shown in Fig. 3-18.

The two-dimensional string and its vacuum housing form a tube-like structure around the activated open string while it is attached to the activated open string on one end and activated looped string on the other. The reason I have termed this a "Tubular Monopole" is that the orbital rotation of the two extra dimension cylinders produces a tube-like structure with a capillary inside. This capillary acts to convey forces that are experienced deep inside the spacetime cavity to distances that are far away from the nucleus of the cavity.

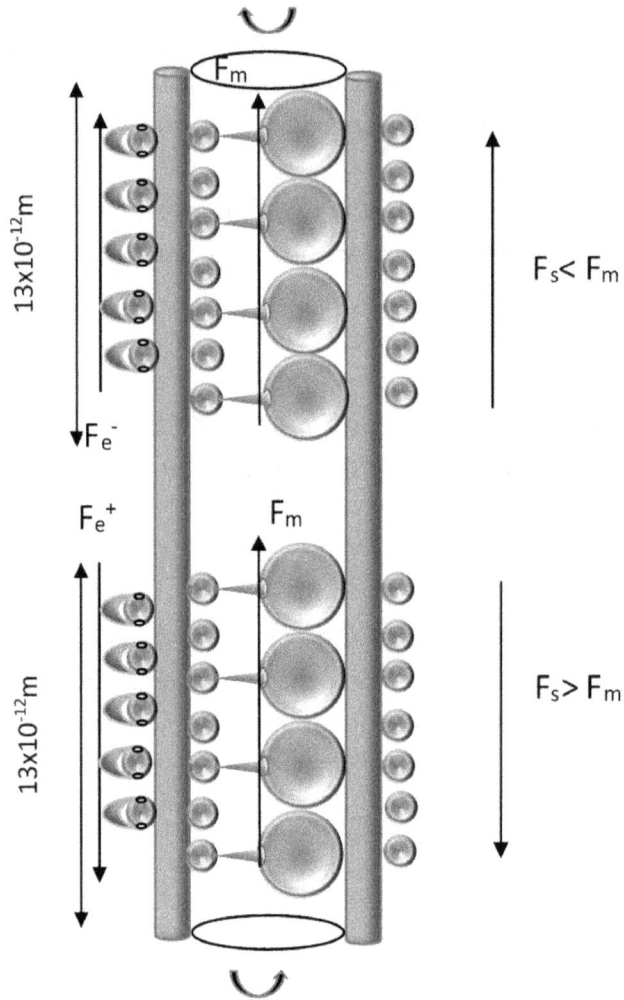

Fig. 3-18 – Conceptual representation of a Tubular monopole, consisting
of two segments each with opposite rotation of activated open strings
designating a positive and negative "charge", the electrical force F_e, and
magnetic forces F_m. The extra dimension cylinder on the left with
activated open strings represents a photon. F_s is the strong force inside
the spacetime cavity

A unique attribute of the Tubular Monopole is that the activated open strings inside
the tubular sleeve around the extra dimension cylinder rotate predominantly in the
same direction. As such, activated open strings flowing into the Tubular Monopole do

not annihilate each other, instead they are pushed by the torque of its orbital rotation creating the charge of the Tubular Monopole. Since activated open strings do not decay, the Tubular Monopole is not exposed to the electromagnetic field which is created by the open strings. This means that the activated looped strings and the Tubular Monopole itself are only subject to the force field of activated looped strings.

However, to make up for the lack of the electromagnetic field, the Tubular Monopole will draw on the strong force to equalize the forces inside the cavity. I will explain this in more detail in Chapter 11.

The Tubular Monopole extra dimension object is the basic structure of a "charge". As mentioned above, the Tubular Monopole is comprised of magnetic monopoles attached to the vacuum of two extra orbiting dimension cylinders.

Yet another unique attribute of a charge is that activated open strings forming the length of the tube can rotate clockwise or counterclockwise. This will be indeed the case when the Tubular Monopole extends the entire length of the spacetime cavity. It will have two segments, one segment with open strings rotating clockwise, the other counterclockwise, forming" positive and negative" charges respectively.

A "charge" has a few important characteristics; mass, which is determined by the presence of the activated looped strings in the Tubular Monopole's structure and a force, with a direction determined by the rotation (spin) of the activated open strings around the extra dimension cylinder. Let us designate this as F_e, the force associated with the electrical component of the charge, Fig. 3-18.

Recall, in the previous section, a magnetic monopole exerted a force F_m on the monopole assembly. When the magnetic monopoles attach to the extra dimension cylinder of a spacetime cavity, a new force is created which is the torque of its orbital rotation of activated open strings around the extra dimension cylinder, F_e.

Therefore, the total Force of the charge will be:

$$F_{emf} = F_e + F_m$$

F_{emf} is the sum of the electrical force as a result of movement of the charge (torque of open strings around the extra dimension cylinder) or the elementary charge and the magnetic force, F_m. When there is no change in F_m, we are left with only the force of the Elementary charge.

As you will see in the next chapter, this tubular extra dimension object plays an important role in the formation of matter as it is the housing for the charge of fundamental particles. I will discuss the mass of the charge in the next chapter.

The magnitude of the force F_m outside the spacetime cavity as shown above is:

$$F_m = (2.18 \times 10^{41} \times 2.2 \times 10^{42})^{1/2} = 6.9 \times 10^{41} N$$

This force must balance out with the forces created by the activated looped strings if there were no change in the field of open strings in the charge (all open strings rotating in the same direction causing no annihilation or decay therefore no electromagnetic force on the Tubular Monopole):

$$51 \times 10^{-60} \times 0.96k/R_a^2 = 6.9 \times 10^{41} \qquad\qquad (3\text{-}33)$$

Where $51 \times 10^{-60} m^3$ is the volume of an activated looped string with a radius of $2.3 \times 10^{-20} m$ outside the spacetime cavity and:

$\rho g = 0.96k/R_a^2$ for activated looped strings (chapter 6),

R_a from the above equation is calculated to be:

$$R_a = 13 \times 10^{-12} m,$$

This is the length of the Tubular Monopole encasing activated open strings around a cylindrical extra dimension object. This means that the length of this tube is repeated at about $13 \times 10^{-12} m$ intervals. Therefore, one of the dimensions of a charge has a length of about $13 \times 10^{-12} m$ as an assembly of activated open and looped strings and two-dimensional open strings in restframe. This distance is about one mole (Avogadro) of activated open strings attached to the extra dimension cylinders:

$$13 \times 10^{-12} m / 2 \times 1.61 \times 10^{-35} = 4 \times 10^{23}$$

The activated open strings inside each 13 x10^{-12}m segment of the Tubular Monopole rotate predominantly in the same direction. This rotation determines if a charge is positive or negative.

If the activated open strings in a Tubular Monopole were to rotate randomly, the open strings will decay, and the decay of the strings will create a field change inside the tube. The force of this field change must also equal the magnetic force, F_m:

$$17.47 \text{x} 10^{-105} \text{ x } 0.25 \text{x} 10^{114} \, (1/R_a) = 6.9 \text{x} 10^{41} \qquad\qquad (3\text{-}34)$$

$$R_a = 6.3 \text{ x} 10^{-33} \text{m}$$

This is the second length for a charge which is about the same as the length of the two-dimensional open strings. The term $\rho g = 0.25 \text{x} 10^{114} \, (1/R_a)$ for activated open strings is borrowed from Chapter 7.

Therefore, a charge can have two dimensions, one is about 13x10^{-12}m, which contains a grouping of activated open strings predominantly rotating in the same direction, and the other, about 6.3 x10^{-33}m, which is an individual charge which contains one activated open string with a random spin. The smaller charge is essentially made up of a single magnetic monopole whose activated looped and open strings are each attached to an extra dimension cylinder.

Since the smaller charge is made up of one magnetic monopole, it can form smaller groupings of the two-dimensional strings with the same rotation and, as a result, it can have a more randomized distribution along the length of an extra dimension cylinder to which they attach. As you will see later, this type of charge makes up the structure of a neutron whereas the charges with a length of 13 x10^{-12}m makes up the structure of a proton and electron.

If we place a Tubular Monopole as shown in Fig. 3-12 inside the spacetime cavity, it will become an extra dimension object which forms the fundamental structure of a charged particle we know as a Proton or Electron. The lower segment of the Tubular Monopole will have a length of 13 x10^{-12}m, followed by another segment of 13 x10^{-12}m in the upper region of the cavity. In the lower segment of the cavity the activated open strings in the Tubular Monopole structure rotate predominantly clockwise around the extra dimension cylinders to which they are attached, and in the upper segment predominantly counterclockwise. The entire assembly of the Tubular Monopole by the virtue of the attachment to orbiting extra dimension cylinders has an orbital rotation inside the spacetime cavity.

The direction of the rotation of the activated open strings is dictated by the arrow of the force F_m. Inside the spacetime cavity, this direction is always up towards the mouth of the cavity due to the direction of the rotation of the two-dimensional strings attached to the monopole. In the upper region of a spacetime cavity, F_m is the dominating force pointing upward. This means that the monopole's movement is upward, as such, all activated strings attached to the magnetic monopole must also create a force that matches the movement of the monopole when attached to the extra dimension cylinder. Therefore, in the upper region of the cavity, starting at about 13×10^{-12}m, all activated open strings spin in the same direction to match the movement of its corresponding monopole. In the lower region of the spacetime cavity, the dominating force is the strong force which I will cover in Chapter 10. This force is the highest force inside the spacetime cavity and its direction is towards the nucleus of the cavity. As a result, the monopoles will be forced to move downwards towards the nucleus. Activated strings attached to the monopole will then all rotate in the same direction creating a force that is downward. The two forces created on the extra dimension cylinder of the Tubular Monopole to which activated strings are attached are equal but in opposite direction which constitutes the positive and negative charge of the force F_e.

Let us now look at the angular momentum of the activated open strings as it spins around the extra dimension cylinder in a Tubular Monopole. This rotation creates what we know as the "charge" and the force associated with it, F_e. This force takes place in the restframe and is the summation of the forces created by the rotation of all activated open strings rotating in the same direction in a Tubular Monopole.

The momentum of a single activated string rotating around an extra dimension cylinder is:

$$p = h/\lambda$$

$$p = h/2\pi(2r)$$

The angular momentum of a single string will be:

$$L_i = p(r) = h/4\pi = \hbar/2$$

In which "r" is the radius of the extra dimension cylinder.

This means that the angular momentum of a single activated open string is the same as the reduced Planck constant, or in other words Planck constant is a measure of the value of the angular momentum of an activated open string around an extra dimension cylinder. Since angular momentum is additive, and in each segment of the Tubular Monopole described above we have about 4×10^{23} activated open strings attached to the vacuum of the extra dimension cylinders:

$$\sum L_i = \sum \hbar_i/2 = \hbar_i/2 \times 4 \times 10^{23} = 2.1 \times 10^{-11} \text{ kg m}^2/\text{s}$$

The energy of the angular momentum is converted to displacement of the mass (torque):

$$E = mgR_i = h_i f \tag{3-35}$$

Substituting for $g = c^2/r_p$, and $f = c/2\pi r_p$

$$mR_i = h/2\pi c$$

or

$$mR_i = \hbar_i/c$$

The summation of the distance created by the angular momentum of 4×10^{23} strings attached to an orbiting extra dimension cylinder is:

$$\sum mR_i = \sum \hbar_i/c = 2\sum L_i/c = 4.2 \times 10^{-11}/3 \times 10^8 = 1.4 \times 10^{-19} \text{ kgm} \tag{3-36}$$

This will be the "Elementary Charge". Which means the distance that one kg of mass moves as a result of this torque. In classical physics this is represented by 1.6×10^{-19} Coulombs as the charge of a particle. Therefore, a charge is a measure of the displacement force of activated open strings in restframe, all rotating in the same direction. This force is shown in Fig. 3-12 as F_e.

If we substitute 2.18×10^{-8}kg, the mass of one activated open string in the above equation:

$$\sum 2.18 \times 10^{-8} \, r_i = 2 \sum L_i/c = 4.2 \times 10^{-11}/3 \times 10^8 = 1.4 \times 10^{-19} \text{ kgm}$$

$$\sum r_i = 6.5 \times 10^{-12} \text{m}$$

Which means that the cumulative angular momentum of one Avogadro of activated strings, will move one activated open string a distance of 6.5×10^{-12}m which is half the length of the Tubular Monopole or "charge".

The force (torque) associated with this angular momentum can be up or down depending on the direction of the spin or rotation of the activated open strings attached to the extra dimension cylinders. If we use the right-hand rule, and the activated strings are rotating counterclockwise, the force will be upward and vice versa. This force occurs in the restframe1 and is exerted on the extra dimension cylinder to which the string is attached. When the number of strings attached to an extra dimension cylinder create a net force that is downward (towards the nucleus), the charge is designated as "positive charge" and when the force is pointing upwards as "negative charge".

Inside the spacetime cavity the forces created by the "charge" on the orbiting extra dimension cylinder must balance out or be net zero. This means that if the cumulative force of charge is upward, the other must be downward to create a net zero torque on the extra dimension cylinder. What happens if the forces of charges do not cancel each other? The extra dimension cylinder to which the strings are attached will curl under the strain of the torque created one way or the other depending on the direction of the force. As you will see later, this may not be an issue for extra dimension objects that are outside the spacetime cavity of a proton. An extra dimension cylinder that is located inside a spacetime cavity that comprises a fundamental particle must maintain its integrity and must not curl on itself, otherwise the path for transport of the two-dimensional strings out of the spacetime cavity will be disrupted and the cavity will turn into a "blackhole". I will cover the nature of forces of the charges and magnetism more extensively in the following chapters after I have introduced the structure of fundamental particles.

Activated looped strings are distributed with random rotations along the length of the extra dimension cylinder. This random distribution means that the rotations will most likely occur with 50/50 left and right rotations. As such, they do not create a torque on its extra dimension cylinder.

Let us now calculate the charge associated with the magnetic monopole, F_m:

$$F_m \, xR = mg_m R = E = 4 \times 10^{23} \times 1.96 \times 10^9 \qquad\qquad (3\text{-}37)$$

In which R is the total distance the torque F_m moves the strings which must equal the total energy of about one Avogadro of strings. If we substitute for g_m calculated in the previous section,

$$g_m = 5.29 \times 10^{50} \text{ m/s}^2$$

$$mR = 1.5 \times 10^{-18} \text{ kg-m}$$

This is "Planck charge". The mass distance relationship of a charge attributed to the magnetic force, F_m as shown in Fig. 3-12. In classical physics, this value is about 1.87×10^{-18} Coulomb.

Since Planck charge is the dominating charge as a component of the electromotive force, it is responsible for movement of the charge in an electromagnetic system. When there is no change in the magnetic field of a system, the force of the elementary charge is the dominant force.

$$F_{emf} = F_e + F_m$$

Alternatively, the elementary charge can also be calculated by substituting for g in equation (3-37) as:

$$g = c^2/r = 9 \times 10^{16}/1.61 \times 10^{-35} = 5.6 \times 10^{51} \text{m/s}$$

c. Photon

Attachment of both ends of a two-dimensional string to one activated open string

In another case, the two-dimensional string with a length of 4.4x10^{-33}m described earlier will attach to the second E-pole of a three-dimensional open string on both ends (Fig. 3-19). The extra dimension cylinder of a photon is the second extra dimension cylinder in a Tubular Monopole without activated looped strings as shown in Fig. 3-10. Therefore, the structure of a photon inside the spacetime cavity similar to the Tubular Monopole only experiences the field force of the activated looped strings (and the strong force which is shown in Chapter 11).

Since the activated open strings do not decay in a Tubular Monopole, the two-dimensional open strings of 4.4x10^{-33}m have sufficient residence time to attach to the same string from both ends. This extra dimension object forms the basic structure of a photon which will be discussed further in Chapter 4.

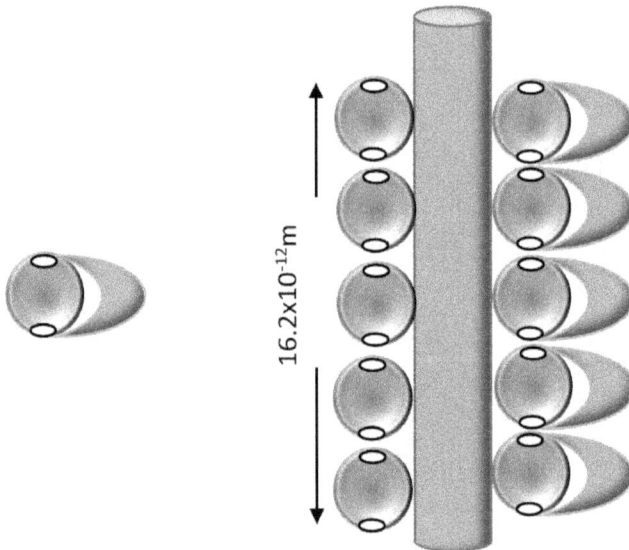

Fig. 3-19 – Conceptual representation of attachment of 2D open string to activated open string on both ends, and structure of a photon.

While attached to the activated open string at both ends, as seen in Fig. 3-19, the vacuum capsule of this extra dimension object will be about $17.47 \times 10^{-105} m^3$ (Planck volume), which stretches out to a length of $4.4 \times 10^{-33} m$. Its energy density on a string level will be:

$$\rho = 2.18 \times 10^{-8} / 17.47 \times 10^{-105} = 0.12 \times 10^{97} kg/m^3$$

Which is the same as an activated open string. Collectively, attachment of multiple strings of two-dimensional open strings to activated open strings which are in turn attached to the vacuum of an orbiting extra dimension object creates an extra dimension object we know as a photon. Since the activated open strings (with its two-dimensional attachment) rotate in the same direction, they do not annihilate each other and as such do not decay. Outside the spacetime cavity as an individual entity, other activated open strings which attach to the extra dimension cylinder of a photon comprise activated strings with random opposite rotations (strings shown on the left of the cylinder). These strings will decay which, as you will see in Chapter 11, provides the energy for the photon's wave function. The photon extra dimension object has a certain length which we can calculate as follows.

When a two-dimensional open string attaches to an activated open or looped string, there will be a change of energy of the extra dimension object. This change of energy will become a mass that along with the constant acceleration, creates a force. The two-dimensional open strings attached to activated open strings create a longer object as a photon which only sees the force field of the activated looped string in a Tubular Monopole. As such, it must be able to sustain the forces generated by the activated looped strings in a Tubular Monopole. But before we write a force balancing equation, we need to calculate the mass of a single string attached to the activated open string. The length of the two-dimensional open string is $4.4 \times 10^{-33} m$. When quantized, this is 273 Planck-size strings:

$$4.4 \times 10^{-33} / 1.61 \times 10^{-35} = 273$$

The mass of the single string attached to the activated open string will then be:

$$2.18 \times 10^{-8} / 273 = 0.8 \times 10^{-10} kg$$

The orbital radius of the open strings attached to the activated open string with respect to the center of the extra dimension cylinder is "2r". Therefore, the g_{rf} for the two-dimensional open string of a photon is:

$$g_{rf} = c^2/2r = 9 \times 10^{16}/2 \times 1.61 \times 10^{-35} = 0.28 \times 10^{52} \text{ m/s}^2$$

The force of the activated looped string in the field must equal the force created by the single string of mass 0.8×10^{-10}kg with an acceleration of 0.28×10^{52}m/s^2 inside the vacuum capsule of the open string:

$$51 \times 10^{-60} \times 0.96k/R_a^2 = 0.28 \times 10^{52} \times 2 \times 0.8 \times 10^{-10} \qquad (3\text{-}32)$$

$$R_a = 16.2 \times 10^{-12}\text{m}$$

This extra dimension object with a length of 16.2×10^{-12}m is the basic structure of a photon which will be discussed in more detail in Chapter 4. This distance is also about one mole (Avogadro) of activated strings attached to the extra dimension cylinders:

$$16.2 \times 10^{-12}/2 \times 1.61 \times 10^{-35} = 5 \times 10^{23}$$

In essence, the two-dimensional open string forms a sleeve around the assembly of the extra dimension cylinder and activated open strings attached to its vacuum.

Note that since a Photon is the second extra dimension cylinder in a Tubular Monopole with no activated looped strings (Fig. 3-10), it has about the same length as a charge. Furthermore, since the mass of the string, 0.8×10^{-10}kg, that attaches to the activated open string is smaller than the mass of the string itself, 2.18×10^{-8}kg, this change in energy will not become measurable to present itself as baryonic mass. As such a photon without activated looped strings will remain a massless particle.

Since the activated strings attached to the above structure comprising a photon can spin right-handed or left-handed, a photon will have two states of polarization. The photon will have a spin corresponding to the spin of the activated strings comprising its structure when it is formed as a component of a Tubular Monopole.

d.Attachment of a two-dimensional open string to two activated looped strings

Unlike activated open strings, activated looped strings have one E-pole. When two oppositely rotating activated looped strings attach to a two-dimensional open string inside the spacetime cavity where the dimensions of the looped string are in the same range as the Planck length, they will exert a force on each other's plane of E-pole causing it to stretch and expand. This will subsequently lead to the expansion of the sphere of the activated looped string increasing its radius as it moves up inside the spacetime cavity. The activated looped string will expand until its radius is increased from 1.61×10^{-35}m to 2.3×10^{-20}m, a factor of 1.4×10^{15}. The two expanded activated looped strings detach and become two individual strings as the centrifugal force overcomes the force of the attachment. Activated looped strings can also expand when they are attached to other extra dimension objects that exert a force on its E-loop pole as shown in Fig. 3-20. There are two instances in our universe that an E-loop expansion from its initial radius of 2.3×10^{-51}m to about Planck length is a common occurrence in addition to formation of two-dimensional Planck strings as will be discussed later. One is a pulsating string cylinder as discussed earlier; the other is expansion of a looped string as discussed here. In both cases, the forces applied on the E-loop string must be uniform and symmetric from all sides to stretch the E-loop to a Planck length. If the forces are unequal, the E-loop will collapse into a narrow structure such as a two-dimensional open string.

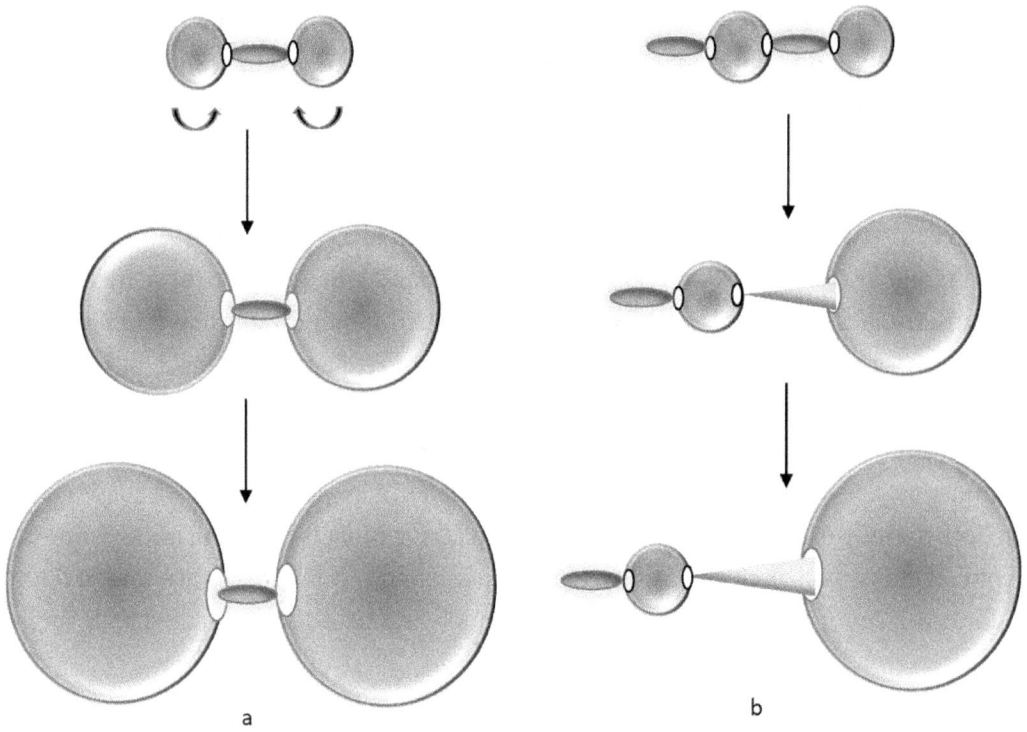

Fig. 3- 20 – Sequential expansion of an activated looped string by the forces of attachments exerted on its pole, a) two looped strings with opposite rotations, b) stretched by other extra dimension objects.

CHAPTER 4
Fundamental Particles and Standard Model

Background

IN THE PREVIOUS THREE CHAPTERS, I introduced activated looped, and open strings, spacetime cavity, spacetime "well", and extra dimension objects that result from the decay of activated strings. In this chapter, we will put these theories into practice and compare the results with known and proven science and theories in classical physics.

As I mentioned earlier, experimental measurement of extra dimension objects and activated three-dimensional string species, and the energy that creates the mass of what appears to be "empty space" as discussed in this theory are currently not possible. However, we can measure their signature effects on the universe as baryonic mass, momentum and the forces created by these objects. The number of subatomic particles identified by the Standard Model and experimental measurements at CERN are too numerous to cover in this book. As such, I will keep my focus on the main fundamental particles, proton, neutron, electron, photon, and Higgs Boson, and their corresponding properties that are well established in classical physics. Perhaps in future publications, other subatomic particles can be correlated to the string species and extra dimensions of the theory presented in this book.

Let us start by revisiting the graph I presented in the previous chapters to describe the "spacetime cavity" or the "cradle of matter". Perhaps the closest approximation of this graph in classical physics, is the "Mexican hat" representing Higgs energy field which is significantly limited in its scope.

One of the differences between the graph of "spacetime cavity" and the "Mexican hat" is the distance and depth of the energy field in which the fundamental particles can be studied. The Standard Model is typically limited to dimensions greater than 10^{-20}m, whereas the model presented in this book allows one to calculate and study quantum particles down to the string level at 1.61×10^{-35}m where many actions are taking place at high energies. And later, it is further expanded to dimensions of 2.3×10^{-51}m, and 8.9×10^{-160}m.

Since I am going to refer to these graphs often in this section, I will recall two of the previously posted figures below and briefly discuss their properties again before moving into quantifying fundamental particles. From Chapter 2, Fig 2-6, shows the dimensions of activated looped strings inside the spacetime cavity on the left side and the depth of the spacetime cavity on the right. From Chapter 3, Fig 3-10, shows the dimensions of open or looped strings in restframe2 as a two-dimensional string on the left side and the depth of the spacetime cavity on the right.

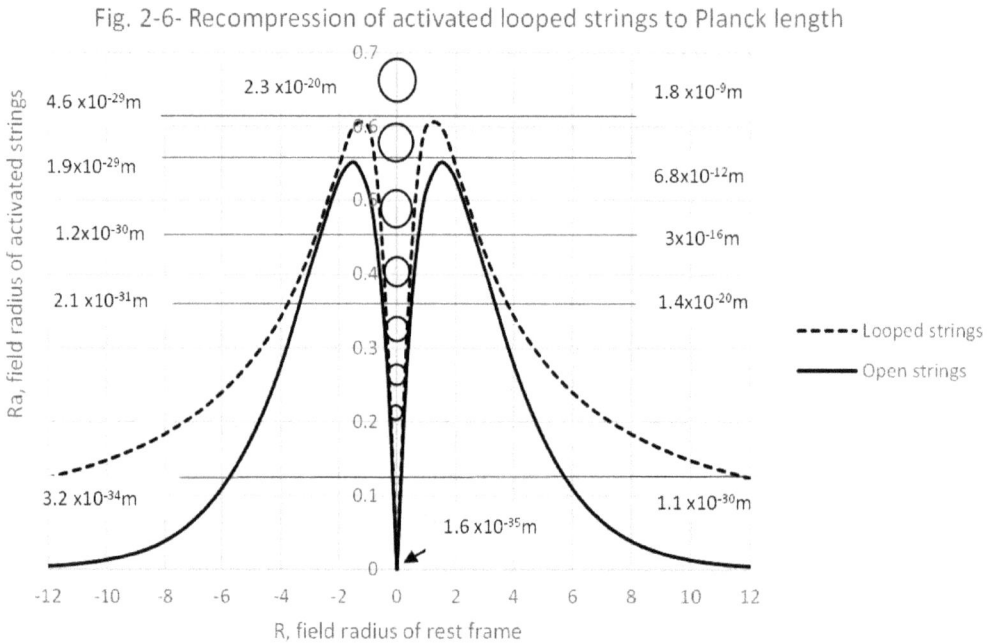

Fig. 2-6- Recompression of activated looped strings to Planck length

Fig. 3-10- Radius of looped or open strings in restframe2 vs cavity depth

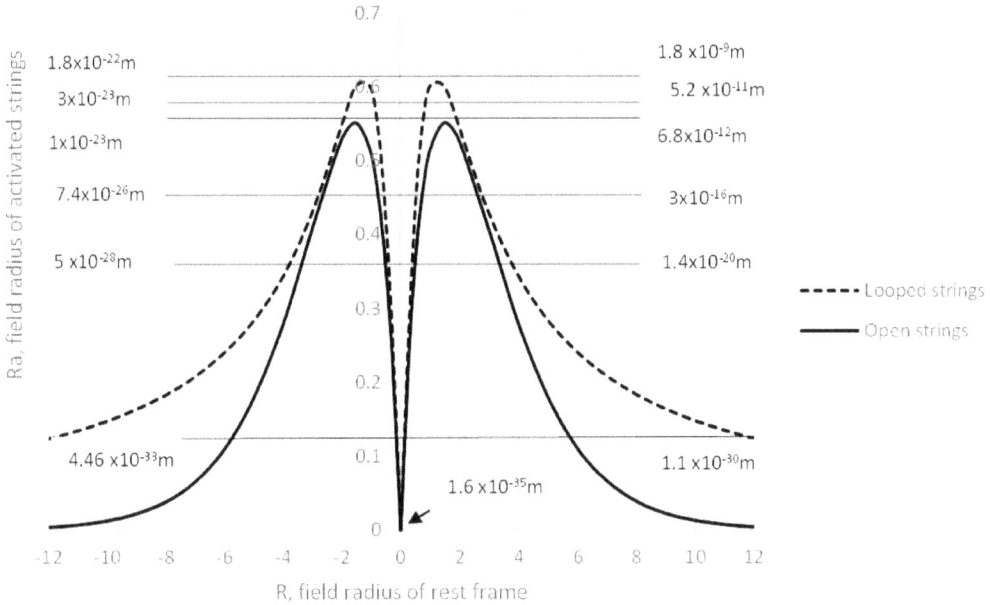

Inside the spacetime cavity, activated strings exists in infinite momentum frame, restframe1, and restframe2 as they would anywhere outside the spacetime cavity. However, something special happens below the dimension of 1.4×10^{-20}m inside the cavity.

Recall from Chapters 1 and 2 that the point of intersection of restframe energy density and the curve of the energy fields, at about 1.4×10^{-20}m, represents the point where all activated strings only exist in restframe1 state or fluctuating between restframe1 and infinite momentum frame.

The second important point to recall is at about 1.1×10^{-30}m. Below this point the field of open strings do not exist in restframe2, meaning that after decaying from a three-dimensional state to a two-dimensional state, it will recombine with its vacuum to recreate the activated open string (via attachment to pulsating extra dimension cylinders). Therefore, two-dimensional open strings as a stable object first appear at this depth and will have a dimension of about 4.46×10^{-33}m.

At this depth, the radius of the two-dimensional strings in restframe calculated from equation (3-19), is about 4.4×10^{-33}m. This means that we have 137 high-vac orbiting extra dimension cylinders in the first orbit of restframe.

$n = 4.4 \times 10^{-33} / 2 \times 1.61 \times 10^{-35} \sim 137$

The total number of possible orbiting extra dimension cylinders in all orbits at the depth 1.1×10^{-30}m are:

$R = 1.6 \times 1.1 \times 10^{-30}$

$N = 1.6 \times 1.1 \times 10^{-30} / 2 \times 1.61 \times 10^{-35} = 5.4 \times 10^4$

The 137 extra dimension cylinders in the first orbit have a special significance in the construction of the atom because it contains the highest amount of vacuum energy of all the extra dimension cylinders. Activated strings attached to these orbiting extra dimension cylinders annihilate each other which produces new high and low-vac extra dimension cylinders as discussed in Chapter 3. The low-vac extra dimension cylinders propagate inside the spacetime cavity with less vacuum energy. In construction of an atom, these 137 extra dimension cylinders will be consumed first by converting into Tubular Monopoles. These 137 extra dimension cylinders also establish the electromagnetic force of the spacetime cavity and, as you will see in Chapter 11, play an important role in the distribution of the strong force inside the spacetime cavity of atoms. In classical physics, the ratio of $1/137$, which is used as an adjustment for the force exerted on the Tubular Monopoles and the resulting dimensional change, is known as the fine structure constant (α).

As we move away from the core of the spacetime cavity into other orbits at the same depth inside the cavity, and above the depth of 1.1×10^{-30}m, the number of orbiting extra dimension cylinders will continue to increase but with slightly less vacuum energy than the 137 in the first orbit.

Using equation (3-13):

$R_s = 1.88 \times 10^{-9} (R_a)^{3/4}$

The radius of the pulsating looped strings at restframe at 1.1×10^{-30}m will be about 6.38×10^{-32}m and the number of possible high-vac pulsating extra dimension cylinders will be about 6.5×10^3:

$n = 6.38 \times 10^{-32} / 2 \times 1.61 \times 10^{-35} = 1981$

Total number of pulsating strings at $R_a = 1.1 \times 10^{-30}$m,

$n_t = (1.1 \times 1.6 \times 10^{-30} / 6.38 \times 10^{-32}) \times 1981 = 5.4 \times 10^4$

Therefore, at the depth of 1.1×10^{-30}m inside the spacetime cavity, there are 137 high-vac orbiting extra dimension cylinders, about 1981 pulsating extra dimension cylinders in the first orbit and a total of 5.4×10^4 orbiting and pulsating extra dimension cylinders in all orbits. As we move up inside the spacetime cavity towards the mouth of the cavity and out in open infinite momentum frame, the number of orbiting, and pulsating extra dimensioncylinders continues to increase and can be calculated from the corresponding equations.

There is a constant flow of the activated looped and open strings towards the orbiting extra dimension cylinders that are inside the spacetime cavity and arise from it. Activated strings that are attracted to the vacuum of extra dimension cylinders must have at least one rotation before they decay into the two-dimensional phase satisfying the Heisenberg uncertainty principle as described in Chapter1:

$L_z = P_z \cdot l_p \geq \hbar / 2$

One may describe the process of replenishment of depleted activated strings to sustain the state of matter in restframe1 as analogous to the spontaneous braking of the supersymmetry in classical physics.

How many activated strings are attached to each orbiting extra dimension cylinder inside the spacetime cavity? Let us first start by calculating the number of activated looped strings attached to an extra dimension cylinder inside the space time cavity. The geometric average radius of the three-dimensional activated looped strings ranging in size from 4.6×10^{-29}m at the mouth of the opening to 1.61×10^{-35}m at the nucleus of the spacetime cavity, is 2.72×10^{-32}m. If we divide the depth of the cavity by the average size of the activated looped strings:

$$1.8 \times 10^{-9} / 2.72 \times 10^{-32} = 0.7 \times 10^{23}$$

Which is in the order of about one mole of looped strings (6.03×10^{23}).

How about activated open strings? Recall from the previous chapter, the length of a Tubular Monopole which constitutes the structure of a charge, was about 13×10^{-12}m. This is also the equivalent of one mole of activated open strings attaching to the vacuum of the extra dimension cylinder:

$$13 \times 10^{-12} / 2 \times 1.61 \times 10^{-35} = 4 \times 10^{23}$$

With the above background we can now look at the structure of the fundamental particles.

Proton

Let us start by having only one Tubular Monopole extra dimension, the "charge", inside the spacetime cavity. The Tubular Monopole occupies two orbiting extra dimension cylinders, one occupied with activated open strings, the other activated looped (and open) strings. All other extra dimension cylinders inside the cavity are occupied by activated open strings. Recall that the Tubular Monopole, has a length of 13×10^{-12}m. The spacetime cavity has a depth of 1.8×10^{-9}m, therefore there will be multiple segments of the Tubular Monopole extending to the upper region of the cavity. The first segment of the Tubular Monopole which starts from about a depth of 1.1×10^{-30}m to 13×10^{-12}m, constitutes the structure of a proton (Fig. 4-1).

The segment of the Tubular Monopole beyond 13×10^{-12}m (which is also 13×10^{-12}m in length) represents the structure of an electron which I will discuss next.

The Tubular Monopole inside the spacetime cavity will have an extra dimension cylinder occupied by activated open strings whose dimension and energy remains constant along the length of the cavity and an extra dimension cylinder occupied by activated looped strings whose dimension and energy change inside the cavity. The radius of the looped strings will be 1.61×10^{-35}m at the nucleus, and about 1.9×10^{-29}m at around

13 x10^{-12}m. The Tubular Monopole orbits using the orbital rotation generated by the pulsating extra dimension cylinders. This includes all other orbiting extra dimension cylinders, two-dimensional strings, and any other extra dimension object.

The direction of the rotation of the extra dimension cylinders (Tubular Monopoles) inside the spacetime cavity is determined by the spin of one set of two-dimensional open strings which creates a force (and angular momentum) pointing towards the nucleus of the spacetime cavity. The direction of the force (and angular momentum) of two-dimensional looped strings inside the Tubular Monopole will be towards the opening of the spacetime cavity. While the rotation of one set of two-dimensional string creates a downward field towards the nucleus, the rotation of the second set of two-dimensional open strings (with opposite spin) creates the force of magnetic monopoles that is upward.

More specifically, inside the proton spacetime cavity, the direction of the movement of the magnetic monopoles is upward in the upper segment of the Tubular Monopole and is forced downward in the lower segment of the Tubular Monopole (Chapter 3).

Therefore, the activated <u>open strings</u> attached to the extra dimension cylinder of the Tubular Monopole will have predominantly clockwise rotation in the lower segment of the Tubular Monopole which comprises the proton, and a counterclockwise rotation in the upper segments of the Tubular Monopole comprising the electron (Chapter 3). Recall that the forces associated with the angular momentum of the strings rotating around the extra dimension cylinder will have an arrow pointing away from each other. This creates a net zero force on the structure of the orbiting extra dimension cylinder in the electron zone (upper) and the proton zone (lower) section of the cavity. The direction of the rotation of the activated <u>open strings</u> in the lower segment of the Tubular Monopole is designated as "positive charge" and the upper segment as "negative charge". Recall that the direction of the string rotation was determined by the direction of the movement of the magnetic monopole. In the previous chapter, I calculated this to be the "elementary charge" of about 1.4x10^{-19} Coulomb. Therefore, there is a chirality in the way the open strings rotate to create the positive and negative charges in the Tubular Monopole structures in the lower and upper sections of the spacetime cavity.

The activated looped strings will be comprised of a random 50/50 mixture of strings of opposite rotations attached to the extra dimension cylinder of the Tubular Monopole.

Therefore, they do not create a force on the extra dimension cylinder as they cancel out.

As a refresher, let us look at spins and rotations created by the decay of activated looped strings inside the spacetime cavity. When two activated looped strings with opposite rotations annihilate each other, two extra dimension objects with opposite rotations are created. The activated looped strings that are not orbiting, create the pulsating extra dimension cylinders that become the source for the orbital rotation of all other extra dimension objects in the universe, including inside the spacetime cavity. The activated looped strings that orbit create the high- and low-vac orbiting extra dimension objects. The direction of the rotation of the low-vac looped strings matches that of the open string which creates the field force towards the nucleus of the cavity. The direction of the strings inside the high-vac extra dimension cylinder will be towards the opening of the cavity which becomes a path for high-speed transport of two-dimensional looped strings out of the spacetime cavity.

Inside the spacetime cavity, there are five forces not including gravitational and electromagnetic forces. Two associated with the Tubular Monopole as shown in Fig. 3-18, which relate to the force F_m and F_e. The third force relates to the field orbital rotation creating a force downward. Therefore, in the lower segment of the cavity, where the proton lies there are two forces pointing downward, one up, and the upper section of the cavity where the electron lies, two forces pointing upward, one down. The other two forces relate to the weak and strong interactions which I will cover in Chapter 11. These forces will be in opposite directions for a neutron as you will see in the next section under atom.

Recall that there are 137 high-vac extra dimension cylinders inside the spacetime cavity. The significance of this is that it creates the electromagnetic force to balance out the forces of the activated open strings inside the spacetime cavity and the same forces exerted by the universe on the spacetime cavity (see Chapter 11).

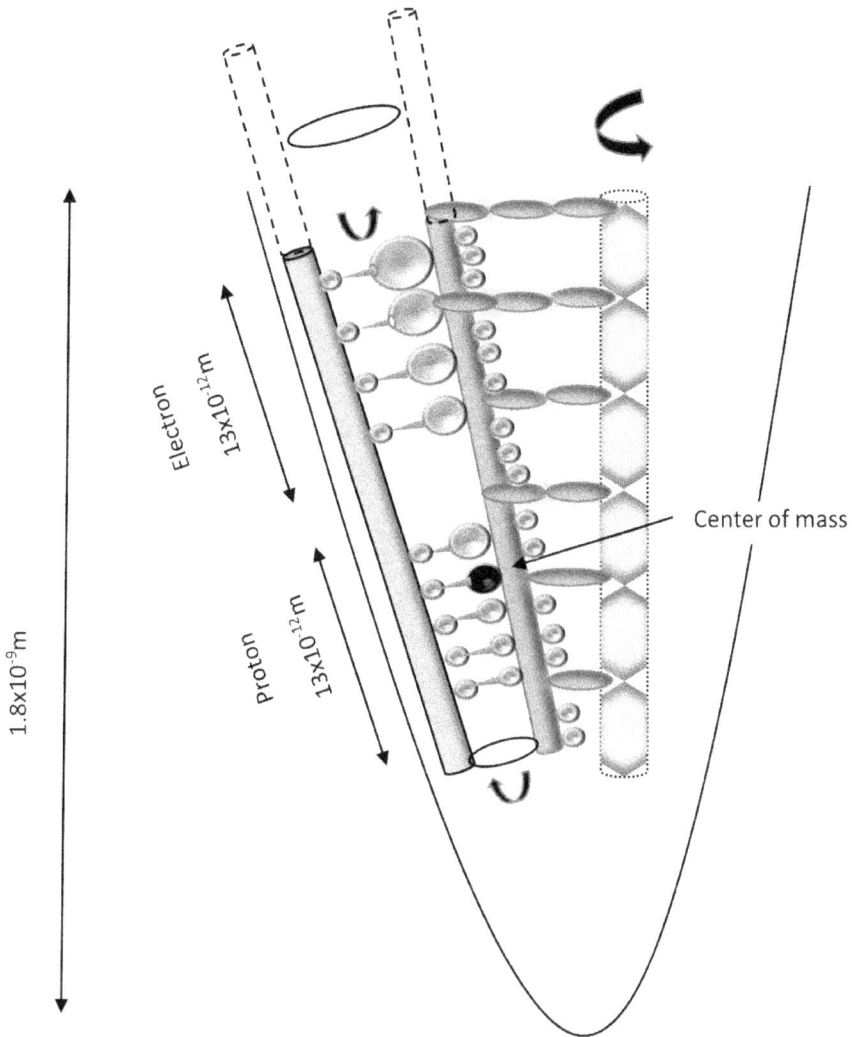

Fig. 4-1 – Conceptual representation of a tubular monopole inside the spacetime cavity. A) Dimensions are not drawn to scale. B) The size of activated looped string reduces with the depth of the cavity. C) The size of activated open string remains the same. D) Direction of rotation of activated open strings in the lower segment is opposite the direction of the rotation in the upper segment. E) Each segment contains ~ 4×10^{23} strings. F) All objects rotate via pulsating string rotational energy.

Fig. 4-1 is not drawn to scale and is merely to demonstrate the coupled configuration of two extra dimension cylinders in a Tubular Monopole. In this example, there will also be many other extra dimension cylinders orbiting inside the spacetime cavity not shown in Fig. 4-1. The remaining extra dimension cylinders contain only activated open strings as their attachments which are randomly distributed 50/50 with opposite rotation along the extra dimension cylinder and as such are charge neutral.

What is the mass of a proton?

The mass of activated looped and open strings is 2.18×10^{-8} kg at the nucleus of the space-time cavity. While the mass of an activated open string remains the same, the mass of an activated looped string becomes smaller as the string becomes larger moving away from the nucleus of the cavity. We do not observe the mass of the activated strings in our universe because as you will see in Chapter 8, the change in its energy is very small in infinite momentum frame. We can only observe its mass when there is a substantial change in energy and when that energy interacts with our three-dimensional universe. This change in energy occurs when the activated strings become attached to the extra dimension cylinder in the <u>restframe</u> state and <u>decay</u>.

The decay of activated open strings attached to the extra dimension cylinders does not create a change in energy because that energy will remain in our three-dimensional universe as an attachment to other activated strings. Recall this was the condition setting up the fundamental equations for open strings. As a result, we do not observe a mass due to the decay of activated open strings. As you will soon see, this will be the case with the structure of a photon which contains only activated open strings.

The decay of activated looped strings attached to the extra dimension cylinders does create a change in energy because the energy of the three-dimensional string is converted to a two-dimensional string that exists our universe (the Hilbert space) via orbiting extra dimension cylinders and enters the vector space.

To understand what baryonic mass is, I must expand a little more on the propagation mechanism of two-dimensional looped strings which are disclosed later in greater detail in Chapter 8. I will first address the change of energy manifesting the mass of proton when the strings decay.

When two activated looped strings with opposing spin annihilate each other, one becomes a high-vac two-dimensional string and the other, a low-vac two-dimensional string. The mass (energy) of the two-dimensional high-vac string is housed within the Planck vacuum of the string which exits our universe at high speeds via the orbiting extra dimension cylinder it forms. The high-vac two-dimensional string exits our universe without any interaction with our three-dimensional universe because the two-dimensional looped strings do not attach to other two-dimensional open strings. There are two reasons that looped strings do not attach to other objects, one is incompatibility of its speed with objects in our universe. The other is that there is only one attachment pole on the string as compared to an open string.

However, the propagation mechanism of the low-vac looped strings involves a momentary attachment of the low-vac string to a two-dimensional open string which is part of our three-dimensional universe (see Chapter 8). After this momentary attachment, the low-vac string also exits our universe via the orbiting extra dimensions that it forms. This momentary interaction between a propagated two-dimensional looped string and a two-dimensional open string links the mass of the string to our universe before it disappears. The change in energy which is the propagated mass of the string which interacted with our three-dimensional universe is the baryonic mass we observe.

The structure of a proton which is a Tubular Monopole with a length of about 13×10^{-12} m contains many activated looped strings with opposite spins that decay upon attachment to the extra dimension cylinder. Which one of these activated strings constitutes the mass of a proton?

The answer is that it will be the activated string precisely at the center of mass of the Tubular Monopole which is orbiting inside the spacetime cavity in infinite momentum frame state.

Why is the string at the center of mass of the Tubular Monopole so important? Because the orbital rotation of an object which behaves like a wave is secured at its center of mass which will have two-dimensional open strings attaching it to pulsating strings. Thus, the interaction of the two-dimensional looped string described above with our three-dimensional universe which makes it observable. This occurs at the string that is located at the center of the mass of the object. All other locations along the length of the Tubular Monopoles will also have two-dimensional string attachments; however, away from the center of mass these attachments are more random and infrequent. As you will see in the next segment, the attachment at the center of mass for an electron is accomplished by a photon which has a fixed dimension and also exists in our universe.

As shown in Fig. 3-7, up to about 1.4×10^{-20}m, the strings exist in restframe1 state. Therefore, the center of mass of the portion of the extra dimension cylinder orbiting freely in infinite momentum frame would be the geometric mean of the two distances; 1.4×10^{-20}m and 1.8×10^{-9}m, the length of the spacetime cavity:

$$R_a = (1.4 \times 10^{-20} \times 1.8 \times 10^{-9})^{1/2} = 5.1 \times 10^{-15} \text{m.}$$

Therefore, the mass of a proton is the change of energy that occurs when activated looped strings decay at $R_a = 5.1 \times 10^{-15}$m inside the spacetime cavity. However, since the decay is caused by two strings with opposing spins which annihilate each other, the change in energy we observe is the mass of the low-vac two-dimensional string propagated at about $R = 5.1 \times 10^{-15}$m where the two activated strings decay. As discussed above, we do not observe the mass of the high-vac string which is calculated below as 2.4×10^{-12}kg. We only observe the propagated mass of the low-vac two-dimensional looped string which has a momentary interaction with our universe before exiting. Recall that we can calculate the propagated mass of the two-dimensional strings as a function of R_a (3-26B):

$$m_s = 3.1 \times 10^{-43}/R_a$$

$$m_s = 3.1 \times 10^{-43}/5.15 \times 10^{-15}$$

$$m_s = 0.6 \times 10^{-28} \text{kg}$$

This is the mass of a proton in the absence of any other forces. However, as you will see in Chapter 11, there are two forces in a spacetime cavity in addition to electromagnetic, gravitational, and other forces described. These forces are weak and strong forces. Of the two forces, the strong force is the dominating one which has a magnitude of 137 times higher than the electromagnetic force. The strong force impacts the dimension of activated looped strings attached to the Tubular Monopole inside the spacetime cavity which causes them to shrink and increase its mass. The reduction in the radius of the activated looped string and the increase in its mass are proportional to the amount of force it experiences. We can adjust the mass of the proton calculated above by multi-plying by 137 to account for the additional force applied by the strong nuclear force:

$m_s = 0.6 \times 10^{-28} \times 137 = 8.2 \times 10^{-27} kg$

Which compares to the mass of a proton in classical physics as $1.7 \times 10^{-27} kg$.

The mass of the two-dimensional string calculated as the observed mass of a Proton is the energy that exits our three-dimensional universe (the Hilbert space) and becomes a component of resframe2 which is essentially invisible to our universe.

Let us now look at the change in energy that represents the mass of proton before the string decays. Using equation (2-4) we can calculate the mass of activated looed string at the center of mass of the Tubular Monopole at $R_a = 5.1 \times 10^{-15} m$ which will be $1.9 \times 10^{-13} kg$:

$m = 0.5 \times 10^{-16} / R_a^{1/4}$

$m = 0.5 \times 10^{-16} / (5.1 \times 10^{-15})^{1/4} = 1.9 \times 10^{-13} kg$

Correcting for the effect of the strong force;

$m = 1.9 \times 10^{-13} \times 137 = 2.6 \times 10^{-11} kg$

Recall in Chapter 3, I demonstrated the change in the mass of an activated string when it attaches to the vacuum of an extra dimension cylinder. Its mass is reduced by a factor of 4.3×10^{16}. This reduction in mass was the energy used for the string to make one rotation around the extra dimension cylinder. The mass of the string attached to the extra dimension cylinder of the Tubular monopole will then be:

$m_p = 2.6 \times 10^{-11} / 4.3 \times 10^{16} = 0.6 \times 10^{-27} kg$

Given that there will be two strings of opposite rotations that decay, the mass of the proton before decay of the string will be $1.2 \times 10^{-27} kg$, which compares to $1.7 \times 10^{-27} kg$ as reported in classical physics.

As you can see above, the change in energy we observe in both cases, before the string decays and after the string decays are about the same. Again, in this case, the string attached to the center of mass makes a momentary contact with the two-dimensional open string which forms the radius arm of the Tubular monopole's orbital rotation before it decays. This contact connects the string with its mass to Hilbert space, our universe. When it disappears we see this as "baryonic mass". All other strings on the same extra dimension cylinder make rare or a random connection to an open string which makes them invisible to our universe.

We will see later that outside the spacetime cavity, the mass we observe is the change of energy of the whole of activated looped strings which comprise the mass of dark matter.

Let us now look at the two-dimensional strings inside the extra dimension cylinder to which the activated looped strings are attached or the mass of high-vac two-dimensional looped strings originating from the nucleus.

The mass of the two-dimensional string inside the extra dimension cylinder at the nucleus starts with 2.18×10^{-8}kg and changes with R_a (equation 3-24):

$m = 2.3 \times 10^{-43}/R_a$

At $R_a = 5.15 \times 10^{-15}$m,

$m = 0.44 \times 10^{-28}$kg

$m = 0.44 \times 10^{-28} \times 137 = 6 \times 10^{-27}$ kg

This means that the mass of a two-dimensional string traveling inside the extra dimension cylinder at precisely the same location where the two strings annihilate each other is the same as what we obtained as the mass of a propagated looped string that interacts with our universe. It seems as if the universe has a backup plan to ensure that manifestation of the mass of fundamental particles is not left to chance.

The propagation velocity of the string inside the high-vac extra dimension cylinder at the center of the mass of the proton will be:

$1.96 \times 10^{9} = 1/2 mV^2 = \frac{1}{2}(0.44 \times 10^{-28})V^2$

V=2.98x10^{19}m/s

Of the two activated looped strings that annihilate each other at the center of the proton's mass, one is converted to a high-vac extra dimension cylinder. The mass of the two-dimensional string will be equal to the mass of its parent three-dimensional string, 2.4x10^{-12}kg. The mass of an un-propagated low-vac two-dimensional string is calculated from equation (3-26A):

$$m = 0.82x10^{-25}/R_a^{1/2}$$

$$m = 0.82x10^{-25}/(5.15x10^{-15})^{1/2} \hspace{3cm} (3\text{-}26A)$$

$$m = 2.8x10^{-17}kg$$

You can see, the mass of the un-propagated low-vac string is significantly lower than the mass of its parent string, 2.4x10^{-12}kg. As you will see in Chapter 11, the substantial portion of this mass (energy) is used as the displacement energy of the proton as a wave. In classical physics, this discrepancy will be the Hamiltonian of the wave function.

The difference between the mass of fundamental particles vs. atomic and larger species is that the mass of fundamental particles is not a summation of individual species making them. For example, a proton is made of about an Avogadro (one mole) of activated looped strings (0.7x10^{23}) having an average mass of about 5.7x10^{-16} kg, yet the mass of a proton is only in the order of 10^{-27}kg.

The magnitude of the charge created by a proton (1.6x10^{-19}C) is the torque force x distance generated due to the rotation of activated open strings around the extra dimension cylinder of the Tubular Monopole. As shown in Chapter 3, this force is generated because the activated open strings in a Tubular Monopole predominantly rotate in one direction making its forces additive.

Since the structure of a proton has two activated strings, looped and open strings spinning around the extra dimension cylinder, it requires two 2π spins for a complete rotation. In classical physics, this is designated as spin1/2 which, as you will see, is the same for electrons and neutrons.

As you saw above, a proton is a Tubular Monopole extra dimension object that is comprised of about one Avogadro of strings which is about 13 x10^{-12}m long. Therefore, in reality, a proton behaves like a wave having a wavelength of 13 x10^{-12}m, although its mass is determined by the energy change at the center of the mass of this wave. As you will see in this chapter, all other fundamental particles have the same property, i.e., they are waves in nature constructed of about one (mole) Avogadro of strings. This confirms de Broglie's theory that particles in the universe behave as a wave. As you will see shortly, these waves under the right circumstance curl up into a donut and act like a point particle.

Electron

Referring to Fig. 3-7, the segment of the Tubular Monopole extra dimension assembly which also has a length of 13 x10^{-12}m and resides in the region between the two peaks of the looped and open strings inside the spacetime cavity, makes up the electron. The electron orbits begin at about 5.2x10^{-11}m from the center of the spacetime cavity which is designated as Bohr radius. In a spacetime cavity, for every Tubular Monopole that is a proton in the lower section with a positive charge, there is Tubular Monopole in its upper section with a negative charge to counter the charge of the proton.

The activated open strings inside the monopole extra dimension and attached to the orbiting extra dimension cylinder rotate counterclockwise or opposite the spin of the activated open strings making up the charge of the proton. This creates an angular momentum and a force pointing downward towards the opening of the cavity. If you recall from the previous chapter, the force of each charge must result in a net zero force on the extra dimension cylinder which is shared by both charges. Maintaining the integrity of the extra dimension cylinder extending out of the cavity is critical to the transport of the two-dimensional strings out of the cavity.

The structure of the electron segment of the Tubular Monopole is the same as that of a proton, i.e., it will have activated looped and open strings attached to the vacuum of their corresponding extra dimension cylinders, Fig 4-2. Except, as you will see below, the change in energy of the strings that establishes the mass of the electron is significantly less than that of proton.

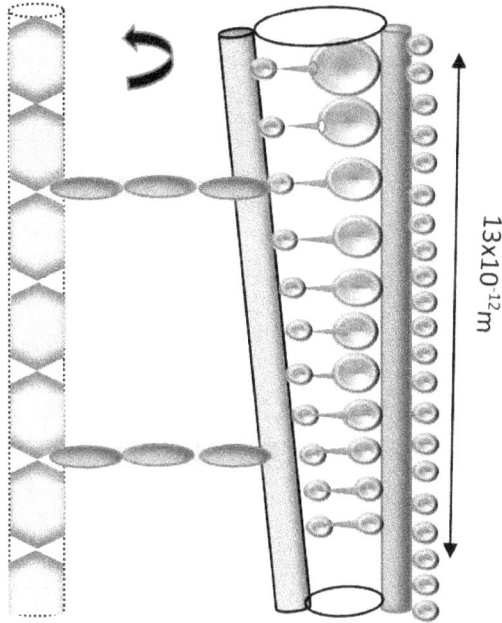

Fig. 4-2 – Conceptual representation of an electron inside the spacetime cavity as a wave, with its axis of orbital rotation.

We can calculate the mass of the electron using the same equation as above by calculating the energy change at the center of mass of the Tubular Monopole where two activated looped strings annihilate each other. The proton segment of the Tubular Monopole ends at 1.3×10^{-11}m, and the electron segment begins at about 5.2×10^{-11}m which is the Bohr radius. This means that the center of mass of the electron's Tubular Monopole in the infinite momentum frame falls at about $R_a = 5.85 \times 10^{-11}$m:

$m = 0.5 \times 10^{-16}/R_a^{1/4}$

$m = 0.5 \times 10^{-16}/(5.85 \times 10^{-11})^{1/4} = 1.8 \times 10^{-14}kg$

The mass of activated looped string at the center of the electron segment of the Tubular monopole is 1.8×10^{-14}kg. When the string is attached to the extra dimension cylinder, its mass is reduced by 4.3×10^{16} as shown in Chapter 3:

$m_e = 1.8 \times 10^{-14}/4.3 \times 10^{16} = 4.1 \times 10^{-31}$ kg

Since we have two strings that decay, the mass we observe is 8.2×10^{-31}kg.

Which compares to the mass of an electron reported in classical physics as 9.6×10^{-31}kg.

Similar to a proton, this mass is also equal to the mass of a two-dimensional string originating from the nucleus and inside the extra dimension cylinder precisely at the location where the two activated looped strings annihilate each other at the center of the mass of the electron.

We observe the location of the mass of electron to be different from one moment to the next. This is because the decay of the two three-dimensional parent strings occurs on the orbiting extra dimension cylinder of the Tubular Monopole that is rotating at the speed of light, c. Therefore, the location of the resultant two-dimensional string which gives us the baryonic mass of the electron changes randomly from one moment to the next. This makes the electron appear to be in many locations at the same time, which is also the case with a proton and neutron. The difference is that the orbital rotation of the location of the mass of protons and neutrons is significantly smaller, at $R_a = 5.1 \times 10^{-15}$m vs 5.8×10^{-11}m for electrons making it appear as more concentrated in a smaller area.

However, regardless of where the three-dimensional strings attach to the extra dimension cylinder of the Tubular Monopole and decay, the activated strings must make at least one rotation around the extra dimension cylinder in restframe1 state before decay or as discussed earlier:

$$L = pr \geq h/4\pi$$

In which,

$$p = h/\lambda$$

$$p = h/2\pi(2r)$$

$$pr = h/4\pi$$

and L is the angular momentum of the activated string around the extra dimension cylinder which points to the Heisenberg uncertainty principle.

The magnitude of the charge of an electron will be the same as that of a proton ($\sim 1.6 \times 10^{-19}$ C or kgm) because it contains the same number of activated open strings in its structure, however the direction of angular momentum and the force created will be opposite (see Chapter 3).

In atoms with multiple protons, the spacetime cavity will have the corresponding number of Tubular Monopoles which will increase the number of electrons orbiting around the center axis creating the electron shells.

In this case, to maintain balance of the torque created due to the orbital rotation of the mass of electrons, several orbits will be created depending on the number of protons and electrons. The distance between each orbit (or shell) of an electron is about equal to the length of the electron itself at 13×10^{-12}m. To maintain a fixed distance between each electron in its orbit, electrons are separated by a photon, which also has about the same length in its wave form. A conceptual diagram of more than one electron orbit is shown in Fig. 4-3.

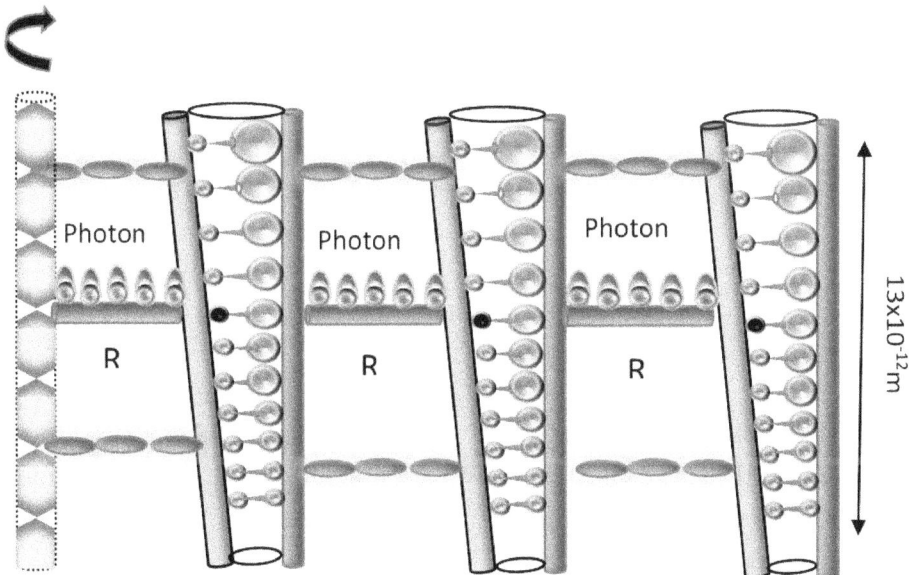

Fig. 4-3 – Conceptual representation of multiple electron orbits rotating around the center axis of a spacetime cavity. Each electron is orbiting around the center of its mass and separated by the length of a photon which is about the same as the electron.

The torque generated by all the orbiting electrons around the center of the mass of the Tubular Monopole has an energy of:

$$E = Mg(R/2) \qquad\qquad (4\text{-}1)$$

Where M is the mass of all orbiting electrons in an orbit (shell), and R/2 is half the distance of the length of the electron. This means that if the torque increases more than half the length of the electron, its tubular structure of length R will collapse. The torque energy of the total mass of electrons in orbit "n" must equal the energy produced by each electron with a mass "m" in "n" number of orbits.

$$E = nmc^2 \qquad\qquad (4\text{-}2)$$

The acceleration of the electrons with "n" number of orbits is:

$$g = c^2/nR$$

The distance between each orbit is constant and equals R, the length of a photon. Substituting for g in equation (4-1) and setting the two energies equal:

$$E = M(c^2/nR)\,(1/2R) = nmc^2$$

$$M/m = 2n^2$$

M/m is the number of electrons in each shell (total mass/individual electron mass)

$$N = 2n^2$$

Therefore, the total number of electrons in each orbit must be distributed according to $2n^2$ to create a balance caused by the torque applied to the Tubular Monopole of the electron. This is the same as the number of electrons calculated for each shell as described in classical physics.

The Tubular Monopole comprising the electron structure can break off and exist outside the spacetime cavity, Fig. 4-4. When this happens, it will be essentially a wave with a length of about 13×10^{-12}m. Electrons outside the spacetime cavity can curl up into a spherical shape and act like a point particle. While inside the spacetime cavity, electrons and protons share the same extra dimension cylinder which is the path for the exit of the two-dimensional strings out of the spacetime cavity. As you will see Chapter 10, if this path is disrupted, the spacetime cavity can turn into a blackhole. As such, electron structure inside the spacetime cavity remains in its wave form.

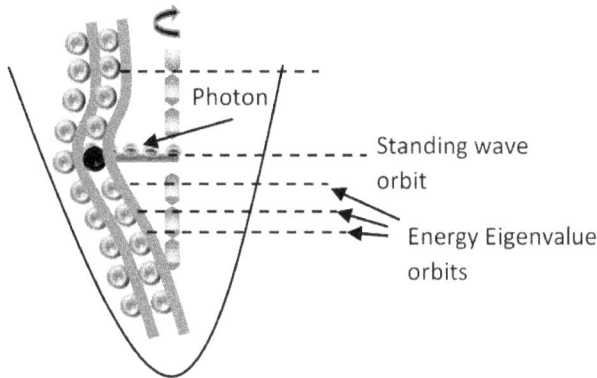

Fig. 4-4 – Conceptual representation of an electron as a wave outside the spacetime cavity of a proton with its axis of orbital rotation. Bold sphere represents center of mass. Each activated looped string represents the energy eigenvalue of the wave.

Electrons outside the spacetime cavity of a proton still retain their orbital rotation around the axis created by the pulsating extra dimension cylinders (Fig. 4-5). Recall that without this orbital rotation, the orbiting extra dimension cylinders of an electron structure will collapse into a pulsating string.

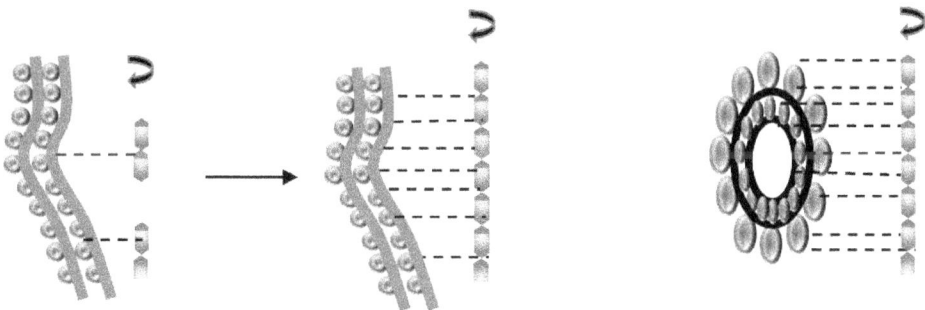

Fig. 4-5– Proposed mechanism of transformation of an electron from wave form to a torus shape representing a point particle.

An electron in its natural form is a wave-like object comprised of 4×10^{23} open strings and slightly less activated looped strings attached to two extra dimension cylinders. Under certain circumstances, it can behave like a point particle.

A proposed mechanism is when there is an increased number of pulsating extra dimension cylinders in the environment of the electron which act as the axis for the orbital rotation of extra dimension cylinders of the electron.

As seen in Fig. 4-5, a wave-like particle in the presence of an increased amount of contact points for orbital rotation will experience an increase in its angular momentum which, as a vector, is additive along the length of the electron. The increased angular momentum then creates a larger torque. When the number of contact points increases to a level that the torque generated will overcome the centrifugal force of the orbital rotation, the wave-like strings curl up into a round torus-like object.

$F_1 = mg$

F_1 = centrifugal force of the wave-like electron

$g = c^2/r_s$

r_s = radius of the orbital rotation = $3.37 \times 10^{-18}(R)^{1/2}$

$F_1 = 2.96 \times 10^{17} mc^2/R^{1/2}$

F_2 = torque of the angular momentum = $E/R = mc^2/R$

$R = 13 \times 10^{-12} m$

Where "m" is the mass of a single activated open string.

In a wave form, F_1 is greater than F_2, meaning that the centrifugal force of the orbital rotation is greater than the torque of the angular momentum. The centrifugal force will keep the Tubular Monopole in its wave form, orbiting only in one plane of motion.

The ratio of the two forces is:

$F_1 = 1.6 \times 10^{32} N$

$F_2 = 1.5 \times 10^{20} N$

$F_1/F_2 = 0.58 \times 10^{32}/0.2 \times 10^{20} = 1 \times 10^{12}$

$F_1 > F_2$ Wave

However, as connections to the axis of rotation increase (more attachments of the activated open strings to the pulsating strings), so does the angular momentum and the associated torque. The centrifugal force created by the individual strings are parallel vectors, whereas the torque from individual open strings is in the same direction along the length of the extra dimension cylinder and are additive. If the number of attachments increase to more than approximately 1×10^{12}, the force of the torque will exceed the centrifugal force:

$F_1 < F_2$ Curled (point particle)

Thus, the Tubular Monopole will curl up into a round torus-like object under the force of the torque. Theoretically, the maximum number of attachments a Tubular Monopole can make along its length of $13 \times 10^{-12} m$ is about 4×10^{23}.

What are the conditions that will cause an increase in the amount of pulsating extra dimension cylinders in the environment of an electron, photon, and other particles?

There are likely many ways that pulsating strings, which are the result of the decay of two three-dimensional looped strings, are introduced into an environment. An example may be when an energy source is aimed at a particle, such as when conducting an experiment for the purpose of measuring a particle's property or observing its behavior. An energy source, regardless of how benign it may appear, likely carries more pulsating extra dimension cylinders directed at the particle under the study. The increased amount of pulsating extra dimension cylinders introduced into the particle during the experiment creates the angular momentum and the torque necessary to change the properties of a wave-like particle including conversion into a point like particle. In classical physics this is referred to as "the Observer Effect" on the quantum behavior of a particle.

An electron becomes a positron if the orbital rotation of the extra dimension cylinders of its cavity is opposite that of an electron, or clockwise. In which case the arrow of the angular momentum of its cavity will be opposite that of an electron. A Positron and electron can annihilate each other by negating each other's orbital rotation.

Since the structure of an electron is similar to a proton, having two activated strings, looped and open strings spinning around the extra dimension cylinder, it requires two 2π rotations for a complete rotation, or have a spin ½ as described in classical physics.

Referring to Fig. 4-4, you can see that the Tubular Monopole forming the structure of an electron orbits around an axis at the speed of light. As a wave, its energy will change across the length of the Tubular Monopole based on the radius of the activated looped string which changes with R_a inside the spacetime cavity. The total energy of the wave will be the summation of the energy of the activated strings attached to the Tubular Monopole. The energy of each string can be calculated from equation (2-5):

$$E_i = 4.45/R_a^{1/4}$$

The total energy will be the summation of the energies of the activated looped strings along the 13×10^{-12}m length of the Tubular Monopole.

$$E = \sum E_i$$

In classical physics, the energy of each activated looped string attached to the wave are described as the energy eigenvalues of the wave. The location designated at the center of mass is where attachment of a photon is permanent. This location behaves as a standing wave in Schrodinger's famous equation. As described earlier, all other locations will contain orbital attachments via two-dimensional open strings which are dynamic and random. The energy of the "stationary wave" can be calculated based on the orbital radius of the Tubular Monopole using equation (3-19):

$$r_s = 4.26 \times 10^{-18}(R_a)^{1/2}$$

The energy of the standing wave will then be:

$$E = hc/2\pi r_s$$

I will cover more on this quantum mechanical property of the wave in Chapter 11.

Neutron

The structure of a neutron is a Tubular Monopole similar to that of proton with a few exceptions. The spacetime cavity of a neutron is opposite that of a proton which, as you will see in the next section, will make the structure of an atom, Fig. 4-6A.

Recall, in Chapter 3, I calculated two different lengths for the Tubular Monopole. One was 13×10^{-12}m which makes the structure of a proton and electron, the other 6×10^{-33}m. The smaller Tubular Monopole makes up the fundamental charge of a neutron.

One of the features of a Tubular Monopole is that the activated open strings comprising it rotate in one direction creating the electrical charge of the particle. This charge in a neutron is comprised of smaller sections of the Tubular Monopole which are randomly distributed along the length of the two extra dimension cylinders rotating left- or right-handed. The end result must be the same, that the force generated on the extra dimension cylinder to which the activated open strings attach must create a net zero force to prevent the extra dimension cylinder from distortion and maintain its integrity.

When a proton is formed, the unidirectional flow of activated open strings with opposite rotation in the two long segments of the Tubular Monopole reduces the rate of decay of open strings attached to the Tubular Monopole inside the proton cavity. As a result, a surplus of magnetic monopoles is produced creating a field of magnetic monopoles outside the spacetime cavity of the proton. This field begins to flow towards the neutron cavity where the monopoles flip changing the direction of its forces and entering the neutron cavity which is located opposite the proton cavity (see Magnetism). The infusion of the magnetic monopoles from the proton cavity disrupts the natural formation of the neutron structure which would have mirrored that of a proton. As a result, instead of forming Tubular Monopoles with each having a segment of 13×10^{-12}m comprising its charge, the magnetic monopoles distribute randomly along two high-vac extra dimension cylinders creating charges that are comprised of a single monopole or a smaller grouping of monopoles, Fig. 4-6B.

The orbital rotation of the neutron spacetime cavity will be opposite that of a proton, with the general orbital rotation of the extra dimension cylinders all counterclockwise, creating a force towards the nucleus of the cavity. All other rotations and extra dimension objects described for the proton will be the same but have opposite rotations. The forces created by the neutron will be opposite that of a proton.

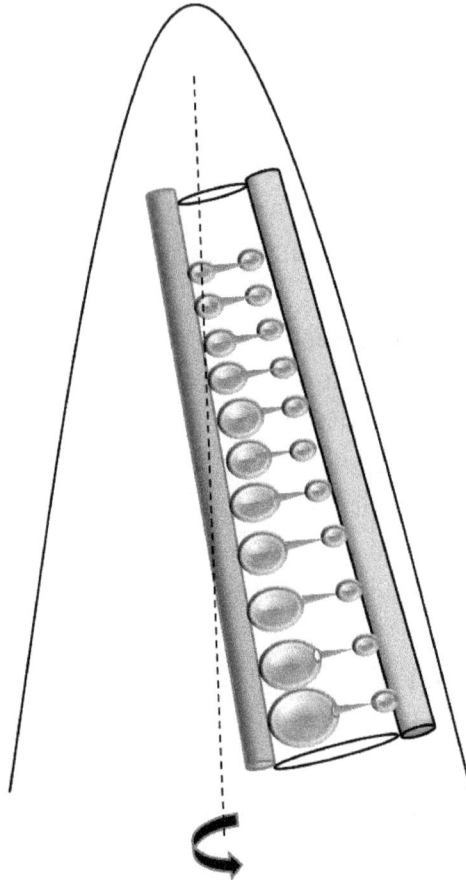

Fig. 4-6A – Conceptual representation of Neutron inside the south cavity. The charge in Neutron is the alternating magnetic monopoles rotating in opposite direction around the orbiting extra dimension cylinder.

Activated looped strings

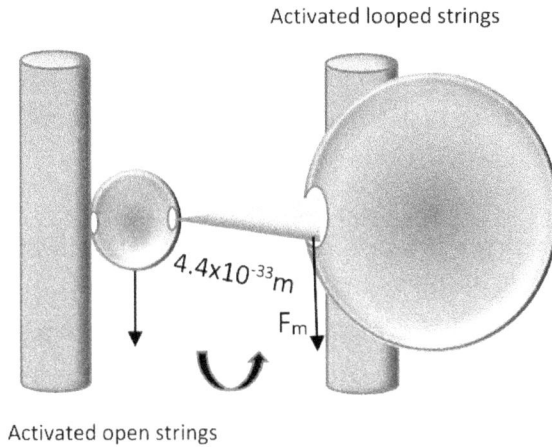

4.4x10^{-33}m

F$_m$

Activated open strings

Fig. 4-6B – Conceptual representation of the charge of a neutron.
F$_m$ is the magnetic force. F$_e$ is the force of the electrical charge.
The charge orientation alternates throughout the length of the
Tubular Monopole as one or a grouping of the magnetic
monopoles.

The mass of a neutron will be approximately equivalent to the mass of a proton. Like a proton, a neutron has two activated strings, looped and open strings spinning around the extra dimension cylinder, and requires two 2π rotations for a complete rotation, or spin 1/2.

Photon

The structure of a photon was described in the previous chapter as an extra dimension object created by two-dimensional open strings with a length of 4.4x10^{-33}m attached to both E-poles of the sphere of an activated open string. It was also shown that a photon similar to an electron, is comprised of about 5 x10^{23} activated open strings attached to an orbiting extra dimension cylinder having a length of 16.2x10^{-12}m, Fig. 4-7.

A photon comprises the structure of a second extra dimension cylinder in a Tubular Monopole which makes up a "charge". This structure can be separated as an entity and become a standalone object on its own. Upon separation, a Tubular Monopole can readily replenish and reproduce this structure with the extra dimension cylinders near its field. As such Tubular Monopoles are the source of an endless production of photons upon excitation.

Since a photon is attached to an orbiting extra dimension cylinder, it is also an orbiting object. Therefore, its structure similar to an electron, will have two rotations. One rotation is the orbit of the extra dimension cylinder around a pulsating cylinder and the other is the orbit of the string and its attachment around the orbiting extra dimension cylinder.

The activated strings with the two-dimensional open string attachment all rotate in the same direction hence they do not annihilate each other. When a photon is not attached to the Tubular Monopoles, other activated open strings not containing the attachment will have 50/50 strings of opposite spin distributed along the length of the cylinder. These strings, upon decay, provide the energy source for the photon's wave function.

Similar to the electron, inside the spacetime cavity, a photon is in its wave form. Outside the spacetime cavity, a photon can also curl up into a torus and act as a point particle under similar circumstances described for an electron, Fig. 4-8.

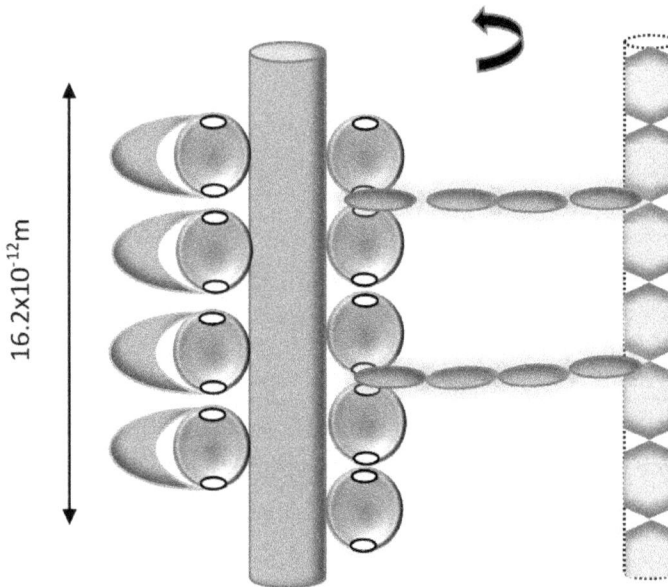

Fig. 4-7 – Conceptual representation of a photon as a wave with its axis of rotation.

As seen in Fig. 4-7, a photon has activated open strings and no activated looped strings in its structure. When its activated open strings attached to the extra dimension cylinder decay, we will not observe a change in energy as the energy of its two-dimensional string remains in our universe. As such, a photon does not exhibit a baryonic mass. In addition, as will be shown in Chapter 7, the net effect of the acceleration of the activated open strings creating the electromagnetic force on a mass is zero with no effect on the curvature of spacetime. However, photons will follow the curvature of spacetime created by the activated looped strings.

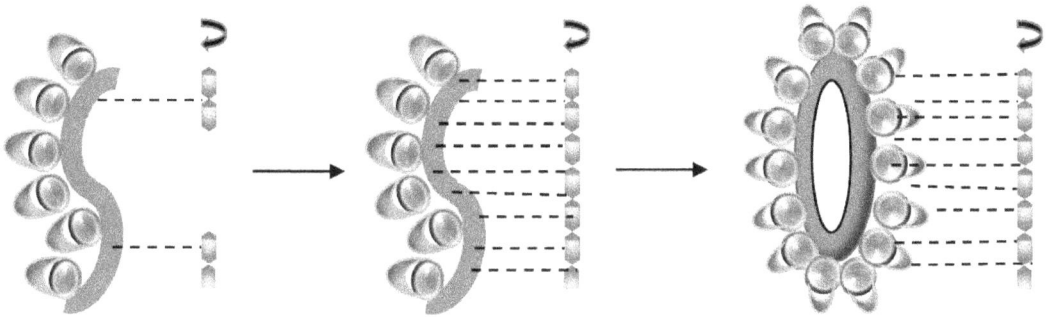

Fig. 4-8 – Conceptual representation of the mechanism of transformation of a photon from a wave form to a torus representing a point particle.

The activated strings comprising the structure of the photon can spin right- or left-handed around the extra dimension cylinder, as well as the orbital rotation of the entire assembly. The latter gives the photon a right- or left-handed polarization.

The structure of a photon contains an activated open string rotating around one extra dimension cylinder. Therefore, it requires one 2π rotation of the activated open string to complete a rotation, or spin 1.

Atom

So far, I have introduced a conceptual representation of how a proton, neutron and electron may look as a composite structure of a three-dimensional looped and open string and their two-dimensional vacuum states inside a single spacetime cavity.

In a three-dimensional space, the spacetime cavity resembles an inverted cone with its base at the mouth of the cavity and its apex at the nucleus of the matter where the radius of both looped and open three-dimensional strings is equivalent to Planck length, 1.61×10^{-35}m. The spacetime cavity is created as a result of the gravitational force due to the disappearance of three-dimensional looped strings when they decay upon annihilation of two strings with opposing spins. As mentioned before, the spacetime cavity has a general orbital rotation that is determined by the rotation of the two-dimensional open strings that attached and are coupled with the pulsating strings of the same spin. This orbital rotation is clockwise creating an angular momentum and a torque that is downward towards the nucleus of the spacetime cavity. Outside spacetime cavity, the orbital rotation of the two-dimensional strings with opposite rotation (the entangled partner) dominates the field of infinite momentum as a whole and is partly responsible for the expansion of the universe. This is very similar to the angular momentum created by the rotation of the strings around the extra dimension cylinders creating the force of the "charge" except that this angular momentum and force affects the entire field inside and outside spacetime cavity.

Without a counter force (similar to a charge that creates a balance for force exerted on an individual extra dimension cylinder), a one-sided angular momentum and the torque created by all extra dimension objects will make the spacetime cavity unstable and cause it to collapse and the strings inside it to quickly decay into more stable particles. Therefore, in order to create symmetry in force and stability, two mirror image spacetime cavities are necessary to counterbalance the forces of each other.

Fig. 4-9 shows the symmetric structure of the two spacetime cavities on both sides of the nucleus, each a mirror image of the other. The two spacetime cavities are connected to each other at the smallest dimension or at a depth of 1.61×10^{-35}m.

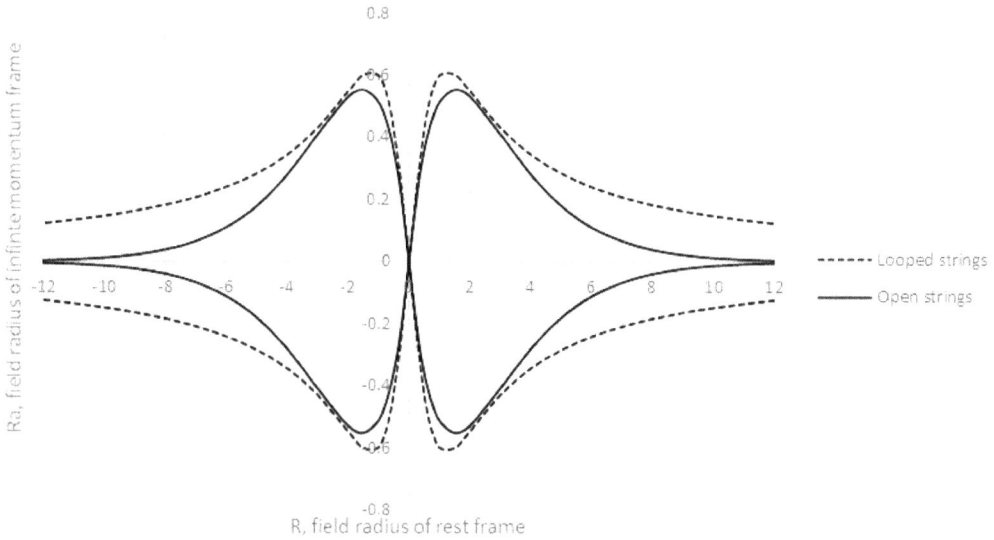

Fig. 4-9 - Spacetime cavity of an Atom comprising North and South cavities

The two spacetime cavities, one the mirror image of its counterpart, will have the same properties and mathematical descriptions as I covered in previous chapters. The main difference is that the orbital rotation of the extra dimension cylinders inside one will be opposite the other. This is a necessity because, as I explained earlier, the number of activated three-dimensional strings rotating clockwise and counterclockwise are equal in the universe for both activated open and looped strings. Stable matter cannot choose one orbital rotation vs. the other. The two cavities adjoined at the nucleus create an equilibrium of forces between the cavities and the infinite momentum frame on both sides. This dual cavity configuration constitutes the stable structure of an atom. To further create a balance of masses, the protons and neutrons are each housed in opposite cavities. Let us call the cavity that protons reside in the "North" cavity and the cavity that neutrons reside in the "South" cavity.

In the north cavity, we have protons (13×10^{-12}m) in the lower region and electrons (13×10^{-12}m) orbiting in the upper region. In the south cavity, neutrons with smaller positive and negative charges (6×10^{-33}m) are distributed along the length of the cavity. Both cavities have almost equal masses.

Let us now describe the structure of the simplest atom, Hydrogen. The Hydrogen atom is an exception with no neutrons in the south cavity. It contains a Tubular Monopole with the positive charge in the lower segment and an electron in the upper segment of the north cavity.

All other extra dimension cylinders inside the spacetime cavity will be occupied by activated open strings with 50/50 left and right spin distribution. Activated open strings do not leave a signature effect inside the spacetime cavity other than the electromagnetic force.

As the number of protons increase in the north cavity forming atoms with a higher mass than hydrogen, neutrons of about equal mass are created in the south cavity to balance out the forces of the north cavity. In an atom, the number of Tubular Monopoles in the north and south cavity are equal. All other extra dimension cylinders in both cavities are occupied by three-dimensional open strings which create the electromagnetic force inside the spacetime cavity. Of these, 137 extra dimension cylinders will constitute the dominating portion of the electromagnetic force needed to balance out the electromagnetic force created by the universe (see Chapter 11).

Higgs Boson

In the above examples given for construction of the atoms, at least one Tubular Monopole is needed inside the spacetime cavity. Each Tubular Monopole carries an extra dimension cylinder occupied by activated looped strings which determine the mass of the fundamental particle and atoms. As the number of Tubular Monopoles are increased in the north cavity, there will be an equal number in the south cavity to balance out the mass. In construction of atoms with higher masses than hydrogen, the 137 extra dimension cylinders with the highest vacuum energy are the first to be occupied by activated looped strings. Atoms that require more than 137 protons and neutrons will employ extra dimension cylinders with slightly less vacuum energy from other orbits. These will be nearest to the first 137 extra dimension cylinders at the depth of about 1.1×10^{-30}m.

How about a scenario where all 137 high-vac extra dimension cylinders in one single spacetime cavity are occupied by the activated looped strings without a south cavity to balance it out? Is this a possibility?

To answer this question, I must now briefly address the expansion of the activated looped strings. As you have seen in previous chapters, activated looped strings expand in size as a function of the radius of the field of a mass in infinite momentum frame. At the nucleus of the spacetime cavity, its dimension, mass, and energy are the same as activated open strings with a Planck radius of 1.61×10^{-35}m. This radius expands to 2.3×10^{-20}m in infinite momentum frame outside the spacetime cavity, an increase of 1.4×10^{15} in size.

The decay mechanism of the expanded activated looped strings in infinite momentum frame is not the same as those described earlier inside the spacetime cavity. Recall that activated looped strings attach to the vacuum of the extra dimension cylinders which are then annihilated by a similar activated looped string with opposing spin rotation. The result of the annihilation is the formation of two extra dimension objects entangled with opposing spin. The annihilation of the expanded activated looped strings with a much larger dimension of 2.3×10^{-20}m results in instantaneous regeneration of a vast number of Planck-size activated looped strings with a radius of 1.61×10^{-35}m. This is the first step in annihilation of the expanded activated looped strings.

I will cover the regeneration mechanism of activated looped strings in Chapter 8 in more detail. Expansion of the activated looped string results in a reduction of its core mass from 2.18×10^{-8} kg to about 1.5×10^{-23}kg. It also results in a substantial increase in the number of E-loop strings forming its pole surface area by a factor of 10^{30}. When two expanded activated looped strings decay, its rather large structure collapses and implodes. The collapse of its structure induces a kinetic energy in the stretched E-loops at its pole (1.6×10^{-36}m) and the stretched E-strings forming the gauges of the two-dimensional surface of its core plane. This energy, in addition to availability of just the right amount of bulk vacuum released ($\sim 18 \times 10^{-40}$m^3), provides the energy needed for creation of about 0.5×10^{77} Planck-size activated looped strings instantaneously. This is indeed the mechanism for recreation of activated looped strings in the universe as the existing ones are constantly being consumed by "matter". I will discuss this concept further under string regeneration mechanisms in Chapter 8.

The sudden generation of about 0.5×10^{77} activated looped strings with Planck radius and energy, creates its own spacetime cavity because of the immediate decay of some of the newly generated activated strings and creation of extra dimension cylinders. Indeed, about 100-250 spacetime cavities will be generated as a result of the decay of each expanded three-dimensional looped string. The spacetime cavities will have the same properties as described earlier, i.e., it will have 137 extra dimension cylinders with the highest level of vacuum energy. Given the sudden burst of 0.5×10^{77} activated looped strings, the extra dimension cylinders in these cavities will be occupied by activated looped strings because of the overwhelming and instantaneous generation of such a huge number of looped strings.

The resulting spacetime cavities at the maximum will have all 137 extra dimension cylinders occupied by activated looped strings with no south cavity to balance it out as in an atom. The reason for this is that there will not be sufficient residence time for the sudden generation and dispersion of some 100-250 spacetime cavities to form north and south cavities as in an atom.

Let's now look at the properties of the fundamental particle that is created from these spacetime cavities. The mass of the particle that will represent the above scenario for each spacetime cavity will have a maximum mass about 137 times that of a proton which only occupies one extra dimension cylinder with activated looped strings. Recall that the mass of a proton was created by the mass of a single two-dimensional looped string located at the center of the mass of an orbiting extra dimension cylinder occupied by activated looped strings and is about 1.67×10^{-27}kg. If all 137 high-vac extra dimensions were to be occupied by the activated looped strings, the maximum mass of the particle in each spacetime cavity will be:

137 x mass of proton:

$137 \times 1.67 \times 10^{-27}$kg $= 2.28 \times 10^{-25}$ kg

Which when converted to Gev/c², it will be 128.5 GeV/c². This means that the fundamental particle that was created by the implosion of an expanded activated looped string in infinite momentum frame is a Higgs Boson. It will have a maximum mass of 128.5 GeV/c², but it will be slightly lower depending on how many of the 137 extra dimension cylinders are occupied by the activated open strings. The experimental value reported by CERN is about 125 GeV/c², which means that about 133 extra dimension cylinders are occupied by activated looped strings and about 4 are occupied by activated open strings. Presence of open strings are required because without them, the resultant two-dimensional looped strings which manifest the mass of the particle will not propagate. Therefore, the mass of Higgs Boson will always be slightly smaller than the maximum shown above.

Since the mass of the decayed activated looped string which is about $2x10^{-23}$kg is converted to $2.2x10^{-25}$kg Higgs Bosons, it means about 100 spacetime cavities are created which means the decay of an activated looped string in infinite momentum frame leads to about 100 Higgs Bosons, each with a mass of about $2.2x10^{-25}$kg.

The particle created by three-dimensional activated looped strings will have spin 0 because it is spherical and rotates symmetrically. It will have no charge because there are no tubular monopole extra dimensions which form the structure of a charge. It is scalar and will have no charged field, and orientation preference. It will have a very short life because, as mentioned before, particles that are housed in one spacetime cavity, with such a heavy mass, will be unstable as the spacetime cavity will collapse without the dual cavity to stabilize it like an atom. When it decays, it will decay into a soup of particles which are likely heavy because of its mass and the overwhelming number of activated looped strings that are present in its structure. The decay will include photons due to a few extra dimension cylinders that are occupied with activated open strings.

The decay of activated looped strings in infinite momentum frame and the creation of Higgs Boson occur in a very short period of time. This event is contained inside the spacetime cavity that is formed immediately shielding it from the medium of infinite momentum frame.

As I mentioned earlier, the mass of a Higgs Boson is observable to our universe and has been experimentally measured at CERN. Creation of Higgs Bosons, due to the decay of the activated looped strings outside the spacetime cavity and in infinite momentum frame, suggests the existence of a substantial mass in the universe with particles that have a very short half-life. As you will see later, these particles represent the decay of activated looped strings forming the fundamental mass of dark matter in the universe.

You have now learned a new concept. While activated looped strings with opposite spin annihilate each other and propagate as high and low-vac two-dimensional extra dimension objects inside the spacetime cavity, their annihilation as expanded looped strings result in generation of new Planck -size three-dimensional looped strings in infinite momentum frame. This process leads to Higgs Boson particles, which in turn produce the two-dimensional high, and low-vac two-dimensional looped strings inside their short-lived spacetime cavity.

One may ask what is the purpose of Higgs Boson in the universe if it is a short-lived particle?

As you will see in Chapter 8, the generation of Higgs Boson is in reality a side product of regeneration of new activated looped strings in the universe. Without constant regeneration of activated looped strings our universe will consume its reservoir of activated looped strings which was created during the Big Bang. Expansion of activated looped strings in infinite momentum frame and subsequent decay and regeneration of 0.5×10^{77} new activated looped strings with every decay event ensures proper supply of activated looped strings for about 100B years.

The decay and regeneration of activated looped strings are accompanied by interim creation of Higgs Boson as a particle. This process is taking place everywhere in the universe outside the spacetime cavity of neutrons and protons. Therefore, it is important to understand the underlying benefit of Higgs Boson as a short-lived particle that evolves and disappears during a regeneration burst of about 0.5×10^{77} activated looped strings.

The second important aspect of the regeneration process of activated looped strings described above is creation of Planck-size activated looped strings. As you have seen earlier in this chapter, particles such as protons and neutrons which form the atoms contain Planck-size activated looped strings that are significantly smaller in size and higher in mass and energy than what is available in infinite momentum frame. Immediately after the Big Bang, when the infinite momentum frame begins to establish, the activated looped strings begin to expand with the expansion of the universe resulting in larger looped strings with less mass and energy. The Planck-size activated looped strings which are generated as a result of the decay of the expanded looped strings may be viewed as the seed for formation of new matter in the universe wherever this need arises. Therefore, one may conclude that Higgs Boson is a side product of a process that creates the ingredient needed to create new baryonic mass and particles.

Other sub-atomic particles

This topic is beyond the scope of this book. As you have seen earlier, due to the existence of many rotations, forces, angular momentums, torques, and the number of extra dimension objects inside a spacetime cavity, defragmentation of a spacetime cavity will lead to creation of many sub-atomic particles that are short lived and cannot be isolated.

My purpose in this book is to establish a relation between the main fundamental particles of the Standard Model and the extra dimension objects described because of various string configurations.

Magnetism

The presence of a dual cavity in atoms produces a new property for the atoms, its magnetic property. Recall that a magnetic monopole was an extra dimension object with an activated looped string (which is significantly larger than the activated open string in infinite momentum frame) attached to an activated open string via two-dimensional open strings with a length of 4.4×10^{-33}m. Also recall that a magnetic monopole extra dimension and a tubular monopole extra dimension (the structure of a charge) are intimately related and only differ by virtue of attachment of a magnetic monopole to two extra dimension cylinders, Fig. 4-10.

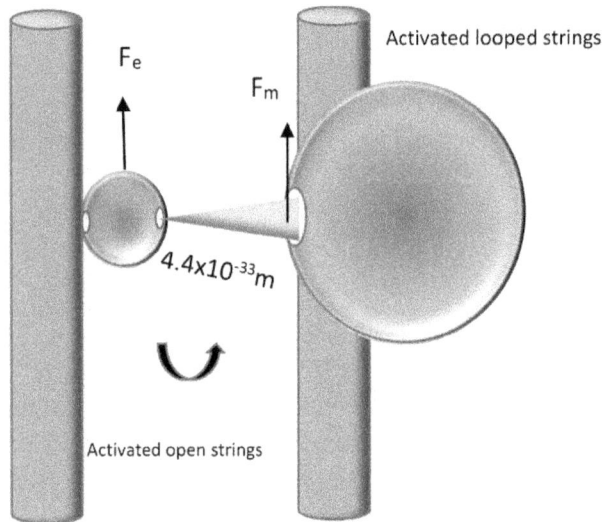

Fig. 4-10 – Conceptual representation of a magnetic monopole and a segment of the tubular monopole. F_e is the electrical force imposed on the activated open string. F_m is the magnetic force on the looped string.

Magnetic monopoles are produced inside the north and south spacetime cavities of an atom and are released to its surrounding environment. As such, there will be a field of magnetic monopoles created at the north and south spacetime cavity of an atom.

In the north cavity, the proton and electron comprise the two segments of the tubular monopole. Recall that in these segments, activated open strings with the same spin make up the structure of the Tubular Monopole. The direction of the movement of the monopole and the spin of the activated open string are determined by the vector of the force F_m which dominates the monopole. If F_m is pointing up, the monopole moves up and the direction of the spin of the activated open string will be counterclockwise creating a force that is upward matching the movement of the monopole, and vice versa.

In the proton cavity, because there exist two large segments of extra dimension cylinders with activated opens strings rotating in the same direction, there will be significantly less decay of the open strings. This means that there will be a surplus of monopoles that are produced with the negative charge. Note that monopoles do not have a charge in infinite momentum frame until they attach to two extra dimension cylinders. I am referring to it as a negative charge because it is simpler to remember than the orientation of the spin of the activated open strings.

With the surplus of the negative monopoles at the north cavity, a flow of monopoles is established into the south cavity where the neutron is located. At the south cavity, the arriving monopoles can readily switch their activated open strings to match that of the field rotation and become positive.

Without the surplus production of negative monopoles coming from the north cavity, the cavity of the neutron may have been identical to that of the proton, and we would have had a tubular monopole with two long segments (13×10^{-12}m) comprising negative and positive charges in the lower and upper segment of the cavity. However, the surplus production of the negative monopoles flows to the neutron cavity changing the symmetry that would have otherwise existed, creating a mixture of randomly positioned positive and negative monopoles with a dimension of 6×10^{-33}m as charges making up the neutron.

To maintain the balance of energy in the neutron's cavity, a flow of positive charges is established out of the south cavity which then become negative at the north cavity matching its field rotation.

Therefore, there will be two counter current flows of the magnetic monopoles with the same sign at each cavity, due to the direction of their forces F_m, one pointing inward, the other outward, but both having the same spin of the activated open strings.

This flow pattern will result in the monopoles leaving and arriving at the north and south cavity having activated open strings of the same spin (counterclockwise in the north, and clockwise in the south).

As shown in Fig. 4-11, monopoles with counterclockwise spinning open strings designated by a "negative sign" exit the north cavity and flow toward the south cavity where it will accept activated open strings with opposite rotation designated by the "plus" sign. Likewise, magnetic monopoles with a "positive" sign exit the south cavity and flow toward the north cavity and accept activated open strings to become "negative". As a result, there will be an internal current produced inside the north and south cavities of the atom where the monopoles attached to the extra dimension cylinders become an electric current and flow inside the atom with the net current equaling zero.

When the spacetime cavities of two atoms are in close proximity and face each other, the extra dimension cylinders exiting them will connect and bridge the two cavities. Monopoles then attach to two extra dimension cylinders of Tubular Monopoles and create the force and angular momentum in restframe to attract or repel each other. Cavities having positive or negative signs will repel and the ones with opposite signs attract each other as a result of the magnetic fields created by the magnetic monopoles, Fig. 4-12. Note that the diagram is by no means drawn to scale. In reality, the extra dimension cylinders are at Planck scale and the opening of the cavity, about 1.8×10^{-9}m.

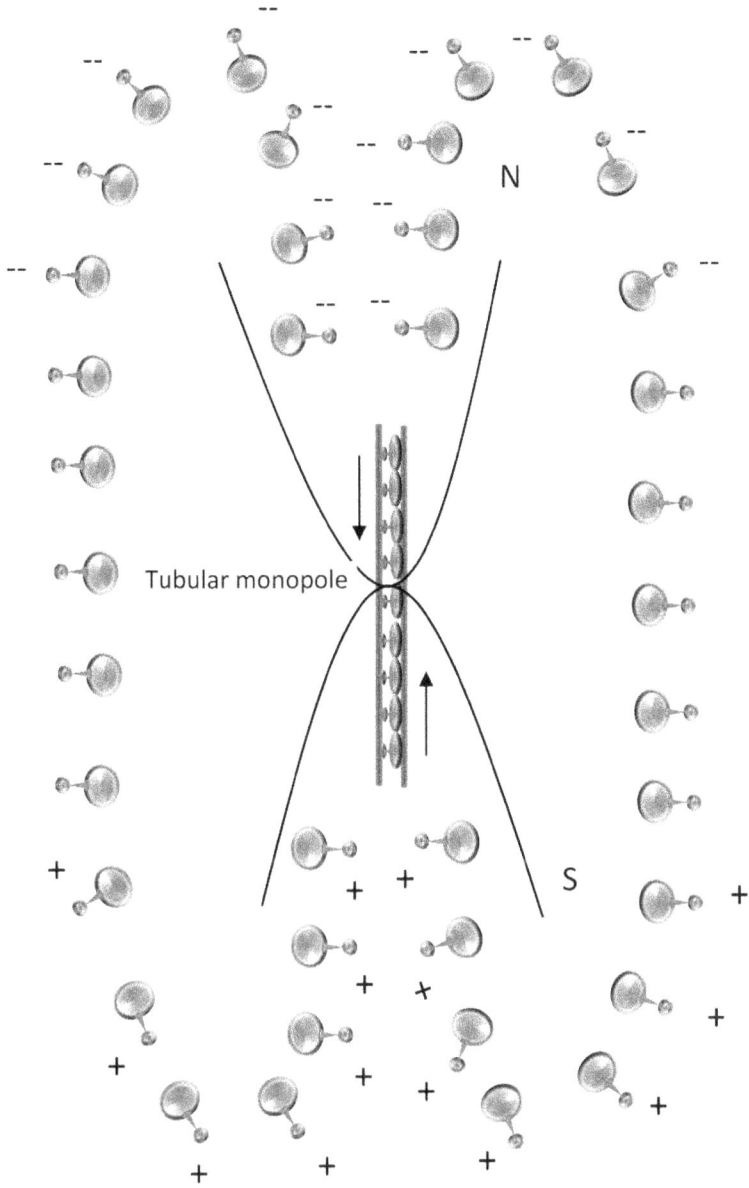

Fig. 4-11 – Flow pattern of magnetic monopoles and magnetic field leading to an internal current inside the spacetime cavity.

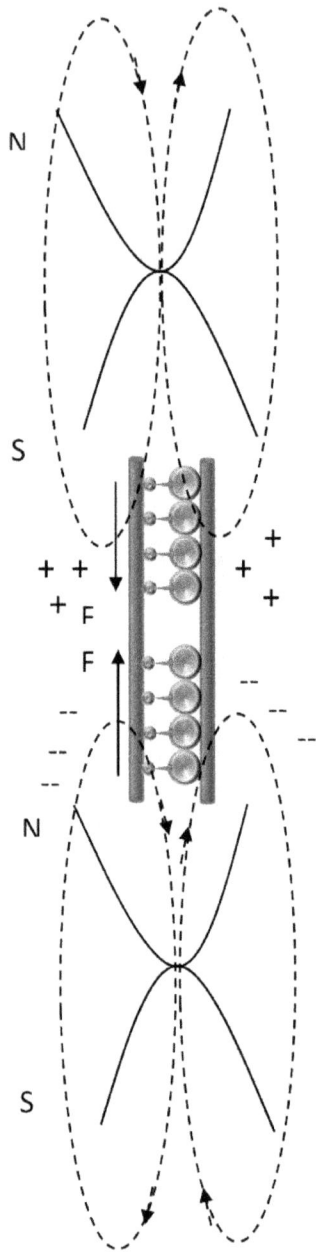

Fig. 4-12 – Conceptual representation of two opposing spacetime cavities bridged by extra dimension cylinders. Forces created by the monopoles attract or reject two opposing cavities of an atom.

The field and the forces associated with the monopoles arising from the north and south cavities create a moment between them which allows an atom and its spacetime cavities to pivot in either direction around the nucleus. This moment allows certain atoms to align themselves towards a specific direction making their forces additive.

In general, all atoms produce the magnetic monopole extra dimensions that create the magnetic field. Therefore, magnetism is an integral part of an atom. However, certain atoms such as Iron, Cobalt and Nickle can pivot and orient all of their north and south cavities in the same direction. When this happens, the forces of the monopoles from individual atoms add together and create a much stronger force that we observe as magnetism. A permanent magnet is one that its atoms are permanently oriented with regard to north and south cavities. Metals with magnetic properties can become magnetic by exposing them to the field of a permanent magnet and causing the north and south cavities to line up in the same direction. Alignment of the high and low-vac extra dimension cylinders exiting the spacetime cavities of certain atoms also means that the vacuum energy of these extra dimension objects become concentrated further magnifying the field effects they create.

Relationship between a charge and magnetic monopoles

When the field of the magnetic monopole is exposed to the tubular extra dimension (the charge), the magnetic monopoles will attach to the orbiting extra dimension cylinders of a Tubular Monopole of a spacetime cavity and can then flow and create an electric current as electrons which have a dimension of 13×10^{-12}m, Fig. 4-13. In the infinite momentum frame, magnetic monopoles attach to random extra dimension cylinders and create the smaller charge of 6×10^{-33}m, hence establishing the lines of the electro-magnetic field and creating a charge field.

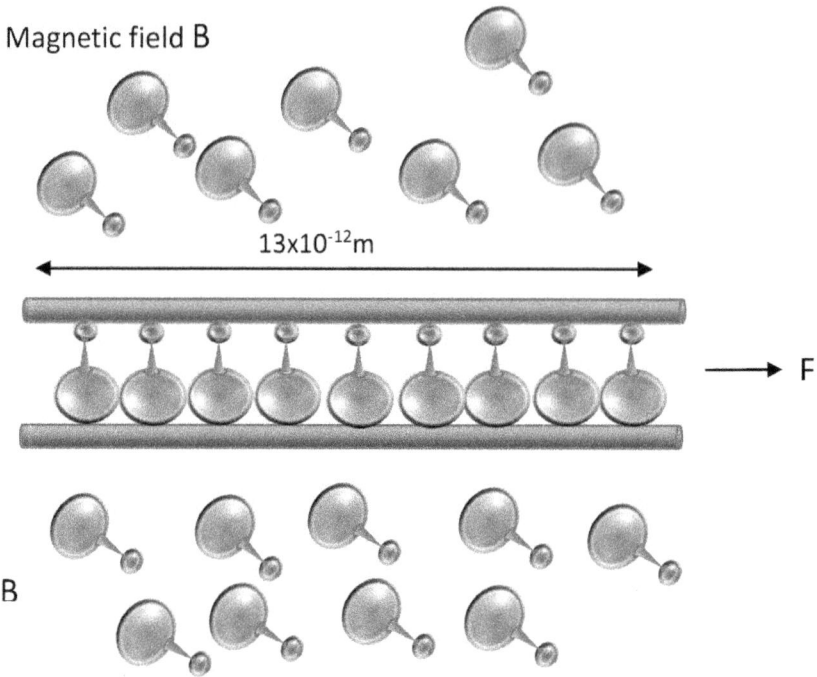

Fig. 4-13– Relationship between a field of magnetic monopoles (B) and tubular monopole (Charge). a) Attachment of moving magnetic monopoles (changing field) to extra dimension cylinders creating electrons (current). b) Attachment of magnetic monopoles to moving extra dimension cylinders creating electrons (current). c) Or creating a magnetic field from a moving electron.

The optimal orientation of a magnetic field is when the field is perpendicular to the extra dimension cylinders as shown in Fig. 4-14. The force generated by the magnetic monopoles will be perpendicular to the field of the monopoles and in the direction of the flow of the charge produced. The least optimal orientation is when the field is parallel to the extra dimension cylinders, resulting in no force in the direction of the flow of the charge.

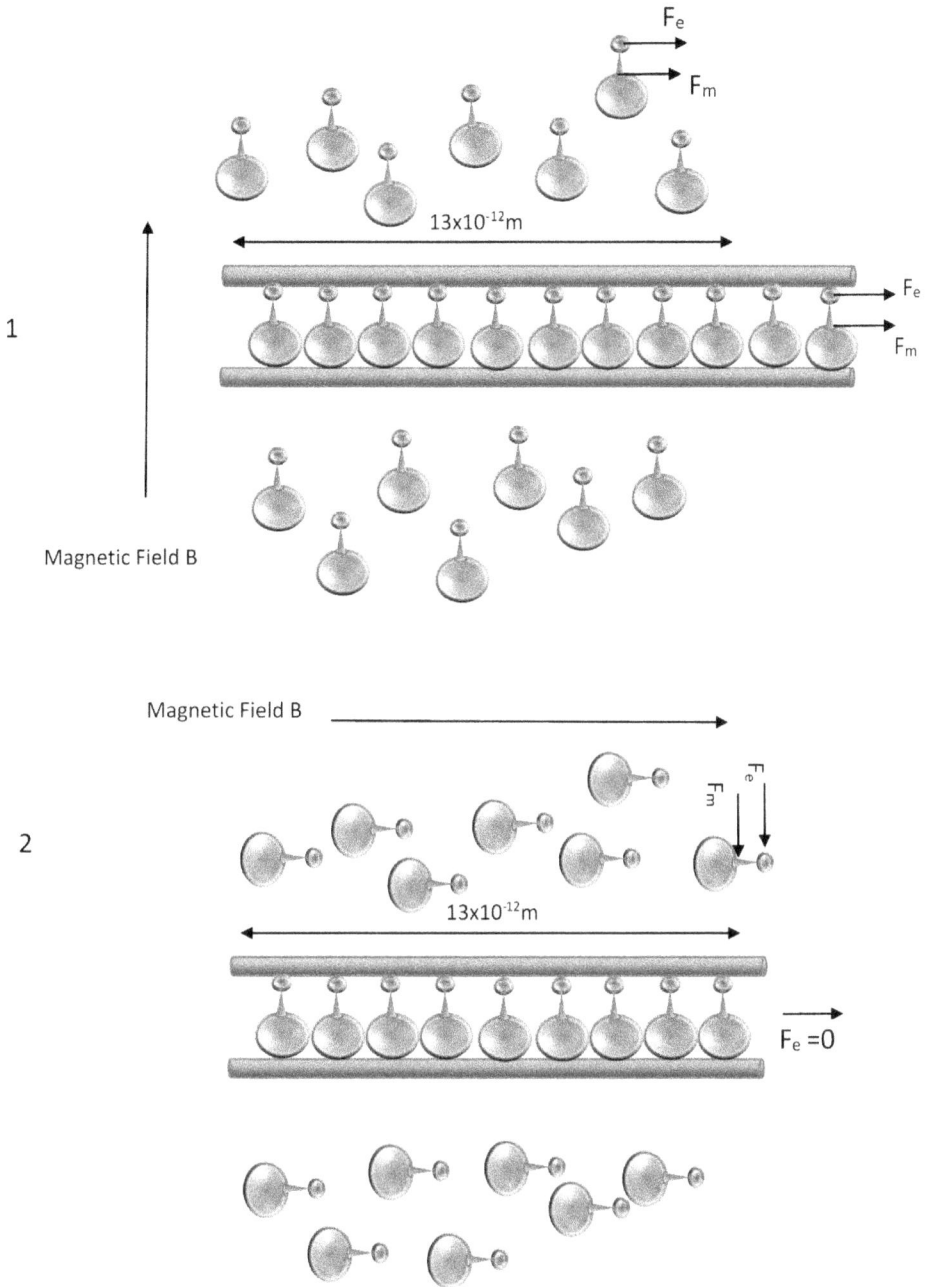

Fig. 4-14 – 1) Magnetic field perpendicular to the extra dimension cylinders, maximum effect. Electric Force is the summation of forces of the monopoles. 2) Magnetic field parallel to the extra dimension cylinders. No magnetic Force in the direction of the flow of charge, no current.

In order to establish a flow of current, there must be a change in the magnetic field B of the monopoles which creates the force F_m:

$$F_{emf} = F_e + F_m \qquad\qquad (4\text{-}3)$$

Since F_m produces the dominant charge (Planck charge), it must change to induce a ΔF in the field. Recall that F_m changes with the acceleration of the string at the E-pole of the activated looped string which changes with the diameter of the E-pole which in turn changes by expansion and contraction of the activated looped string.

Activated looped strings arising from the spacetime cavity of the atoms of a conductor material will have an expanding activated looped string as it enters the open field of infinite momentum frame outside the cavity. Therefore, inducing a motion in the field will allow a change in the field of magnetic monopoles with expanding looped strings. This motion then induces the ΔF in the field which becomes the driving force needed to make the elementary charge attached to the Tubular Monopole move in the direction of ΔF_m, and create a flow of current, Fig. 4-14.

The forces in equation (4-3) are the principle forces of the Lorentz Force equation:

Electromagnetic Force = Electrical force + Magnetic force

$$F_{em} = qE + qvxB \qquad\qquad (4\text{-}4)$$

When the monopoles attach to the extra dimension cylinders outside the spacetime cavity, they create a field of charges with driving force of qE. This force translates into the elementary charges in restframe (F_e). The change in the magnetic field creates ΔF_m. This is the driving force creating the flow of charges in the restframe.

The change in ΔF_m is the principle of Farady's law, where the electromotive force is the change of the field flux of the magnetic field:

$$\mathcal{E} = - d\phi_B/dt \qquad\qquad (4\text{-}5)$$

Where ϕ_B is the flux of the magnetic field.

The motion introduced in the field can be either by the field itself or the movement of conductors carrying the Tubular Monopoles. In either case, a gradient in the field will establish a ΔF which will then induce F_e and a flow of current.

CHAPTER 5
Dark Matter

Activated Looped strings

LET US NOW LOOK AT the overall energy-matter picture in the universe. Let us start with Fig. 1-10 introduced in Chapter 1 which demonstrates the fields of activated open and looped strings in infinite momentum frame from the time of the Big Bang until the fields are dissipated.

Fig. 1-10 - Radius of activated looped and open strings in infinte momentum frame, energy density in restframe vs The radius of the universe in restframe

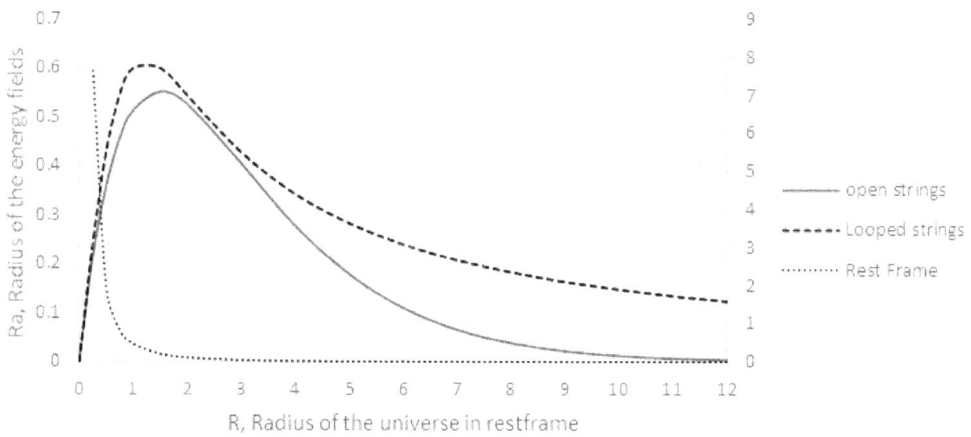

There are two important features to note on the graph of looped and open strings. One is the linear nature of both graphs from the time of the Big Bang when R=R$_a$, until the peak of the curves at around R $\approx$ 1.5. The linearity of the curves indicate that the three-dimensional activated open and looped strings follow the ρ =KR rule. Meaning that this portion of the field of the three-dimensional strings is in restframe1 and in equilibrium with the strings in restframe2. In other words, this is the portion of the three-dimensional strings that will decay creating the total mass of "matter" in the universe. The second important feature on the graph is the point where the field energy density of the strings in restframe equals that of open strings or Planck scale, at R=0.56 (see Chapter 1). This condition only occurs at the nucleus of the spacetime cavity where the energy density of strings in restframe equals the Planck energy density. Therefore, the segment between R=0 and R=0.56 represents the energy of strings comprising the mass inside the spacetime cavity of fundamental particles and the segment between R=0 and R =1.5 represents the energy of strings comprising the mass outside the spacetime cavity.

The mass created inside the spacetime cavity of fundamental particles is the baryonic mass and the mass created outside the spacetime cavity is "dark matter".

We can calculate the area corresponding to these two points to obtain the relative portion of the total energy in the universe that corresponds to dark matter (R=1.5) and baryonic matter (R=0.56). We will do this calculation for activated open strings and as seen from the graph, the relative proportion of energy is about the same for both looped and open strings.

First, the area under the graph of the activated open strings from R=0 to R=12 represents the total energy of the activated open strings in the universe. If we scale the graphs to the dimension of the mass of the universe:

$$\int_0^{12} Re^{-2/3R}\, dR = -3/2\ Re^{-2/3R} - 9/4\ e^{-2/3R} + C = 22.5 \text{ (units of R}_a{}^2)$$

The area under the graph from R=0 to R=0.56 represents the baryonic mass:

$$\int_0^{0.56} Re^{-2/3R}\, dR = -3/2\ Re^{-2/3R} - 9/4\ e^{-2/3R} + C = 1.24 \text{ (units of R}_a{}^2)$$

The area under the graph from R=0 to R=1.5 represents dark matter:

$$\int_0^{1.5} Re^{-2/3R}\, dR = -3/2\ Re^{-2/3R} - 9/4\ e^{-2/3R} + C = 6 \text{ (units of } R_a^2)$$

As such the percentage of each component as a ratio of the total energy in the universe will be:

Baryonic mass: $1.24/22.5 = 5.5\%$

Dark matter mass: $6/22.5 = 26.6\%$

Activated open strings in infinite momentum frame (Dark Energy): 67.9%

Baryonic mass/total mass $= 17.1\%$

Dark matter/total mass $= 82.9\%$

Let us now look at the mass of dark matter related to activated looped strings. This is the mass that is measurable in our universe just as the baryonic mass. From Fig. 1-10, this means the mass of all activated looped strings in restframe, i.e., attached to the vacuum of extra dimension cylinders.

In Chapter 3, I described the formation of the orbiting extra dimension cylinder and its role in establishing the baryonic mass of the matter inside the spacetime cavity.

Outside the spacetime cavity, the size of activated looped strings increases substantially from its dimension at the cavity nucleus (1.61×10^{-35}m, to 2.3×10^{-20}m). As I discussed in the previous chapter, the decay mechanism of the expanded activated looped string in infinite momentum frame is different from inside the spacetime cavity where the activated looped string is much smaller.

In Chapter 3, I described the formation of 2×10^{30} high-vac extra dimension cylinders that extend a distance of $R = 2.5 \times 10^{26}$ m or the restframe radius of the universe. It is likely that these extra dimension cylinders are formed as part of the structure of the singularity of a blackhole. I will discuss this in more depth in Chapter 10. The length and the vacuum energy of these extra dimension cylinders are unique in a sense that it can stretch the entire length of the universe and still attract and attach to activate looped strings despite the enormous dimensional mismatch between the looped string

(2.3x10^{-20}m) and the extra dimension cylinder (1.61x10^{-35}m). The structure created by the attachment of activated looped strings to select extra dimension cylinders (~2x10^{30}) constitute dark matter in the universe, Fig. 5-1.

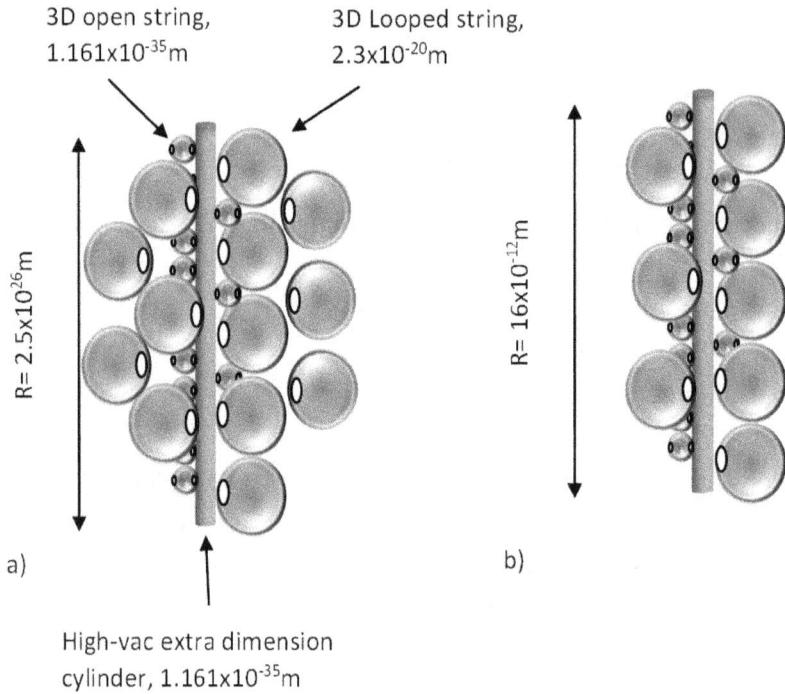

Fig. 5-1 - Conceptual representation of a) structural dark matter, b) distributed dark matter

Outside the spacetime cavity, the decay of the activated looped strings attached to these extra dimension cylinders results in the generation of new activated strings in a burst of sudden release which leads to a momentary formation of new particles as Higgs Bosons (see Chapters 4 and 8). In a way, Higgs Bosons which are a side product of the regeneration process for activated looped strings, are also a manifestation of an enormous mass in the universe that is "Dark Matter". The mass created due to the decay of the activated looped strings is the mass of "dark matter" as has been referred to in classical physics. This mass is, in large part, organized as a structure that extends the length of the universe and has 2x10^{30} high-vac extra dimension cylinders extending the entire length of the universe with activated looped and open strings as its attachments, Fig. 5-1.

This does not mean that the decay of activated looped strings does not occur away from these extra dimension cylinders. In fact, activated looped strings decay anywhere in the infinite momentum frame where two strings with opposing spin rotation can stay in contact long enough (at least one revolution). As such, Higgs Bosons are created all around us and anywhere in the infinite momentum frame. However, the substantial amount of the mass dominating the mass of "dark matter" in the universe in an organized, structured way is associated with decay of three-dimensional looped strings attached to the high-vac extra dimension cylinders.

In addition to the structural "dark matter" discussed above, activated looped strings also attach to the **"vacuum threads"** I described in Chapter 3, with a length of about $16x10^{-12}$m. These vacuum threads create the distributed mass of "dark matter" which forms the backbone structure of the restframe in the universe. However, the mass of this type of dark matter is significantly less than the structural dark matter because the mass of the strings attached to the "vacuum threads" drops by a factor of $4.3x10^{16}$. The mass of a strand of distributed dark matter is calculated as follows:

Number of activated looped strings attached to a strand of the wave:

$16x10^{-12}/2.3x10^{-20} = 6.95x10^8$

Mass of single activated looped sting attached to extra dimension cylinder:

$2x10^{-23}x0.23x10^{-16} = 0.46x10^{-39}$kg

The total mass of the wave:

$0.46x10^{-39}$ x $6.95x10^8 = 3.19 x10^{-31}$kg

As you can see above the mass of a strand of distributed "dark matter" is about the same as an electron!

Because distributed dark matter makes up the backbone restframe structure of space-time in the universe, it will be subject to the "String Resistance Force" I have described in detail in Chapter 11. "String Resistance Force" is the force in the universe that causes all the objects in the restframe to move and is created by the movement of activated and open strings in infinite momentum frame flowing past the strings in the restframe. As such, this force acting on the distributed dark matter will make the galaxies spin faster than you would expect from gravitational force of masses. Perhaps, one may consider this as one of the existing proofs of both "String Resistance Force" and "Dark Matter". Without the String Resistance Force, dark matter will not create this effect by itself.

The mass of the Higgs Boson particles that represent the decayed looped strings of "dark matter" is observable to our universe in the same way as I described the baryonic mass of fundamental particles such as a proton and a neutron. The only difference is that it has a significantly shorter half-life than stable particles with a baryonic mass. As such, one may view "dark matter" as a transitory state between "dark energy" which represents the mass of space and "baryonic mass".

The mass of the open strings attached to the dark matter structure is not measurable as a particle because, as stated earlier, open strings attach to other strings in our universe with no measurable change in the energy of the system. However, as you will shortly see, this mass is calculable as the mass contributing to the electromagnetic mass of the universe.

Recall that the baryonic mass of matter is a phenomenon observed when a three-dimensional looped string decays into a two-dimensional string by the virtue of annihilation of two strings with opposing spins inside the spacetime cavity. The mass observed is the change in energy that exits our three-dimensional universe (Hilbert space) via extra dimension cylinders corresponding to the string with the highest change in energy (at the center of mass). The change of energy in a baryonic mass is observable to our universe because it interacts with a two-dimensional open string which is the arm of the orbital rotation to which the three-dimensional string is attached. Two-dimensional open strings reside in Hilbert space, our universe, as such we can observe and locate this loss of energy as mass. The change of energy representing "dark matter" mass has no interaction with our universe where it exits Hilbert apace, as such it will exit our universe without a trace. As you will see shortly where it does manifest this energy loss as a mass is randomly scattered across the space and away from the original location of the string decay because its structure is moving at the speed of light. As such it will remain elusive.

A special case where a strand of vacuum thread does interact with our universe at its center of mass or wave, it manifests itself as a Neutrino with a mass of 0.46×10^{-39}kg and a wave of about 16×10^{-12}m as shown above.

The mass of dark matter in the universe is a cumulative addition of the masses of all activated looped strings that decay as a component of the fabric of spacetime in rest-frame1 (Fig. 5-1) and converted to Higgs Bosons with a mass of about 2×10^{-25}kg. Since the mass of an activated looped string in infinite momentum frame is about 2×10^{-23}kg, it converts to about 100 Higgs Bosons upon decay. In essence, the mass of "dark matter" is the energy of activated looped strings that exits our universe through the decay of 100 Higgs Boson. In the case of baryonic mass, it is the energy of the string that is located at the center of the mass of the specie exiting the three-dimensional universe. In either case, the two-dimensional strings that exit our three-dimensional universe manifesting the "mass" of an object form the two-dimensional circumference of the universe. Recall that this was an important condition in setting up the fundamental equations in Chapter 1.

We can calculate the mass of dark matter as an attachment of the activated looped strings to the 2×10^{30} high energy extra dimension cylinders that extend the length of the universe as follows:

The mass/length of each activated looped string attached to the extra dimension cylinders at restframe1 is:

2×10^{-23}kg$/2 \times 2.3 \times 10^{-20}$m $= 0.435 \times 10^{-3}$ kg/m

$M = 0.435 \times 10^{-3}$ kg/m x 2.3×10^{26}m x $2 \times 10^{30} = 2 \times 10^{53}$kg

As you can see above the mass of activated looped strings forming the structural mass of "dark matter" is significantly heavier and not reduced by a factor of 4.3×10^{16} in the same way as the mass of distributed "dark matter". This is because of crowding due to population of the entangled vacuum strands around the structural extra dimension cylinders which also causes the decay of activated looped strings but not attached to the extra dimension cylinders overshadowing the smaller mass of the strings that do attach to the extra dimension cylinders.

The composite structure of an activated looped string attached to an extra dimension cylinder in infinite momentum frame which constitutes the structure of "dark matter" is less complex than the structures leading to the baryonic mass inside the spacetime cavity.

Since the $2x10^{30}$ high-vac extra dimensions are likely to emanate from the singularity of a blackhole (see Chapter 10), one may suggest that blackholes are contributors to the formation of dark matter in the universe which is the basic structure of the mass of the universe. Higgs Bosons and dark matter are directly involved in the generation of $0.5x10^{77}$ three-dimensional looped strings per decay of one activated looped string (Chapter 8).

Continual generation of new three-dimensional strings in the universe is critical to maintaining the supply of looped strings which are being consumed by the baryonic masses. Therefore, blackholes play an important role in the regeneration process of three-dimensional looped strings and sustainability of the universe.

Energy and momentum in the universe

Earlier, I described the energy momentum equation for a single activated open and looped string which is the same for a two-dimensional string:

$$E^2 = (mc^2)^2 + (P_i c)^2$$

The energy momentum relation for the universe encompassing activated looped and open strings as a system will then be:

$$\sum E^2 = \sum (mc^2)^2 + \sum (P_i c)^2$$

$$E_t = [\sum (mc^2)^2 + \sum (P_i c)^2]^{1/2}$$

$$E_t = P_i c \sum [(mc^2)^2/(P_i c)^2 + 1]^{1/2}$$

Since the mass of matter is a small portion of the momentum energy, we can use the Taylor series approximation and simplify to:

$(1 + \epsilon)^{1/2} = 1+\epsilon/2$

Where ϵ is a very small number.

$\epsilon = (mc^2)^2/(P_i c)^2$

$E_t = \sum P_i c + \sum mc^3/2P_i$

Since we have an equal number of activated strings (open or looped) that rotate in opposite directions creating an equal amount of opposing momentum in opposing directions:

$\sum P_i = 0$

$E_t = \sum m^2\, c^3/2P_i$

Or,

$E_t = M^2 c^3/2P_i$ 　　　　　　　　　　　(4-4)

Where M is the total mass of matter in the universe. Equation (4-4) applies to both activated open and looped strings.

Let us apply this to activated looped strings. The momentum of the activated looped strings varies from $P_i = 6.5$ kgm/s when the string is at Planck length as a regenerated string from the decay of dark matter, to 0.45×10^{-14} kgm/s as a fully expanded looped string in infinite momentum frame. Since the latter is significantly smaller, the dominating momentum of the two will be 6.5 kgm/s. We know that the energy density of looped strings in infinite momentum frame is about 0.28×10^{36} kg/m^3, or 2.52×10^{52} J/m^3. The volume of the infinite momentum frame is:

$V = (4\pi/3R_a^3) = 6.4 \times 10^{78} m^3$

$E_t = 1/2(6.4\times10^{78}) \times 2.52 \times10^{52} = 1.6 \times10^{131}$J

Substituting in equation (4-4):

$M_i = 1.96 \times 10^{53}$kg

This is the total mass of matter in the universe in restframe which is essentially the same as the mass of dark matter earlier calculated by the number of extra dimension cylinders to be 2×10^{53}kg.

Open Strings Contributing to Dark Matter

If we apply equation (4-4) to open strings:

$$E_t = M^2 c^3 / 2 P_i \qquad\qquad (4\text{-}4)$$

The energy density of activated open strings is 0.17×10^{97}kg/m^3 or 1.53×10^{113}J, and the interstitial volume filled by the open strings in the universe is about 2.72×10^{78} m^3.

$$E_t = 1.53 \times 10^{113} \times (2.72 \times 10^{78}) = 4.16 \times 10^{191}\text{J}$$

$$M_o = 0.44 \times 10^{84} \text{ kg}$$

This is the total (space) mass of activated open strings in restframe1 in the universe contributing to electromagnetic forces. As mentioned earlier, this mass is not observable in our universe in the same way as the baryonic mass or the mass of dark matter created by the looped strings because it results in no change of energy in our universe. The energy of a decayed activated open string remains in our universe attached to other strings as a component of extra dimension objects. Unlike looped strings, as you will see in future chapters, this dark matter mass created by the activated open strings and its corresponding gravity leaves no effect on the curvature of spacetime.

Again, we can verify this mass by the virtue of the attachment of the activated open strings to the 2×10^{30} high energy density extra dimension cylinders that span across the radius of the universe.

Mass/ length $= 0.675 \times 10^{27}$kg

Mass$= 0.675 \times 10^{27} \times 1.0 \times 10^{26} \times 2 \times 10^{30} = 0.135 \times 10^{84}$kg

The average of the two above numbers will be about 0.28×10^{84}kg.

CHAPTER 6
Gravity

Activated Looped Strings

IN CHAPTER 1, I DEMONSTRATED the derivation of the equations representing the energy fields of a mass in infinite momentum frame vs. restframe for activated looped and open strings. I will now use the same fundamental equations described in Chapter 1 to derive the gravitational field equations of activated looped strings in our universe. Recall equation (1-11):

$$\sum c(F_i) = (\tfrac{1}{2})\text{k d /dt} \sum \Delta A_i \qquad\qquad (6\text{-}1)$$

where,

$$F_i = M_i g$$

g = gravitational acceleration of activated looped strings

M_i = analog mass (space mass) of activated looped strings

$$M_i = vd\rho$$

ρ= density of activated looped stings

R_a = radius of the field of the activated looped strings, our observable universe

$A = 4 \pi R_a^2$, surface area of the two-dimensional surface of the universe containing activated strings.

$dA = 8\pi R_a\, dR_a$

$v = 4/3\pi R_a^3$ = volume of the universe containing the activated looped strings

Substituting all the above in equation (6-1) and integrating the summations:

$\int -cgvd\rho = \int d[(\tfrac{1}{2})k\, dA]/dt$

Substituting for dA,

$\int -cgvd\rho = \int d[4\pi R_a\, k\, dR_a\, /dt]$

$$\int -4/3cg\pi R_a^3 d\rho = \int d(4k\pi R_a)\, dR_a/dt \qquad (6\text{-}2)$$

The term dR_a/dt is the propagation speed of the field of looped strings which is equal to c, the speed of light;

$\int -4/3cg\pi R_a^3 d\rho = \int d(4k\pi c R_a)$

Simplifying and integrating;

$$\int g d\rho = \int -(3k)dR_a\,/R_a^3 \qquad (6\text{-}3)$$

$g\rho = (3k/2)1/R_a^2 + C$

Using the boundary condition of $\rho = 0$ at $R_a = R$,

$C = - (3k/2)1/R^2$

$$g = 3k/2\rho\, (1/R_a^2 - 1/R^2) \qquad (6\text{-}4)$$

Equation (6-4) represents the gravitational acceleration of looped strings as a function of the radius of the energy field of a mass in infinite momentum frame and restframe for any mass since equation (6-1) is independent of mass "m". In essence, "g" is the acceleration of activated looped strings with a changing density and dimension towards any mass.

If we substitute for $R_a = 0.6R$, the slope of the linear portion of the graph of the field of activated looped strings, we will obtain "g" as a function of the radius of the field of the mass in infinite momentum frame (R_a):

$$g_m = (0.96k/\rho)\,(1/R_a^2) \tag{6-5}$$

As you will see shortly, equation (6-5) is the basis for the classic Newtonian gravitational acceleration of any mass "m" with a radius of R_a.

Substituting for $k=0.24\times10^{79}$ J/m^2, and $\rho = 0.28\times10^{36}$kg/m^3, and $R_a =1.15\times10^{26}$m in equation (6-5):

$g_m =6.2\times10^{-10}$ m/s^2

This is the gravitational acceleration of activated looped strings in infinite momentum frame in the universe because $R_a = 1.15\times10^{26}$m is the radius of the mass of the universe.

If we substitute for $R_a = 1.61\times10^{-35}$m and $\rho = 0.12\times10^{97}$, the gravitational acceleration of activated looped strings at the nucleus of a mass will be about:

$g_m = 7.4\times10^{51}$m/s^2

Let us re-write equation (6-4):

$$g_m = 3k/2\rho\,(1/R_a^2 - 1/R^2)$$

as:

$$g_m = 3k/2\rho\,[(1/R_a - 1/R)(1/R_a + 1/R)]$$

And Substitute for ρ (eq.1-15a):

$$\rho = (3k/c^2)(1/R_a - 1/R)$$

$$g_m = c^2/2 \ (1/R_a + 1/R)$$

$$g_m = g_\perp = c^2/2R_a + c^2/2R \qquad\qquad\qquad (6\text{-}6)$$

This new "g_m" has a different meaning. Since ρ, the density of the strings which represents variability of the dimension of activated looped strings comprising the mass, has been eliminated, the resultant $g_\perp$ in equation (6-6) represents the gravitational acceleration of looped strings in the direction perpendicular to g_m where the dimension of activated looped strings remain the same. In other words, this will be the gravitational acceleration of strings in a direction that is perpendicular to the extra dimension cylinders that make up spacetime in restframe. Whereas g_m in equation (6-5) is the gravitational acceleration of the looped strings comprising the macro-mass "m" in a direction where the dimension of activated looped string changes with R_a (changing ρ), Fig. 6-1.

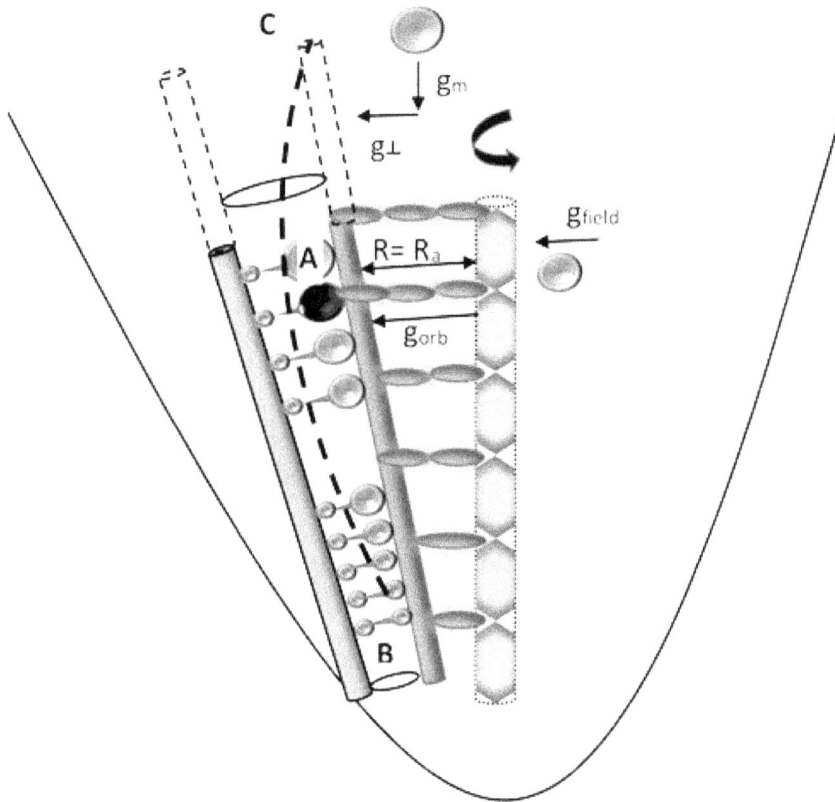

Fig. 6-1 – Conceptual representation of relationship of g_m in the direction of changing looped string dimension, vs. g_{fld} and g_{orb} in the direction of constant dimension. Dashed line BAC represents the curvature of the restframe at quantum scale. Axis of rotation in the center is the line of zero gravity.

Therefore, the gravitational acceleration in the direction where the dimension of activated looped strings remain constant has two components, one is $c^2/2R_a$ which is the field gravitational acceleration as a function of R_a (constant ρ) due to the decay of activated strings and the other is $c^2/2R$ which is the gravitational acceleration due to the orbital rotation of the strings in restframe. The vector of these two accelerations is in the same direction:

$g_\perp = c^2/2R_a + c^2/2R$

$g_\perp = g_{fld} + g_{or \,(orbital\ gravity)}$

where

g_{fld} = field gravitational acceleration (this is the lateral acceleration of looped strings towards the decaying strings attached to the extra dimension cylinders at constant ρ)

g_{or} = gravitational acceleration of looped strings due to orbital rotation (e.g., attached to orbiting extra dimension cylinders or other orbiting extra dimension objects)

If we rearrange equation (6-6), we will obtain:

$$R_a = Rc^2 /(2Rg_\perp - c^2) \qquad\qquad (6\text{-}7A)$$

If we substitute g_m for $g_\perp$, the gravitational acceleration of a mass with the changing dimension of an activated looped string (changing ρ) for macro-masses.

$$R_a = Rc^2 /(2Rg_m - c^2) \qquad\qquad (6\text{-}7B)$$

This is the gravitational field equation of activated three-dimensional looped strings in our universe which describes both the gravitational field of <u>large-scale masses</u> and <u>quantum gravity</u> to the string level.

Equation (6-7B) is the gravitational field equation for three-dimensional looped strings for any mass "m". This equation does not discriminate between large or quantum scale masses. As you will see shortly, how we distinguish quantum scale and large-scale gravity is based on how R_a and R are scaled according to the dimension of space, taking into account the dimensional transformation of space as discussed in Chapter 2.

Choosing natural units, c =1

$$R_a = R /(2Rg_m - 1) \qquad\qquad (6\text{-}8)$$

Recall that R_a and R can be extremely small but never zero.

A plot of the curvature of spacetime, i.e., R_a vs. R for different values of "g_m" is provided in Fig. 6-2. "g_m" represents the gravitational acceleration of a mass "m" which is calculated from Equation (6-5). Therefore, the graph of spacetime as shown in Fig. 6-2 will represent the curvature of spacetime created by the mass "m" which will have different "g_m" values depending on the radius and mass of the object.

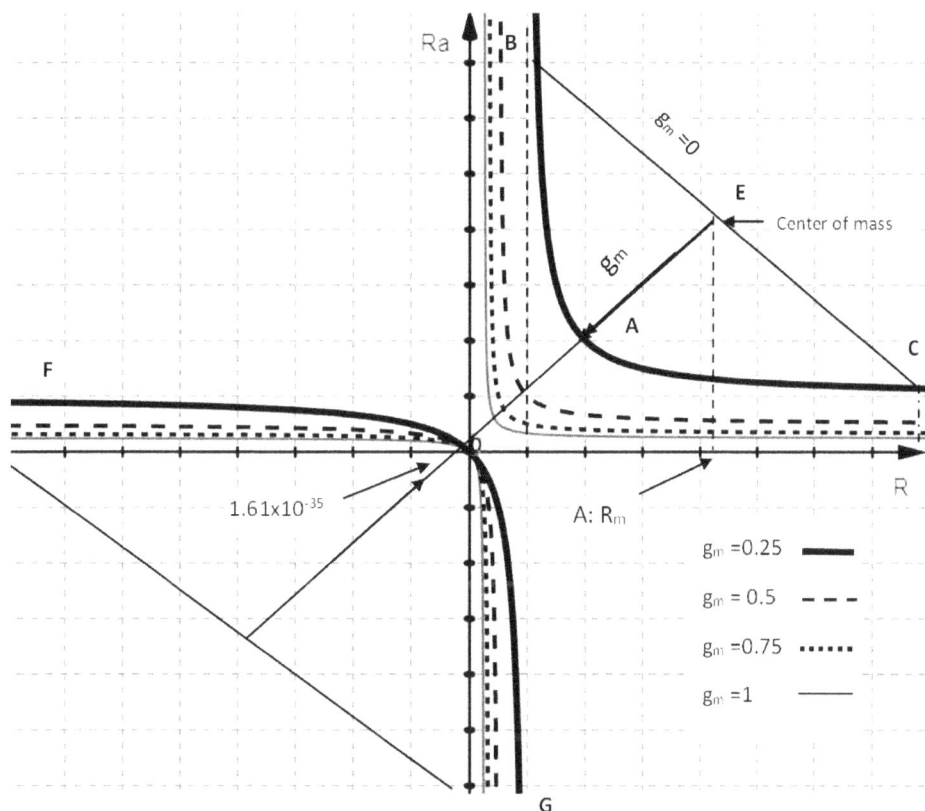

Fig. 6-2 – Spacetime curvature created by gravitational forces of various masses "m", radius R_m, and g. Upper right quadrant represents gravitational field of large scale masses, lower left quadrant represents quantum scale masses.

The graph of the gravitational field of the activated looped strings has two segments along the XY axis, one is the upper right quadrant, and the other is the lower left quadrant. To understand the significance of this graph and equation (6-8), we must now translate the corresponding space transformation with the energy field of looped strings we observed in the previous chapter to the gravitational field of the looped strings in Fig. (6-2).

Recall that from Chapter 2, "space" experiences a transformation at about 1.8×10^{-9}m where the spacetime cavity is created, Fig. 6-3. We also learned that this space transformation coincides with the shrinkage of the dimension of activated looped strings which dominates the space with a dimension of 2.3×10^{-20}m in infinite momentum frame as compared to the activated open strings with a dimension of 1.61×10^{-35}m.

Fig. 6-3 - Two stages of dimensional transformation of space in the universe

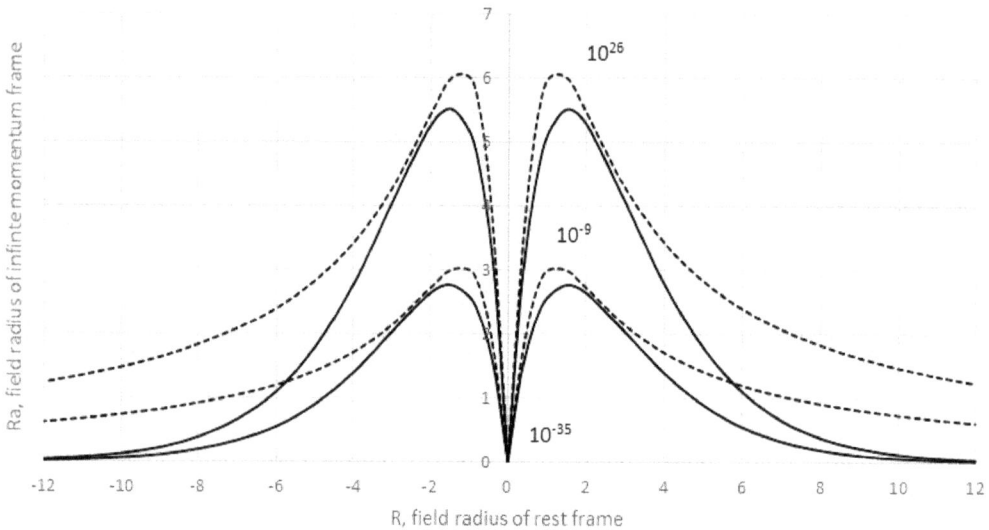

We must account for this space transformation in the gravitational field equations by making the corresponding rescaling of the spatial dimensions of the fields. In Fig. 6-2, the origin 0, similar to the spacetime cavity, represents the nucleus of matter at 1.61×10^{-35}m.

Referring to Fig. 6-2, the graph in the upper right quadrant represents the curvature of spacetime in restframe created by any macro-mass "m" with a gravitational acceleration of its mass "g_m" and a quantum mass with a gravitational acceleration of "$g_\perp$" depending on the scale chosen for R_a and R. The graph on the lower left quadrant of Fig. 6-2 represents the curvature of spacetime in infinite momentum frame created by the changing dimensions of three-dimensional looped strings in quantum field of the mass. Let us first explore the interpretation of the graph of the macro-mass "m".

As you can see from Fig. 6-2, the curvature of spacetime increases with "g_m" meaning that as the mass and radius of the object increases, the curvature of the field in restframe also increases. On the line of zero-gravity, the spacetime is flat (a straight line). On the body of any mass, the radius of the field gravity and orbital gravity is always the same, i.e., R_a = R, Fig. 6-1. This is because looped strings accelerating towards any part of a mass in restframe have the same field radius (R_a) as its orbital rotation. Therefore, the intersection of the line with a 45-degree angle through the origin and the curves represent the points on the mass. The intersection of this line and the line of zero gravity represents the center of the mass "m" with the radius of R_m.

The graph has two asymptotes. The asymptotes represent the limits or the boundaries of the curvature of spacetime in the restframe created by the mass. The limits of each asymptote can be calculated based on the "g_m" value of the mass:

$$R = c^2/2g_m$$

For example, if we substitute for g_m = 6.4x10^{-10}m/s^2, which is the gravitational acceleration of the looped string in the universe as a whole, the value of the asymptote will represent the boundaries of spacetime at restframe, which will be 0.7x10^{26}m. If we substitute for g_m = 5x10^{25}m/s^2, the boundaries will be 0.9x10^{-9}m and the graph in the lower left quadrant of Fig. 6-2 will be reduced to the spacetime cavity of a single particle or atom. Therefore, the graph can be rescaled to any mass by calculating its asymptote based on the "g_m" value of the mass.

However, regardless of the magnitude of mass "m", the corresponding coordinates of the curvature of spacetime for that mass can be calculated precisely and reproducibly. Let us now look at some of these important coordinates. We will do this first for macro-masses in the upper right quadrant of the graph.

Referring to Fig. 6-2, there are four major coordinates for a macro-mass, A, B, C, and E. Coordinate A represents the surface of a mass where the effect of g_m is at its maximum. As you can see, this is where R_a =R. The coordinate of point A is obtained by projecting the coordinate of the center of mass E on the R axis. This will yield the radius of the mass as "R_m". Coordinate B on one boundary represents a point where its radius is about $R = R_m/4$. Coordinate C on the other boundary is projected on the R axis and represents a point where its radius is about $R = 2R_m$.

Let us now calculate the gravitational acceleration of the looped strings in restframe corresponding to the above coordinates using equation (6-6):

$$g_\perp = c^2/2R_a + c^2/2R$$

On the mass itself, $R_a = R = R_m$

$$g_{fld} = g_{or} = c^2/2R_a = c^2/2R = c^2/2R_m$$

$$g_\perp = c^2/2R_a + c^2/2R_a = c^2/R_a = c^2/R = c^2/R_m \qquad (6\text{-}9)$$

Incidentally, if we multiply both sides of equation (6-9) by the mass "m":

$$mg_\perp = mc^2/R_m$$

Or,

$$mg_m R_m = mc^2 \qquad (6\text{-}10)$$

The left side of the equation is the force "$mg_\perp$" applied by the activated looped strings on the mass "m" with the radius of R_m. This is in essence the energy applied to the mass "m" in restframe by the gravitational effect of the activated looped strings in order to prevent the mass "m" from disintegrating into its string components and hence the famous:

$$E = mc^2$$

Let us rearrange equation (6-6) to represent $g_\perp$ only as a function of R in the restframe by substituting for $R_a = R/1.6$, the slope from the energy field graph of the activated looped string:

$$g_\perp = c^2/2R_a + c^2/2R = 1.3\ c^2/R \qquad (6\text{-}11)$$

If we substitute for R $=R_m/4$ in equation (6-11), we will obtain the gravitational acceleration at restframe corresponding to the coordinates of point B:

$$g_\perp = 1.3c^2/R = 5.2\ c^2/R_m \qquad\qquad (6\text{-}12)$$

This is the maximum gravitational field effect on the restframe, not on the body of mass "m".

If we substitute for R $=2R_m$, we will obtain the minimum gravitational effect on the restframe corresponding to points C, not on the body of mass "m".

$$g_{rf} = 1.3c^2/R = 0.65c^2/R_m \qquad\qquad (6\text{-}13)$$

Note that the curvature of spacetime for macro-masses will never cross the nucleus of the matter (the origin of the axis, Fig. 6-2) no matter how high the value of "g_m" is, except as you will see later in a blackhole where the curvature of spacetime for a large macro-mass will cross the nucleus of the matter.

As you can see above, the highest gravitational acceleration corresponds to point B which is 5.2 times the gravitational acceleration of activated looped strings on the mass itself. The lowest gravitational acceleration of activated looped strings corresponds to point C which is 0.65 times that of the mass.

Coordinates of curvature of spacetime of mass "m" inside the energy field of activated looped strings

Let us now apply what we have learned from the gravitational field equation and the graph representing the effect of "g_m" on the restframe (Fig. 6-2) to the energy field of activated looped strings we have studied in previous chapters.

We can obtain the coordinates of any mass "m" on this energy field and the corresponding curvature of spacetime based on the information we obtained from Fig. 6-2 and the equations (6-7B) and (6-8). Referring to Fig. 6-2, the center of the mass of matter "m" for large scale masses or macro-masses with a dimension greater than 1.8×10^{-9}m, falls on the intersection of the axis of R_a and the line representing zero gravity where the spacetime is flat.

On the graph of the energy field, Fig. 6-4, this center of mass falls on the intersection of the line of zero gravity (line CC), and the axis R_a, point E. The center of the mass will be at $R_a = R_m = 0.6$.

Fig. 6-4 - Curvature of spacetime of macro-mass "m" in the field of looped strings

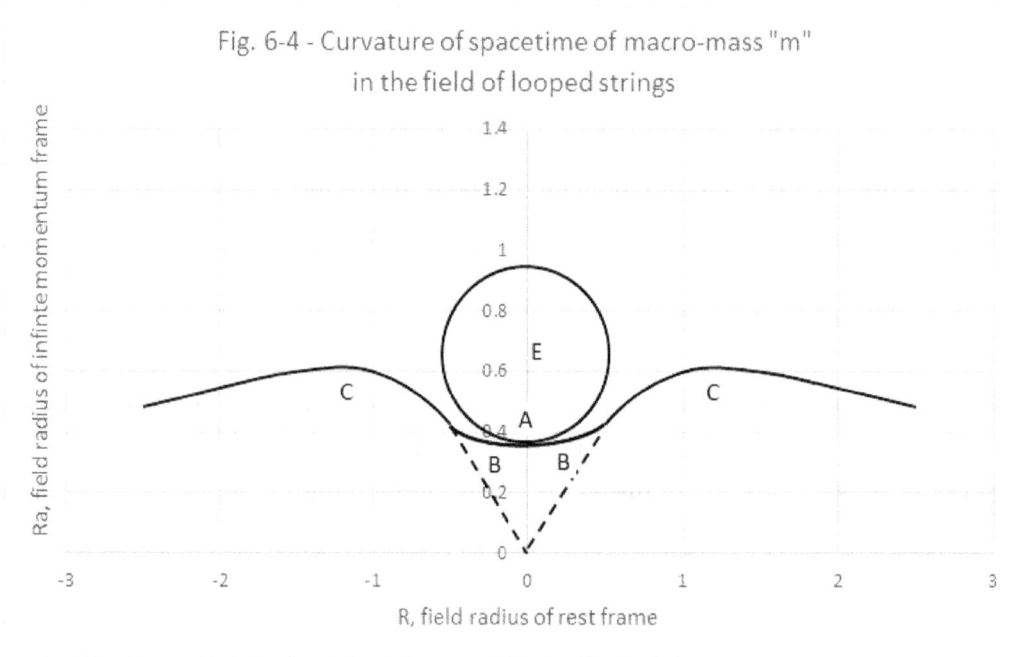

The maximum gravitational field effect of mass "m" on the restframe at one boundary where $R = R_m/4$ was calculated to be:

$g_{rf} = 5.2c^2/R_m$

Referring to Fig. 6-4, this corresponds to the coordinate of spacetime at point B where

$R = R_m/4 = 0.6/4 = 0.15$

The coordinate of point B on the axis of R (0.15) is 1/8 the value of the coordinate of C (1.2) matching the same in Fig, 6-2.

The smaller gravitational field was calculated to be:

$g_\perp = 0.65 \, c^2/R_m$

This corresponds to the coordinate of spacetime at point C where the curvature of spacetime is flat, R_a or R = 2R_m = 1.2.

The gravitational acceleration of activated looped strings on the mass "m" itself, where R_a =R =R_m, was calculated to be:

$$g_{rf} = c^2/R_m$$

Since any mass "m" similar to its smallest constituent has a mirror spacetime cavity, the above coordinates are repeated for the opposite side of the mass, Fig. 6-5.

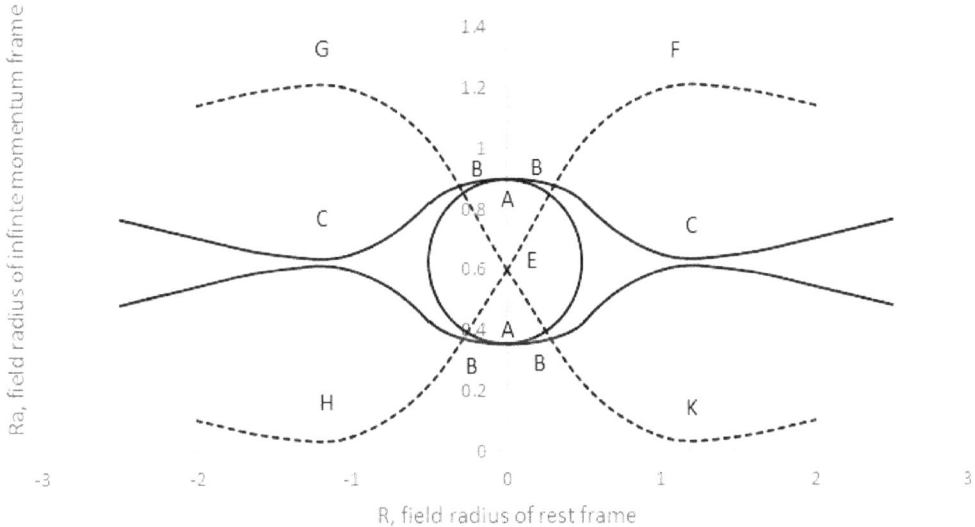

Fig. 6-5 - Symmetric curvature of spacetime of macro-mass "m" and its quantum field

In summary, the curvature of spacetime created by any macro-mass "m" greater than 1.6×10^{-9}m has 6 major inflection points and boundaries. Two coordinates with the weakest gravitational acceleration in rest frame where the orbital radius is twice the radius of the mass. Four coordinates with the highest gravitational acceleration where the orbital radius is one-fourth of the radius of the mass. The above coordinates create what I had termed as the "spacetime well". Note that the above coordinates were given as examples of how to determine the curvature of spacetime at its maximum and minimum gravitational coordinates in restframe. All other points can be precisely calculated from equations (6-5) to (6-8) and their corresponding graphs. In essence, the curvature of spacetime created by a macro-mass is the summation of all the $g_\perp$ vectors of individual spacetime cavities as shown in Fig. 6-1. The effect of the $g_\perp$ vectors on the many orbiting extra dimension cylinders arising from each cavity with its associated curvature are added up to produce the overall spacetime curvature surrounding a macro-mass as shown in Fig. 6-5. This is spacetime curvature produced by a mass as described in general relativity. It is the curvature of restframe structure of the spacetime.

Therefore, the upper right quadrant of Fig. 6-2 represents the spacetime curvature of a macro-mass as described in general relativity, and "quantum gravity" as described in classical physics depending on the scale of R_a, and R. The difference is that the line of zero-gravity for a macro-mass falls outside the energy well of the mass, whereas, it becomes the axis of orbital rotation of the strings inside the spacetime cavity of a quantum mass. In either case, Point A is the deflection point of the curvature of spacetime for both quantum mass and macro-mass.

Let us now address the gravitational field of the activated looped strings corresponding to the same mass at quantum scale as shown in the lower left quadrant of Fig. 6-2. The asymptotes of the graphs in the upper and lower quadrant represent the break point between the space transformation of two distinctively different energy fields.

The graph in the lower quadrant represents a field with an effect on the curvature of spacetime which is independent of the magnitude of the mass. Its maximum curvature of spacetime and gravitational acceleration is located at the nucleus. This point is represented by the origin of the graph of Fig. 6-2 and will translate to coordinate E in Fig. 6-5.

The graph in the lower left quadrant also derived from equation (6-8), demonstrates the gravitational field of the subatomic constituents comprising mass "m". Let us refer to this field as the "quantum gravitational field" of the mass "m". The corresponding graph of the "quantum gravitational field" on the energy field of the looped strings is shown in blue dotted lines in Fig. 6-5.

If mass "m" was a single atom or a subatomic particle, the graph in the lower left quadrant will represent its energy field as shown in Fig. 6-4. However, since a large mass is an aggregate of many particles and spacetime cavities oriented in many directions, it will create a field that is a vector summation of all the fields created by the spacetime cavities of individual atoms.

Therefore, the quantum gravitational field is created as a result of the addition of the forces and energies of all extra dimension objects arising from the spacetime cavities of individual atoms making up the mass of an object. Since the fields inside and arising from the spacetime cavities are vector fields discussed in Chapter 3, the resultant quantum gravitational field will be the net field of the summation of all vectors pointing in different directions. There are two spacetime cavities, north and south, creating the quantum fields, the nucleus of the field will be located at the center of the mass, Point E, Fig. 6-5 which falls on the line crossing through the origin on both upper and lower graphs of Fig. 6-2. Let us refer to the quantum gravitational field as simply the "quantum field" for abbreviation from this point forward.

The quantum field is created as a result of the abundance of pulsating extra dimension cylinders arising from the mass "m". The pulsating strings create and supply the energy for the orbital rotation of all other extra dimension objects I discussed in Chapter 3. The orbiting extra dimensions arising from the many spacetime cavities of all atoms making up a mass converge into creating a more organized field.

In essence, the quantum field of mass "m" is a giant spacetime cavity of vector fields with its nucleus at a dimension of 1.61×10^{-35}m and its boundaries (the asymptotes) extending as far as that of the macro-mass gravitational field in the upper right quadrant.

Referring to Fig. 6-5, the nucleus of the quantum field spacetime cavity will be at point E. The two boundaries of the spacetime cavity represented by the points F, G, H, and K are those shown in Fig. 6-2. Its coordinates on the axis of R are the same as the coordinates of Point C where R_a equals $2R_m$. The coordinates of the quantum field of the mass are shown in both Fig. 6-2 and 6-5.

If we substitute for $g_m = 5 \times 10^{25}$m/s^2, the boundaries of the quantum field in Fig. 6-2 will be 0.9×10^{-9}m and the graph in the lower left quadrant will be reduced to the spacetime cavity of a single particle or atom.

An important significance of this quantum gravitational field is that it behaves similar to a very large spacetime cavity creating the same effect as the gravitational field of the mass. Meaning that the activated looped strings entering this field will be subjected to its gravitational field and begin to shrink in size. Shrinkage of the activated looped string which dominates the space in our universe, means the shrinkage of the space itself, Fig. 6-6.

Simply put, a quantum field is the summation of all vacuums generated inside spacetime cavity of each atom, as a result of the decay of three-dimensional looped and open strings. with its sharp apex at the nucleus of the each spacetime cavity, Fig. 6-1. As you can see, this is the main cause of gravity, disappearance of three-dimensional strings, the main constituents of spacetime and dark energy in our universe, and the vacuum it generates. General relativity does not address this important process. The curvature produced is the shrinkage of space itself in both infinite momentum frame and restframe. One must think of the field of quantum gravity as a giant spacetime cavity.

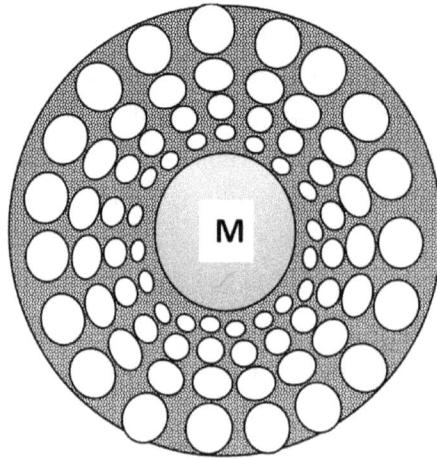

Fig. 6-6 -Shrinkage of 3D looped strings (space) inside the quantum field of mass M. White circles represent activated looped strings, small dots, activated open strings.

The change in the dimension of the activated looped string inside the gravitational and quantum field of a mass follows equation (2-3):

$$r_a = 0.71x10^{-26} (Ra)^{1/4}$$

In which R_a, is the distance from the center of a mass "m", point E, Fig. 6-5.

As you will see in Chapter 11, this has a significant effect on how we measure time. We see this effect as the time dilation effect around any body of mass and it will be the most significant inside the quantum field of a Blackhole.

Classical Newtonian gravity

The forces created by the activated looped strings in any body of mass must balance out the force created by the mass of the universe "M" with a gravitational acceleration of "g =6.2 x10^{-10}m/s^2" as calculated earlier:

The force created by activated looped string is:

$$F= V(\rho g)$$

Where $V = 4/3\pi R_a^3$, substituting for (ρg) from equation (6-5),

$$F= 4/3\pi R_a^3(0.96k/R_a^2) = Mg = Mx6.2x10^{-10}$$

$$M= 0.15x10^{89}R_a \qquad\qquad (6-14)$$

On one extreme, if we substitute for $R_a = 1.61x10^{-35}$m, the forces outside the spacetime cavity must equal that of the highest force inside the nucleus of the cavity. In that case, the mass of the universe in restframe is calculated to be about:

$$M = 2.5x10^{53} \text{ kg}$$

This means that the universe must have a mass of about 2.5x10^{53} kg to maintain the balance of the forces inside and outside the spacetime cavity of an atom. This is about the same as the 2x10^{53}kg calculated in the previous chapter.

On the other extreme, if we substitute for R=1.15x10^{26} in equation (6-14):

$$M=1.7 x10^{114} \text{ kg}$$

This is the mass of activated looped strings creating the infinite momentum frame of space itself. In other words, space itself must have a mass of 1.7x10^{114}kg for the forces of infinite momentum frame and restframe to equal. As I mentioned before, this mass is not observable, because the change in its energy is too small to be detectable in our timescale. As shown earlier, this mass will be consumed over the next 120B years.

As you saw earlier, the gravitational acceleration of activated looped strings for any mass "m_1" was derived from equation (6-5) to be:

$$g_1 = 0.96k/\rho \ (1/R_a^2)$$

The force created and applied to any mass "m_1" with a radius R_a is:

$$m_1 g_1 = 0.96 k m_1 /\rho (1/R_a^2) \tag{6-15}$$

This force must equal the force applied by the universe with mass M (in restframe) and gravitational acceleration of g to mass m_1:

$$Mg = m_1 g_1 = 0.96 k m_1 /\rho (1/R_a^2)$$

$$g = (0.96k/\rho M)(m_1 /R_a^2)$$

This is the gravitational acceleration of the force applied by the mass of the universe to any mass "m_1".

Substituting for the values of $k = 0.24 \times 10^{79} J/m^2$, $\rho = 0.2 \times 10^{36} kg/m^3$, and $M = 1.8 \times 10^{53} kg$, we see that the constant "$0.96k/\rho M$" is in reality G, the universal gravitational constant.

$$0.96k/\rho M = 6.4 \times 10^{-11} m^3 /kgs^2 = G$$

Substituting for the constant $0.96k/\rho M = G$, equation (6-15) becomes:

$$g = Gm_1 /R_a^2 \tag{6-16}$$

Equation 6-16 is Newton's gravitational acceleration of mass "m_1" with a radius of R_a. If we place another mass "m_2" in this field, it will experience the force created by the "g" calculated above, and R_a becomes the distance between the two masses:

$$F = m_2 g = Gm_1 m_2 /R_a^2 \tag{6-17}$$

Equation (6-17), which is essentially a variation of equation (6-5) derived from the "energy balance" equations of the activated looped strings in the universe, is Newton's classical equation of gravity describing the force between two masses.

CHAPTER 7
Electromagnetic Forces

Activated Open strings

ACTIVATED OPEN STRINGS CREATE THE electromagnetic fields and the associated forces in our universe. In conjunction with activated looped strings, they create the extra dimensions responsible for the electric and magnetic fields. Recall an important property of the activated open strings is that unlike activated looped strings, its size, energy, and mass remain the same from the time of the Big Bang and throughout the expansion of the universe, inside and outside the spacetime cavity.

Let us now look at the fundamental equations leading to the acceleration of activated open strings (g) and the corresponding electromagnetic forces.

From Chapter 1, the open string energy balance was given as:

$$\sum c(F_i) + 1/2 \sum c(F_o) = (k/2L_p)\, d/dt \sum \Delta v_i \qquad (7\text{-}1)$$

$$F_i = M_i g$$

g = electromagnetic acceleration of activated open strings

M_i = total analog mass (space mass) of open strings

$M_i = vd\rho$

ρ = density of open strings

R = radius of the universe in restframe

R_a = radius of the universe containing activated open strings, the observable universe

$v = 4/3\pi R_a{}^3$ = volume of the field of activated open strings in the universe

$dv = 4\pi R_a{}^2 dR_a$

Substituting all the above in equation (7-1) and integrating:

$\int -cgvd\rho +1/2 \int cgvd\rho = \int d[(k/2L_p)dv/dt]$

$\int -4/3cg\pi R_a{}^3 d\rho = \int d[(k/L_p) \, 4\pi R_a{}^2 dR_a/dt]$ (7-2)

The term dR_a/dt is the propagation speed of the field of activated open strings which is equal to c, the speed of light. Equation (7-2) then reduces to:

$\int -4/3gR_a{}^3 d\rho = \int (k/L_p)8R_a dR_a$

$\int gd\rho = \int -(6k/L_p)dR_a/R_a{}^2$

$g\rho = (6k/L_p)1/R_a + C$

Using the boundary condition of $\rho = 0$ at $R_a = R$,

$C = - (6k/L_p)1/R$

$g = 6k/\rho L_p (1/R_a - 1/R)$ (7-3)

Since ρ is constant for activated open strings, $\rho = 0.18 \times 10^{97}$ kg/m³ and substituting for $k = 0.24 \times 10^{79}$ N/m, $L_p = 1.61 \times 10^{-35}$

$g = 0.49x10^{18} (1/R_a -1/R)$ (7-4)

Equation (7-4) is the equivalent of the gravitational field equation for activated open strings or the electromagnetic forces.

Substituting for R, R_a/R =0.7 is the slope of the graph of the field of open strings:

$g = 0.147x10^{18} (1/R_a)$ (7-5)

Equation (7-5) is the electromagnetic acceleration of activated open strings for any mass "m" in infinite momentum frame.

Substituting for $R_a = 1 x10^{26}$ m (for open strings):

$g = 14.7 x10^{-10}$ m/s^2

This is the acceleration of activated three-dimensional open strings in infinite momentum frame which is about the same as the gravitational acceleration of activated looped strings in the infinite momentum frame, $6.2x10^{-10}$m/s^2.

Similar to the activated looped strings, on the mass itself, whether it is a macro-mass or inside the spacetime cavity at quantum scale, in restframe, R_a =R. Substituting for R =R_a in equation (7-4):

g=0

This means that activated open strings, unlike the activated looped strings, do not have a net electromagnetic force on the restframe and thus have no effect on the spacetime curvature. In other words, the field acceleration of the open strings and the g-force of the orbital rotation in restframe cancel each other out creating a net zero effect on the curvature of spacetime. However, because of its coexistence in the universe with activated looped strings which define the curvature of spacetime, the field of activated open strings in infinite momentum frame and restframe follow the spacetime curvature created by the much larger looped strings which dominate the space.

Coulomb's law

In Chapter 6, I demonstrated the derivation of Newton's universal law of gravitation between two bodies of masses based on the gravitational acceleration of three-dimensional looped strings:

$$F = mg = GMm/R_a^2$$

Also recall in Chapter 3, I introduced an extra dimension object called a magnetic monopole. A magnetic monopole is composed of one three-dimensional activated looped string attached to a three-dimensional activated open string via two-dimensional open strings having a length of about 4.4×10^{-33}m. When a magnetic monopole attaches to two extra dimension cylinders in restframe, it forms a new extra dimension described as a Tubular Monopole. Tubular Monopoles are the basic structure of a "charge" with a length of about 13×10^{-12}m or 6×10^{-33}m. As I described in Chapter 3, the charge of a Tubular Monopole outside the spacetime cavity is dominated by Planck charge which is larger than the elementary charge by an order of magnitude. This charge is applied to a mass of 2.18×10^{-8} kg which is the mass of a three-dimensional open string. Therefore, using the standard value for the Planck charge, the mass/charge of the charge will be:

$$m/q = 2.18 \times 10^{-8}\text{kg}/1.87 \times 10^{-18}\text{kg-m} = 1.16 \times 10^{10} \text{ m}^{-1} \text{ or kg/Coulomb}$$

In other words, the Planck charge can displace an activated open string 1.16×10^{-10}m along the length of the Tubular Monopole. If we apply Newton's law of gravity to two quantities of charges and use their mass to calculate the forces between the two masses:

$$F = Gm_1 m_2/R_a^2$$

Substituting for m_1, m_2, q_1 and q_2 and multiplying by the factor calculated above, we will obtain:

$$F = Gq_1(1.16 \times 10^{10})q_2(1.16 \times 10^{10})/R_a^2$$

$$F = Gq_1 q_2 (1.16 \times 10^{10})^2/R_a^2$$

Substituting for $G = 6.674 \times 10^{-11} \text{ m}^3/\text{kgs}^2$

$F = 8.98 \times 10^9 \, q_1 q_2 / R_a^2$

The coefficient 8.98×10^9 which has units of kg m/s^2 or Nm^2/ C^2 is k_e or the Coulomb's constant we know in classical physics:

$$F = k_e \, q_1 \, q_2 / R_a^2 \qquad\qquad\qquad (7\text{-}7)$$

Equation (7-7), Coulomb's law is essentially Newton's gravitational law for two bodies of masses converted to the "charge" or the force-distance relationship of the field of the Tubular Monopoles between the two bodies of masses.

Comparison of electromagnetic and gravitational forces

The gravitational acceleration of activated looped strings in restframe is calculated to be about 2.46×10^{-10}m/s^2 and for activated open strings 10.3×10^{-10}m/s^2 which are approximately the same. On the other hand, the energy density of activated looped and open strings in infinite momentum frame are respectively, 0.2×10^{36}kg/m^3 and 0.18×10^{97}kg/m^3.

The total mass associated with activated looped and open strings at restframe in the universe was calculated to be about 2×10^{53}kg, and 0.48×10^{84}kg respectively. The force exerted on the restframe for each component of string outside the spacetime cavity are:

$\sum F \text{ (gravity)} = 2.46 \times 10^{-10} \times 2 \times 10^{53} = 0.49 \times 10^{44}$ N

$\sum F \text{ (electromagnetic)} = 10.3 \times 10^{-10} \times 0.28 \times 10^{84} = 2.9 \times 10^{74}$N

$\sum F \text{ (electromagnetic)} / \sum F \text{ (gravity)} = 0.59 \times 10^{31}$

As you have seen in previous chapters, inside the spacetime cavity, and at the nucleus, the energy density, mass, and gravitational acceleration of activated looped and open strings are about the same. Meaning that the above ratio will be about 1.

CHAPTER 8

Creation and Regeneration of Activated Three-Dimensional Strings

E-loop String

RECALL THAT IN CHAPTER 3, I introduced an extra dimension object which was created when a pulsating string collapses from a Planck radius of 1.61×10^{-35}m to about 2.3×10^{-51}m, named the **E-loop string** or E-loop extra dimension. In the absence of a sweeping orbital motion and orbital gravity, the two-dimensional looped (or open) string with a Planck radius shrinks to the smallest dimension that the universe will allow it to exist as a vacuum entity. This tiny, looped string is the largest vacuum object that can exist by itself as a stable entity in our universe without an orbital rotation. In other words, this tiny, looped string with a surface area of about 1.66×10^{-101}m^2 is the largest space that the universe will allow as a stable space without requiring an orbital energy to sustain its existence. The E-loop is a (substantially) two-dimensional vacuum specie that differentiates itself from the bulk vacuum of the universe by virtue of its spin which exists in the form of its potential energy.

The E-loop strings are a vacuum species that are in turn created from yet a smaller vacuum specie that is (substantially) one dimensional. The mechanism and process for the formation of the E-loop strings will be discussed in Chapter 12. However, to describe the process for the formation of Planck-size strings which form the foundation of our three-dimensional universe, I must briefly describe some of its important characteristics.

The E-loops have an important property, given the required energy they can be stretched from its circular shape to a large extent while maintaining its surface area of 1.66×10^{-101}m^2. Therefore, as the E-loop string is stretched it becomes narrower to maintain its surface area at a constant 1.66×10^{-101}m^2. This is the E-loop vacuum rule that cannot be violated. I'll refer to this as the **E-vac rule** from this point forward. The reason for this property will become evident when the most fundamental string in the universe is identified later.

E-loops are created by a cosmic process right before the Big Bang from yet smaller vacuum species and as you will see later, inherent with an intrinsic spin of a linear velocity of 3×10^8m/s which is the speed of light. The spin of the E-loops is either right-handed or left-handed. Since the energy of an E-loop as a stand-alone object does not change, it is massless. As such, its energy can be viewed as stored potential energy until it is converted to kinetic energy which can then change giving the E-loop its mass.

Another important property of the E-loop is that it is a substantially two-dimensional object with an even smaller thickness than its diameter. Since it has no surface that may cause friction between its layers, they can be stacked on top of each other. Given that it is two-dimensional, a large number of E-loops of the same spin layered on top of each other give rise to a thickness that is still far below the Planck length and would remain (substantially) two-dimensional. We will explore the dimensions and energy-momentum properties of E-loop strings as a vacuum specie in this chapter.

How does the universe create a three-dimensional space such as ours from such tiny two-dimensional objects? To answer this question, we must put these tiny E-loops in motion.

If we induce a motion into the E-loops, its potential energy can be harvested to create larger spaces that are sustainable and grow to become the three-dimensional universe we observe today. The process of inducing kinetic energy into the E-loops began at the onset of the Big Bang which I will cover later. However, let us first explore the mechanism of conversion of an E-loop string to a Planck-size three-dimensional looped or open string.

Activated three-dimensional looped strings

Referring to Fig. 8-1, we will put two E-loops (EL_0, and EL_1) attached to each other by the virtue of its vacuum into a circular (orbital) motion rotating at the speed of light c. Since E-loops have an intrinsic rotational speed of light, two E-loops attached to each other will rotate at the speed of light one acting as anchor (EL_1) spinning the other (EL_0) thus creating an orbital rotation with a centrifugal force. As will be described in Chapter 12, EL_0 and EL_1 are indeed comprised of many layers of E-loop strings that are substantially two-dimensional, each having the same intrinsic potential energy described above. The E-loops as vacuum entities with energy, can be layered because it has no surface to cause decay of its energy, so long as the rotation of the E-loops layered are in the same direction.

When the two adjoined E-loops rotate at the speed of light, the potential energy of the E-loops comprising EL_1 is converted to kinetic energy of the E-loops of EL_0 and stored in an E-portal as shown in Fig. 8-1. The E-portal is an extra dimension cylinder with the same radius as the E-loop string containing the E-loops with kinetic energy. As will be discussed later, the individual E-loops making up the layered EL_0 will become separated under the torque created by the angular momentum of the orbital rotation, hence creating a cylindrical object with E-loops expanding. The potential energy of the E-loops comprising EL_1, will become the source of energy for this orbital rotation. In essence, the potential energy of E-loops comprising EL_1 is converted to kinetic energy of EL_0 E-loops.

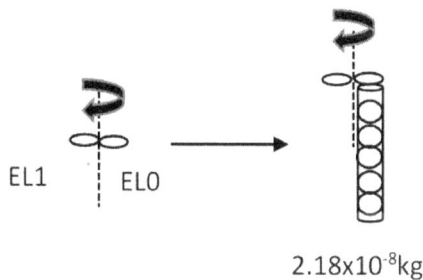

2.18×10^{-8}kg

Fig. 8-1- Transformation of potential energy of E-loop strings into kinetic energy and creation of E-portal energy storage.

The acceleration of the force created by the orbital rotation will be:

$$g = c^2/r$$

In which r is the radius of the E-loop, 2.3×10^{-51}m. This will give rise to an acceleration of

$$g = 9 \times 10^{16}/2.3 \times 10^{-51} = 3.9 \times 10^{67} \text{m/s}^2 \qquad (8\text{-}1)$$

The frequency of the rotating E-loop will be:

$$f = c/2\pi r = 3 \times 10^{8}/2\pi \times 2.3 \times 10^{-51} = 0.207 \times 10^{59} \text{ s}^{-1}$$

And the energy created:

$$E = hf = 0.2 \times 10^{59} \times 6.626 \times 10^{-34} = 1.37 \times 10^{25} \text{J} \qquad (8\text{-}2)$$

The E-loop, EL_0 containing the above energy as stored mass, spins off individual E-loops with kinetic energy. The E-loops are attached to each other and stretched and propagated by the centrifugal force of the rotation until the diameter of the looped string reaches Planck length or 1.61×10^{-35}m, Fig. 8-2. This is the first tier of geometric space transformation from a dimension of about 2.3×10^{-51}m to a dimension of 1.61×10^{-35}m, a factor of 10^{16}. We will call this the base disc of Planck diameter. The new Planck string will have the same thickness as the E-loop strings, also rendering it a substantially two-dimensional string.

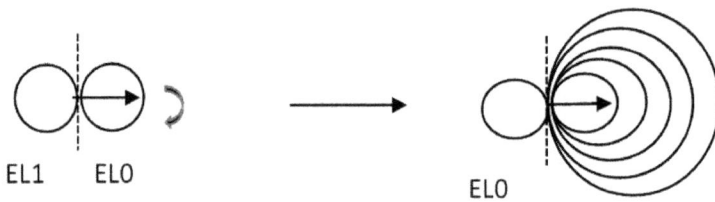

EL1   ELO

ELO

Fig. 8-2- Transformation of a two-dimensional E-loop string to a two-dimensional Planck length looped string.

Recall that I described a geometric transformation of space from 1.61×10^{-35} to 1.8×10^{-9}m which formed the spacetime cavity distinguishing the quantum space from macro-space. This is the geometric transformation of space preceding the earlier one.

Fig. 8-2 demonstrates the transition of an E-loop to a larger two-dimensional looped string with Planck radius. As you can see from this figure, each increment of the E-loop added will have the same surface area of 1.6×10^{-101}m^2. Therefore, the total number of E-loops that will comprise a looped string with Planck radius at any moment of time will be:

$$\pi(1.61 \times 10^{-35})^2 / 1.66 \times 10^{-101} = 5 \times 10^{31}$$

Let us call each segment of the stretched E- looped string propagating and forming a larger string, a "**gauge**". Therefore, 5×10^{31} gauges form a Planck scale looped string at any given moment of time. The gauges are dynamic and move and sweep across the surface of the two-dimensional Planck-size string and are then detached under the centrifugal force of the orbital rotation once it reaches Planck radius. The stretched E-loop string, when detached, collapses back into its original E-loop string releasing its energy as orbital rotation of the Planck disk to sustain its rotation creating a closed looped self-sustaining two-dimensional string.

To sustain a Planck-size two-dimensional string, a supply of a large number of E-loop strings with kinetic energy is then needed as stored energy inside the E-portal.

Referring to Fig. 8-2, since the radius of the original rotating E-loops remain constant, the acceleration of the Planck disk remains constant. As the radius of the looped string increases to r_2 (Planck length), the stored energy or mass of the E-loop, EL_1 will continue to be consumed to maintain the Planck space. We can calculate the kinetic energy of E-loop strings comprising EL_0 and its mass when the disk grows to Planck length:

$$g = c^2/r_1 = 9 \times 10^{16}/2.3 \times 10^{-51} = 3.9 \times 10^{67} \text{ m/s}^2$$

$$g = V^2/2r_2 = 3.9 \times 10^{67} \text{m/s}^2$$

$$V^2 = 3.9 \times 10^{67} \times 2 \times 1.61 \times 10^{-35}$$

$V = 3.54 \times 10^{16}$ m/s

This is the inflation velocity of the E-loops spun from EL_0 to create the two-dimensional surface of the Planck disk. This means that while the Planck-sized disk containing the gauges rotates at the linear speed of 3×10^8 m/s forming the disk, EL_0 where the mass is stored has a kinetic energy with a velocity equivalent of 3.54×10^{16} m/s, which is about c^2. The stored kinetic energy of the E-loop then becomes the mass of the larger two-dimensional Planck-size string:

$$E = hf = 1/2mV^2 \qquad\qquad (8\text{-}3)$$

If we substitute for $E = 1.37 \times 10^{25}$ J, equation (8-2), the energy generated by the rotating E-loop, and $V = 3.54 \times 10^{16}$ m/s the speed of the EL_0 E-loops, the mass of the two-dimensional disk with a radius of 1.61×10^{-35} m will be:

$m = 2hf/V^2 = 2.18 \times 10^{-8}$ kg

Which in classical physics is known as Planck mass. As you can see, from the rotational energy of two looped strings with a radius of 2.3×10^{-51} m, rotating at the speed of light c, a new two-dimensional string with a radius of 1.61×10^{-35} m and a mass of 2.18×10^{-8} kg has been produced. Later, in this section, I will describe how this energy is stored by the E-loop, EL_0, however, as you can see from Figs. 8-1 and 8-2, this mass is stored in a tiny E-loop cylinder and is located at one end of the newly formed looped string as EL_0.

Since the velocity of the two-dimensional looped string with Planck radius is $V = 3 \times 10^8$ m/s, the energy of this mass will be:

$E = mc^2 = 2.18 \times 10^{-8} \times 9 \times 10^{16} = 1.96 \times 10^9$ J

Why is the kinetic velocity of the E-loops stored inside the E-portal critical? This is because the velocity of 3.54×10^{16} m/s is required to create tension energy to stretch the E-loop string from its dimension of 2.3×10^{-51} m to the Planck length:

The tension energy of the initial E-loop rotating at the speed of light c is:

$K_1 = m/2 \times 2.3 \times 10^{-51}$

The tension energy of the stretched string with an inflation velocity of 3.54×10^{16}m/s is:

$$K_2 = m/\, 2r$$

The energy balance for the string before and after stretching is:

$$K_1 c^2 = 1/2 K_2 V^2$$

Substituting in the above equation:

$$r = (2.3 \times 10^{-51})\, V^2/2c^2 = 1.6 \times 10^{-35} m$$

This means that the E-loop strings require a kinetic energy with a propagation velocity of 3.5×10^{16}m/s to stretch to Planck length. When an E-loop string collapses from its Planck length, it will return its energy to the un-stretched anchoring E-loop.

In essence, an E-loop string with potential energy will convert to one with kinetic energy through the process of orbital rotation, gain the velocity it requires to stretch into a Planck-size string, and collapse back into the original E-loop string, releasing its energy for orbital rotation of the Planck disk, a closed loop energy balance.

In this process, two E-loop strings are used and the energy of one returned to the other in the process. The energy of one E-loop string is used to rotate the other, and convert its potential energy to kinetic energy, which in turn creates a larger space and depletes back into an E-loop string. The E-loop with depleted energy exits the Hilbert space as a two-dimensional string forming the circumference of the universe at restframe.

Recall in Chapter 1, in setting up the energy balance equations (1-19) for activated open strings, two activated open strings were also consumed in the process, and the energy of one was returned to the universe as an attachment to other strings creating a volume that is an extra dimension object. In the process of forming a three-dimensional string from E-loop strings, we also create a space volume from a field of E-loop strings. The string for string energy balance of E-loop strings to activated strings also requires two E-loops strings to be consumed but one is returned to the universe as part of the three-dimensional string formed. Therefore, in general, the energy field equation obtained for the activated open strings (1-27) also applies to the E-loop strings in the process of creating the volume of activated strings described above. As you will see in Chapter 12, this will be an important equation describing the universe as a whole.

We began with a (change in the) energy of 1.37×10^{25}J created by two rotating E-loops with a radius of 2.3×10^{-51}m and arrive at a Planck-size disk with a mass of 2.18×10^{-8}kg and an energy of 1.96×10^{9}J. This means that much of this energy was used to create a two-dimensional space that is about 10^{16} times larger. This is the energy required for the transformation from an E-loop space to Planck space with a conversion factor of:

$1.37 \times 10^{25} / 1.96 \times 10^{9} = 0.7 \times 10^{16}$

Recall from Chapter 3, each two-dimensional Planck-size string with a mass of 2.18×10^{-8}kg was converted to 8×10^{60} single looped strings each with a mass of 2.7×10^{-69}kg. This means that the above two-dimensional Planck-sized string created by propagation of E-loops with a mass of 2.18×10^{-8} kg can spin off an equivalent of 8×10^{60} similar single Planck-sized looped strings each with a mass of 2.7×10^{-69}kg, referred to as the base Planck disk above, to sustain its Planck space.

In other words, a two-dimensional Planck-size string with a mass of 2.18×10^{-8}kg and an energy of 1.96×10^{9}J has sufficient energy stored in its E-portal to spin-off 8×10^{60} two-dimensional Planck disks of the same diameter and each with a mass of 2.7×10^{-69}kg. The energy of each Planck disk will be about:

$E = 2.7 \times 10^{-69} \times 9 \times 10^{16} = 24.3 \times 10^{-53}$J

Which is sufficient to advance a Planck disk one Planck length:

$E = mrg = mc^2$

$r = 9 \times 10^{16} / 5.6 \times 10^{51} = 1.61 \times 10^{-35}$m

In which $g = 5.6 \times 10^{51}$m/s^2 is the acceleration of a Planck disk.

Therefore, the mass of the two-dimensional string at 2.18×10^{-8}kg can sustain a two-dimensional Planck disk the entire length of the universe:

$R = 8 \times 10^{60} \times 1.61 \times 10^{-35} = 1.3 \times 10^{26} \text{m}$

When the dimension of the rotating Planck disk of the new two-dimensional looped string of mass 2.18×10^{-8}kg reaches Planck radius, the vectors of the centrifugal force **F1**, and the radius **r**, which are perpendicular to each other with the center of rotation at EL_0, produce the torque **F2** which is perpendicular to the Plane of the two-dimensional string. Once the torque F2 is equal or greater than F1, it forces the disk to flip and also rotate perpendicular to the plane of its original rotation.

Referring to Fig. 8-3A, the rotating disk, with a radius of $r = 1.61 \times 10^{-35}$m and the center of rotation at EL_0, creates a torque that is perpendicular to the plane of the disk's rotation, F_2. The centrifugal force, $F_1 = mg$, keeps the two-dimensional disk in its horizontal position. When the torque generated is equal or exceeds the centrifugal force, the plane of rotation is forced to flip to a vertical orientation.

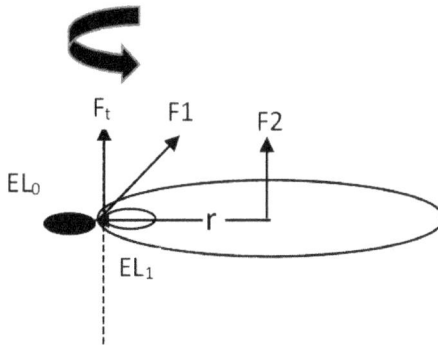

Fig. 8-3A- Balance of forces between a Planck size looped string and an E-loop string. EL_0 is the center of orbital rotation. Vector F_1 is perpendicular to **r**.

The centrifugal force of the mass of the two-dimensional string, 2.18×10^{-8}kg is:

$F_1 = mg = 2.18 \times 10^{-8} \times 3.9 \times 10^{67} = 0.85 \times 10^{60}$ N

The torque generated is:

$$F_t = F_1 \cdot r$$

The magnitude of F_t is:

$$F_t = mg \cdot r \, Sin \, 90 = mgr$$

The force F_2 at the center of the disk required to flip the disk and cause it to also rotate vertically must be equal to or greater than F_1:

$$F_2 \, x1.61x10^{-35} = 1.37 \, x10^{25}$$

The force at the center of the disk:

$$F_2 = 0.85x10^{60} \, N$$

In other words, when the energy of the orbital rotation matches the value of the torque, the disk will also move in the direction of the torque with its angular momentum perpendicular to its horizontal disk.

As you can see above, as E-loop strings propagate and the dimension of the two-dimensional string reaches $1.61x10^{-35}$m, the force F_2 overcomes the centrifugal force causing the disk to flip vertically and the center of rotation to shift to the center of the disk. The new rotation path becomes perpendicular to the original horizontal rotation path of the disks. One can also see above why Planck dimension is so important in our universe. The propagation of the E-loops halts when the dimension of the disk reaches Planck length because the Planck-size disk flips and rotates vertically due to the magnitude of the torque created at this dimension, making Planck dimension the limit of space propagation.

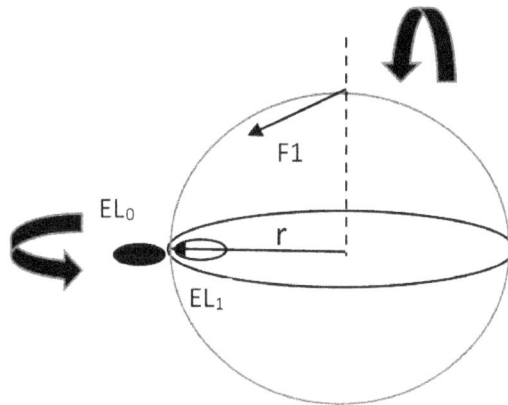

Fig. 8-3B- Simultaneous rotation of the two-dimensional string horizontally and vertically. Center of rotation shifts to the center of sphere. Vectors F_1 and **r** in the same direction.

Referring to Fig. 8-3B, the vectors **F1** and **r** with the center of rotation at EL_0 which are perpendicular in the horizontal plane become parallel and create a zero-degree angle between them. This means that the Planck disk with a mass of 2.18×10^{-8}kg will have two simultaneous rotations; an orbital rotation that is centered at the Planck disk and a spin around the center of the disk. The two simultaneous rotations of the Planck disk sweeping the space at the speed of light create a three-dimensional spherical object with the perimeter (thickness) of the two-dimensional Planck string forming the gauges of its surface. As you will see shortly, the thickness of the Planck disk is about 8.6×10^{-160}m. This will make the three-dimensional object behave as one without a surface.

The inner product of the two vectors **F1** and **r** creates a scalar object which is known as the Hilbert space:

$$F . r = mg . r \cos 0 = mgr$$

Fig. 8-4 demonstrates the sequence of E-loop string transformation into a Planck-size three-dimensional string.

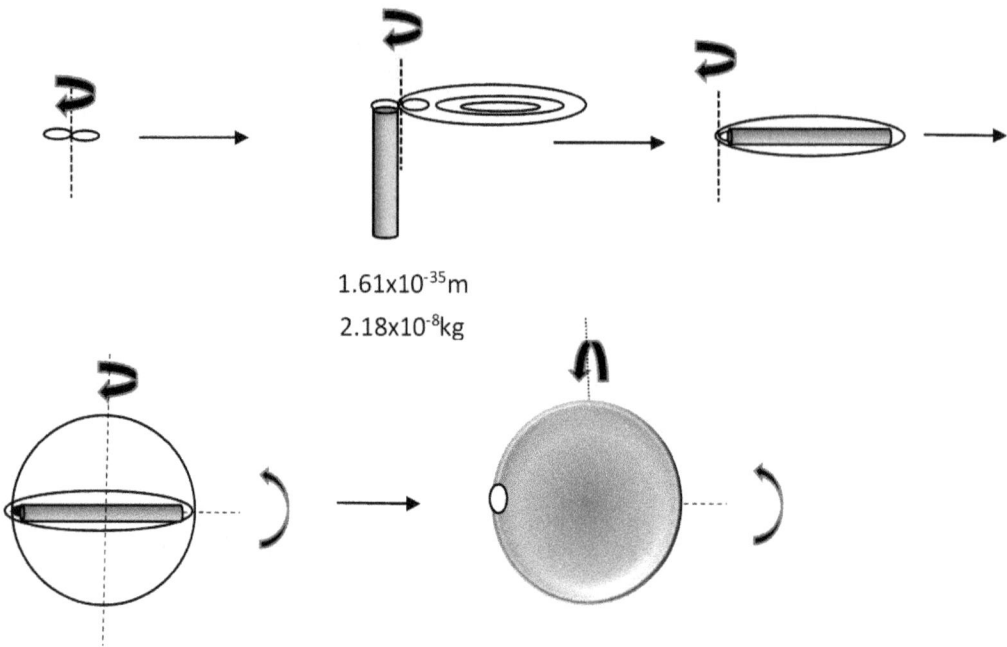

1.61x10^{-35}m

2.18x10^{-8}kg

Fig. 8-4- Transformation of a two-dimensional E-loop string to a two-dimensional Planck length looped string and a three dimensional looped string.

When the Planck disk acquires two simultaneous rotations, the energy stored inside the E-portal attached to EL_0 will become the source of energy for both motions. One motion is the horizontal rotation of the disk with the mass near its end (at EL_0) creating the angular momentum of the sphere and the other is the vertical motion created by the torque which establishes the spin of the sphere. The horizontal motion of the Planck disk sustains the mass of the disk as a two-dimensional string of 2.7x10^{-69}kg and the vertical motion creates the momentum component of the string as a three-dimensional string. The total energy of the scalar three-dimensional object which is the inner product of the two vectors is:

E= mrg

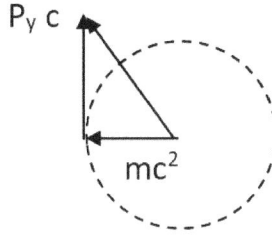

This energy according to Pythagoras' theorem splits into:

$$E^2 = (mc^2)^2 + (P_y c)^2 = (1.96 \times 10^9)^2$$

Since the mass of the two-dimensional Planck string is very small (2.7×10^{-69}kg), the energy consumed to sustain the mass of the Planck disk is also small. As such, the energy of the three-dimensional string essentially is available for its momentum component:

$$P_y = h/\lambda = h/2\pi l_p$$

$$E = P_y c = 1.96 \times 10^9 \text{J}$$

For this reason, I used the term "infinite momentum frame", not because the momentum is infinite, but because the energy of the three-dimensional string is largely available to its momentum component.

I also termed this three-dimensional string object "activated strings" because in a sense, the action of the torque activates a two-dimensional string that is essentially invisible to our universe into a three-dimensional string that is the basis of our existence.

It is important to note that two-dimensional strings of Planck length must be considered as always being under a rotation speed of c, the speed of light, and having an angular momentum that creates the g-force to propagate the gauges or they will not exist, i.e., collapse into an E-loop string.

Let us now review the above model for transformation of a two-dimensional looped string with a radius of 2.3×10^{-51}m into a three-dimensional activated looped string with a radius of 1.61×10^{-35}m.

We started with two adjoining E-loops (comprised of many two-dimensional layers which we will calculate later) which were put into motion subsequent to the Big Bang. The E-loops are rotating at the speed of light c and converting their potential energy into kinetic energy of 1.37×10^{25}J. Using the stored energy, the centrifugal forces stretched the E-loops into 5×10^{31} propagating gauges with a surface area of 1.6×10^{-101}m^2 forming the base disk of a two-dimensional looped string with a radius of 1.61×10^{-35}m. The angular momentum and the associated torque of the stored mass causes the rotating disk to flip and also rotate vertically to its original rotation plane. The simultaneous rotation of the Planck disk in a horizontal (XY) and vertical (XZ) direction creates the structure of a three-dimensional sphere. The stored E-loops with kinetic energy then become the source of energy for the mass and momentum component of the three-dimensional sphere.

Since the surface area of the E-loop remains constant, when stretched to Planck length, it will have a width of 10^{-66}m:

$1.6 \times 10^{-101} / 1.61 \times 10^{-35} = 1 \times 10^{-66}$m

Therefore, the gauges forming the surface of the two-dimensional looped strings are substantially wider than those of the surface of the spheres.

Let us now calculate the mass of each E-loop or gauge with a surface area of 1.6×10^{-101}m^2. As mentioned above, we know that the mass of a single two-dimensional Planck-size base string is 2.7×10^{-69}kg. We also know that it takes 5×10^{31} E-loops or gauges to make a single looped string with a radius of 1.61×10^{-35}m.

$\pi(1.61 \times 10^{-35)2} / 1.6 \times 10^{-101} = 5 \times 10^{31}$

Therefore, the total number of E-loops or gauges making up the mass of a two-dimensional string is:

$5 \times 10^{31} \times 8 \times 10^{60} = 4 \times 10^{92}$

m x $4 \times 10^{92} = 2.18 \times 10^{-8}$

Therefore, the mass of an E-loop with an area of 1.66×10^{-101}m^2 is:

m= 5.4 x10^{-101} kg

This is interesting! The mass of the E-loop and its surface area are about the same.

Therefore, one may propose that the E-loop dimension (E-loop Space) is a dimension in which mass equals space.

Mass = Space

We have known over the last century that mass = energy by the virtue of E=mc^2. However, E-loop space teaches us that:

Space = mass = energy

This has profound implications in understanding what space mass is. This means that space mass is the change in the amount of energy that it takes to create space, but as energy that remains in our universe (Hilbert space). We started with two E-loop strings and in the process of creating space mass we lost one. An E-loop with a spin rotating at the speed of light has no mass and its energy is merely potential energy. However, in the process of two E-loops orbiting one another there is a change in energy, one is lost but mass is created. For an activated three-dimensional string, this change is equivalent to the energy of one E-loop or

E= 5.4x10^{-101}x9x10^{16} = 48.6x10^{-85}J/ per rotation

For this reason, the space mass is unmeasurable for now because the change is too small to measure and the exhausted E-loop exits apparently, leaving no signature behind. Recall that the baryonic mass of say an electron was equivalent to the change in the energy of a two-dimensional looped string of about 6x10^{-31}kg, or of 5.4x10^{-14}J, which is significantly larger and has a momentary interaction with our three-dimensional universe hence detectable. The baryonic mass we measure is the change of energy that exits our three-dimensional universe (Hilbert space). Space mass is the change of energy that remains in our universe.

In Chapter 9, you will learn that the energy of the lost E-loop string converted to space mass manifests itself as thermal energy and entropy of the string. This is the only signature creation of space mass leaves behind. The E-loop string exhausted of its energy, exits our universe carrying nothing more than information about its parent two and three-dimensional strings.

The space mass described above is significantly higher than the baryonic mass. To put this in perspective, the mass density of an activated string with a Planck radius is $0.2 \times 10^{97} \mathrm{kg/m^3}$. This means that the mass of one cubic meter of space is about 10^{44} times heavier than the baryonic mass of the entire universe which is about $(0.1 \times 10^{53} \mathrm{kg})$ yet not detectable because its change is too small to be measured. In fact, the change in the energy of an activated string with a mass of only $2.18 \times 10^{-8} \mathrm{kg}$ is so small that it can last just about the entire life of the universe, roaming around the infinite momentum frame if the string is not decayed by annihilation.

We can calculate the life of an activated open string based on the energy used for the work performed on one string by the electromagnetic force:

F= mg = 2.18x10^{-8} x14.7 x10^{-10} = 3.16x10^{-17}N

F . R = mc² = 1.96x10⁹ J

Substituting for F, R = 0.62 x10²⁶m

t=R/c = 0.2 x10¹⁸s = 6.48 B yrs

Similarly for an activated looped string, the work performed by the gravitational force:

F= mg = 2.x10^{-23} x6.2 x10^{-10} = 1.24x10^{-32}N

F . R = mc² = 2x10^{-23} x9x10¹⁶ = 1.8 x10^{-6}J

R= 1.45 x10²⁶m,

t=R/c = 0.2 x10¹⁸s = 0.48x10¹⁸s = 15.5 B yrs

In describing the process of the formation of a three-dimensional string, three ingredients make up the critical recipe for its creation:

E-Loop strings +orbital rotation (kinetic energy) + bulk vacuum = activated string

Availability of bulk vacuum, which as you will see later has a different property than the vacuum that creates the E-loop string itself, is not an issue at the time of the Big Bang. However, as the universe begins to form in the initial few seconds, the formation of activated strings consumes much of the bulk vacuum which then becomes a valuable commodity in our existing universe.

In our existing universe, crowded with a tremendously large number of activated looped, open strings, and other extra dimension objects, the interference by other strings prevents a two-dimensional string to reconstruct back into a three-dimensional structure. As you will see later, the conditions for string regeneration in our universe are special cases where interference by other objects have been eliminated creating sufficient bulk vacuum space to allow formation of three-dimensional strings.

E-loop extra dimension cylinder or E-portal

Let us now look more closely at the initial conditions that led to creation of a three-dimensional looped string. We started with two attached two-dimensional E-loops creating a rotational plane rotating at the speed of light, c. One E-loop is anchoring at the center of the rotation EL_0, and the other is the orbiting E-loop EL_1. As I showed earlier, the energy created as a result of this rotation is:

$E = hf = 1.37 \times 10^{25} J$

With respect to the orbiting plane of the two E-loops which is orbiting at the speed of light c, this means that the E-loop EL_0, with a dimension of $2.3 \times 10^{-51} m$, has a mass of:

$m = E/c^2 = 1.37 \times 10^{25}/9 \times 10^{16} = 0.15 \times 10^9 \ kg$

The energy for this orbital rotation is supplied by and converted from potential energy of E-loops of EL_1. Since the energy of one E-loop string is:

$E = 5.4 \times 10^{-101} \times 9 \times 10^8 = 48.6 \times 10^{-85} J$

The total number of E-loops layered in EL_0 is:

$1.37 \times 10^{25} / 48.6 \times 10^{-85} = 0.28 \times 10^{109}$

Which matches the total number of E-loops layered in EL_1:

$0.15 \times 10^9 / 5.4 \times 10^{-101} = 0.28 \times 10^{109}$

This means that the two adjoining E-loop strings are each comprised of 0.28×10^{109} individual two-dimensional E-loops, each having an energy of $1.37 \times 10^{25} J$ respectively. The potential energy of EL_1 will be converted to kinetic energy of the E-loops in EL_0 due to the orbital rotation.

In Chapter 12, I will describe the mechanism that creates 0.28×10^{109} layers of individual E-loops.

Since the mass of an E-loop is about 5.4×10^{-101}kg, the kinetic energy induced into the E-loops of EL_0 will result in an inflation velocity of:

$$E = 1/2mV^2$$

$$1.37 \times 10^{25}J = \frac{1}{2}(5.4 \times 10^{-101})V^2$$

$$V = 0.7 \times 10^{63} m/s$$

The orbital rotation of the plane of the two E-loops, rotating at the speed of light c, creates an angular momentum and a torque that is perpendicular to the plane of rotation. The torque applied on the string stretches the individual E-loops from a layered configuration into a cylindrical configuration I described as an **"E-portal"**. In essence, the kinetic energy created causes a thicker multi-layered E-loop to stretch out into a thinner elongated E-loop, converting the kinetic energy into "stored" tension energy.

The magnitude of the torque created by the above energy is:

$$FxR = E$$

$$mgR = 2(1/2mV^2)$$

$$V = (Rg)^{1/2}$$

Substituting for the E-loop inflation velocity of 0.7×10^{63}m/s, and $g=3.9 \times 10^{67}$m/s, the distance the E-loop string will stretch using the above energy is:

$$R = V^2/g = 1.3 \times 10^{58}m$$

In essence, the mass or energy of the E-loop, EL_0, which was calculated above to be 0.15×10^9kg, is stored in an E-portal as tension energy with the length of about 1.3×10^{58}m and a radius of 2.3×10^{-51}m. The torque created by the orbital rotation of the E-loop forces the layered E-loops to stretch and **inflate** to a distance of 1.3×10^{58}m. As mentioned earlier, at the time of the Big Bang since the availability of the bulk vacuum and interference with other string species is not as pronounced, the E-portal using the above energy can stretch to the maximum length allowed by its torque.

We can verify this distance since the E-loops stretched into the E-portal are separated by a length equal to its outside diameter. Since the number of E-loops inside the E-portal is 0.28×10^{109}:

$$R = 0.28 \times 10^{109} \times 2 \times 2.3 \times 10^{-51} = 1.3 \times 10^{58}$$

Which is about the same as the distance calculated above. In other words, an E-loop (EL_0) with a mass of 0.15×10^9kg is equivalent to storing 0.28×10^{109} E-loops of the mass 5.4×10^{-101}kg with a length of 1.3×10^{58}m.

The E-portals make an ideal energy storage extra dimension because there is no energy penalty for this storage. E-loops of the same spin direction with kinetic energy can traverse, expand, and retract into layers depending on its environment and availability of vacuum space. The individual E-loops of the same spin remain connected to each other by virtue of its like vacuum. In the absence of interference, the E-portals will expand to their maximum length. However, as more strings are created and there is less available space and bulk vacuum, the E-portals can retract into smaller E-portals with layered E-loops of the same spin.

If we adjust the above number of E-loops creating 1.37×10^{25}J by a factor of 0.7×10^{16} for the space transformation from an E-loop space to Planck space, or from 2.3×10^{-51}m to 1.61×10^{-35}m, we will have:

$$0.28 \times 10^{109} / 0.7 \times 10^{16} = 4 \times 10^{92}$$

Which as shown earlier is the number of E-loops required for constructing a three-dimensional activated looped string with a mass of 2.18×10^{-8}kg.

The first step of creation of a three-dimensional string is converting the potential energy of the E-loops into kinetic energy and stored tension energy as described above.

The next step is the transformation of space from a dimension of 2.3×10^{-51}m to 1.61×10^{-35}m and creating a surface with 5×10^{30} E-loops and a mass of 2.7×10^{-69}kg. After this step, the inflation velocity of the E-loops of the E-portal is reduced from 0.71×10^{63}m/s to:

$$E = 1/2mV^2$$

$1.37 \times 10^{25} = \frac{1}{2}(2.7 \times 10^{-69})V^2$

$V = 1 \times 10^{47} \text{m/s}$

This means that the mass of the individual stretched E-loop of the E-portal has increased from 5.6×10^{-101}kg to 2.7×10^{-69}kg, and the length of the E-portal has decreased from 1.28×10^{58}m to:

$R = V^2/2g = (1 \times 10^{47})^2/2(3.9 \times 10^{67}) = 1.3 \times 10^{26}$m

As the space goes through a transformation from 2.3×10^{-51}m to 1.16×10^{-35}m and much larger strings are created, there will be less available free space and bulk vacuum coupled with interference from the newly formed strings. As such, the mass of the E-loops of the E-portal become layered and compacted, reducing the length of the E-portal correspondingly.

Compaction of the E-loops means that E-loops of the same spin as two-dimensional objects can be layered without significant increase in their thickness. This is possible because unlike the two-dimensional Planck-size looped strings, E-loops have no surface and as such the layers do not cause friction and loss of energy.

It is also interesting to note that as soon as the first disk of a Planck-size string is formed, the length of the E-portal as shown above becomes equal to the maximum radius of the field of infinite momentum in the universe (see Fig. 1-11, Chapter 1). This means that the maximum radius of the energy field in the universe is established at the time of the Big Bang.

If we divide the total mass created by this energy, 0.15×10^9kg, by the mass of an E-loop string, 2.7×10^{-69}kg, we can calculate the number of compacted E-loops of the E-portal with the new compacted mass:

$0.15 \times 10^9/2.7 \times 10^{-69} = 0.55 \times 10^{77}$

The length of the E-portal R:

$R = 0.55 \times 10^{77} \times 2 \times 2.7 \times 10^{-51} = 2.97 \times 10^{26}$m

Which is about the same as calculated above. This means that an E-loop with a mass of 2.7×10^{-69}kg and an inflation speed of $V = 1 \times 10^{47}$m/s behaves as storing 0.5×10^{77} E-loops of the same mass in an E-portal with a length of 2.97×10^{26}m.

After the space transformation from an E-portal to a Planck-size extra dimension cylinder, as the population of strings grow, the stored energy inside the E-portal is further compacted to a mass of 2.18×10^{-8}kg.

$$E = 1/2mV^2$$

$$1.37 \times 10^{25} = \tfrac{1}{2}(2.18 \times 10^{-8})V^2$$

$$V = 3.5 \times 10^{16}\text{m/s}$$

And at this point the length of the E-portal will shrink down to about the radius of the new Planck-size dimension.

$$R = (3.5 \times 10^{16})^2/2 \times 3.9 \times 10^{67} = 1.6 \times 10^{-35}\text{m}$$

And the kinetic velocity of the new E-loop of the E-portal is reduced to 3.5×10^{16}m. Again, using the above calculations:

$$0.15 \times 10^9/2.18 \times 10^{-8} = 0.7 \times 10^{16}$$

This is equivalent to having 0.7×10^{16} E-loops with a mass of 2.18×10^{-8}kg each stored in an E-portal with the length of 3.2×10^{-35}m. Fig. 8-5 represents creation of E-portal energy storage and the sequential compaction of the E-portal. Fig. 8-6 represents transformation of an E-loop into a Planck-size two and three-dimensional looped string with its core mass stored as an E-portal attached to the two-dimensional string.

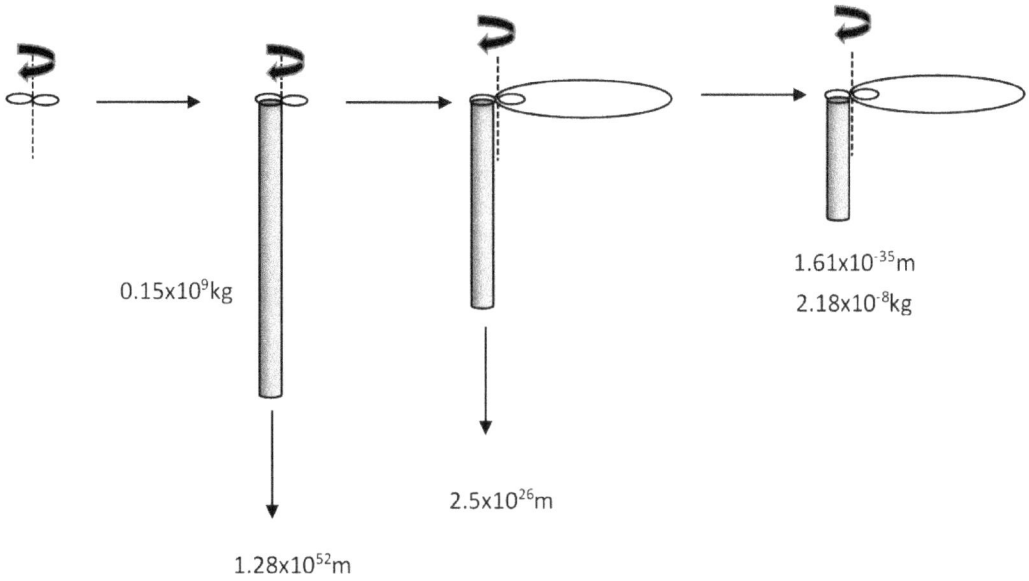

0.15x10^{9}kg

1.28x10^{52}m

2.5x10^{26}m

1.61x10^{-35}m

2.18x10^{-8}kg

Fig. 8-5- Creation of an E-portal as an energy storage extra dimension object and compaction of its mass to a Planck length.

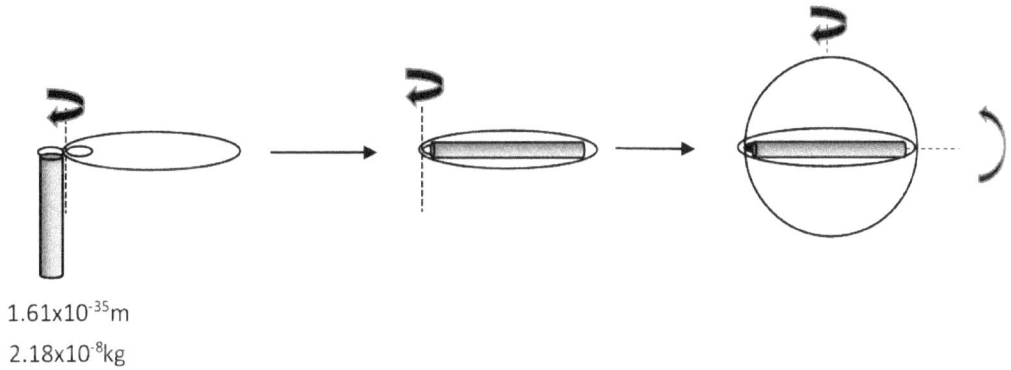

1.61x10^{-35}m

2.18x10^{-8}kg

Fig. 8-6- Transformation of a two-dimensional E-loop string to a two-dimensional Planck length looped string and a three dimensional looped string with core mass stored in an E-portal.

Therefore, at this point we have gone through a space transformation, creating a Planck-size disk with a mass of 2.7×10^{-69}kg and an E-portal attached at one end which also has a length equal to the diameter of the two-dimensional Planck-size disk. The stored energy of the E-portal is equivalent to 0.7×10^{16} E-loops which, after the space transformation (which as shown earlier has the same conversion factor), equals 2.18×10^{-8}kg of "Planck space" becoming the core mass of the Planck-size two and then the three-dimensional activated string.

Recall from earlier, when the two-dimensional Planck-size string reaches a mass of 2.18×10^{-8}kg, it will flip vertically due to the torque created by this mass and become a three-dimensional looped string which was discussed in the previous section. After the formation of the three-dimensional activated string, the E-portal will remain as the stored energy source comprising the core mass of the activated string at 2.18×10^{-8} kg. As such, the activated looped and open string will have an E-portal with a Planck space mass of 2.18×10^{-8}kg inside its spherical three-dimensional structure.

This also means that upon conversion of a three-dimensional activated string to a two-dimensional Planck-size disk of mass 2.7×10^{-69}kg, the attached E-portal has an equivalent stored mass of 2.18×10^{-8}kg of E-loops (or 0.7×10^{16} on E-loop space).

The E-portal can be further compacted from the Planck length above to the length of an E-loop itself, in which case its mass will be 0.9×10^9 kg. This is the initial condition of E-loops as described at the start of this section with E-loops EL_0 and EL_1 each having an energy of 1.37×10^{25}J comprised of 0.28×10^{109} layers of individual E-loops compacted into a dimension of 2.3×10^{-51}m:

$$E = 1/2mV^2$$

$$1.37 \times 10^{25} = \tfrac{1}{2}(0.9 \times 10^9)V^2$$

$$V = 3 \times 10^8 \text{m/s}$$

$$R = V^2/g = (3 \times 10^8)^2/(3.9 \times 10^{67}) = 2.3 \times 10^{-51}\text{m}$$

$$R = 2.3 \times 10^{-51}\text{m}$$

This is the maximum compaction of an E-loop, or the maximum length EL_1 can stretch with a mass of $0.15x10^9$ kg. When the E-portal reaches its maximum compaction, it is equivalent to having one E-loop with a radius of $2.3x10^{-51}$m and a length (thickness) of $2.3x10^{-51}$m (or a volume of $3.7x10^{-152}$m^3), spinning at $3x10^8$m/s. As you will see later, this is the condition of the E-loop strings immediately after the Big Bang and prior to its inflation.

Since we know the mass of an individual E-loop is $5.6x10^{-101}$kg, and the number of E- loops corresponding to this mass is:

$$0.15x10^9/5.6x10^{-101} = 0.26x10^{109}$$

We can calculate the width or the thickness of the two-dimensional E-loop:

$$\text{E-loop thickness} = 2.3x10^{-51}/0.26x10^{109} = 8.9x10^{-160}\text{m}$$

We have now arrived at the fourth transformation of space from $2.3x10^{-51}$m to about $8.9x10^{-160}$m which will be the subject of discussion in the upcoming chapters. However, one can see that the term two-dimensional as I have referred to throughout this book means three-dimensional, but the thickness is extremely small as compared to the length of an E-loop string or Planck string that was treated as two-dimensional. A two-dimensional string with zero thickness is only an abstract mathematical concept that has no physical meaning.

E-portals play an important role in propagation of two-dimensional strings. When activated strings decay into two-dimensional strings, its attached E-portal is no longer confined by the sphere of the three-dimensional string and can immediately expand or inflate into its maximum length given the availability of vacuum space. The E-portals can attach to two-dimensional open strings and the E-loop of the pulsating extra dimension cylinders creating the orbital rotation described above. They can then spin-off two-dimensional looped or open strings with the stored E-loop energy of the E-portal using the same process described in this section. Given the extremely high expanding velocity of the E-portal, the stored mass of the two-dimensional strings can be used to instantaneously spin an array of two-dimensional Planck-size strings as described above.

E-portal and singularity

In construction of a three-dimensional string, we began with two maximally compacted E-loop strings which inflate to their maximum length and retract to Planck length with a mass equivalent to 2.18×10^{-8}kg as bulk vacuum space becomes restricted and confined as the two and then three-dimensional strings are formed.

When three-dimensional strings decay, this process is reversed. First, two-dimensional strings emerge which in combination with the three-dimensional strings make all the extra dimension objects in the universe as discussed in the previous chapters. Two-dimensional Planck strings further decay into E-portals and E-loop strings which exit our universe as two-dimensional strings making up the circumference of the universe.

As you will see in Chapter 10, the rate of activated string decay is significantly higher in a blackhole. This means that a significant number of E-portals are generated inside the Schwarzschild radius of a blackhole. The highest population of E-portals is inside an extra dimension cylinder at the center of the black hole which is known as the singularity of a blackhole. The singularity of a blackhole will contain 5×10^{30} E-portals compacted in an area the size of a Planck disk at the center of the blackhole. The E-portals that exist in our universe as an energy storage component of the two and three-dimensional strings are essentially the same as those that are a constituent of the singularity. The difference between the E-portals is that those of a blackhole are preceded by the event horizon of the Schwarzschild radius and contain the highest population density of E-portals that can be found anywhere in the universe.

When an E-portal propagates as a two-dimensional object, it will maintain a constant energy density because as its thickness becomes smaller, its mass also becomes smaller, maintaining the same mass to volume ratio.

For example, the energy density of the E-portal when the E-loops are fully compacted to a thickness and radius of 2.3×10^{-51}m is:

$$\rho = 0.15 \times 10^9 / 1.6 \times 10^{-101} \times 2.3 \times 10^{-51} = 4.1 \times 10^{159} \text{ kg/m}^3$$

This is the same as the energy density of a single E-loop string when the E-portal expands to its maximum length:

$V = 1.6 \times 10^{-101} \times 8.5 \times 10^{-160} = 1.36 \times 10^{-260} \text{m}^3$

$\rho = m/V = 5.6 \times 10^{-101}/1.36 \times 10^{-260} = 4.1 \times 10^{159} \text{ kg/m}^3$

In Planck space:

$\rho = 4.1 \times 10^{159}/0.7 \times 10^{16} = 5.85 \times 10^{143} \text{ kg/m}^3$

The energy density of about 5.85×10^{143} kg/m^3, along with $g = 3.9 \times 10^{67}$ m/s^2 produces the highest force the universe can offer:

$F = 5.85 \times 10^{143} \times 3.9 \times 10^{67} = 2.28 \times 10^{211} \text{ N/m}^3$

Activated three-dimensional open strings

In constructing an activated looped string, I started with two adjoining E-loops each with a potential energy of 1.37×10^{25}J, one rotating the other at the speed of light c. Let us start with three adjoining E-loops, EL_0, the anchoring string, EL_1, the propagation string, and EL_2 the pulling string. Again, all three E-loops being multilayered with 0.28×10^{92} individual E-loops comprising its potential energy. The mass of the outer E-loop, EL_2, pulls and stretches the middle E-loop, EL_1, while the assembly of El_1 and EL_2 are anchored at EL_0.

Referring to Fig. 8-7, if we put three adjoining E-loops in motion at the speed of light c, the centrifugal force of the outer E-loop EL_2, will pull the string of the middle E-loop, EL_1. The middle E-loop will stretch to a radius of 1.61×10^{-35}m. Similar to the looped strings, the E-portals forming the open strings undergo the same process of inflation and compaction until the E-portal length is at Planck length before a three-dimensional structure is formed.

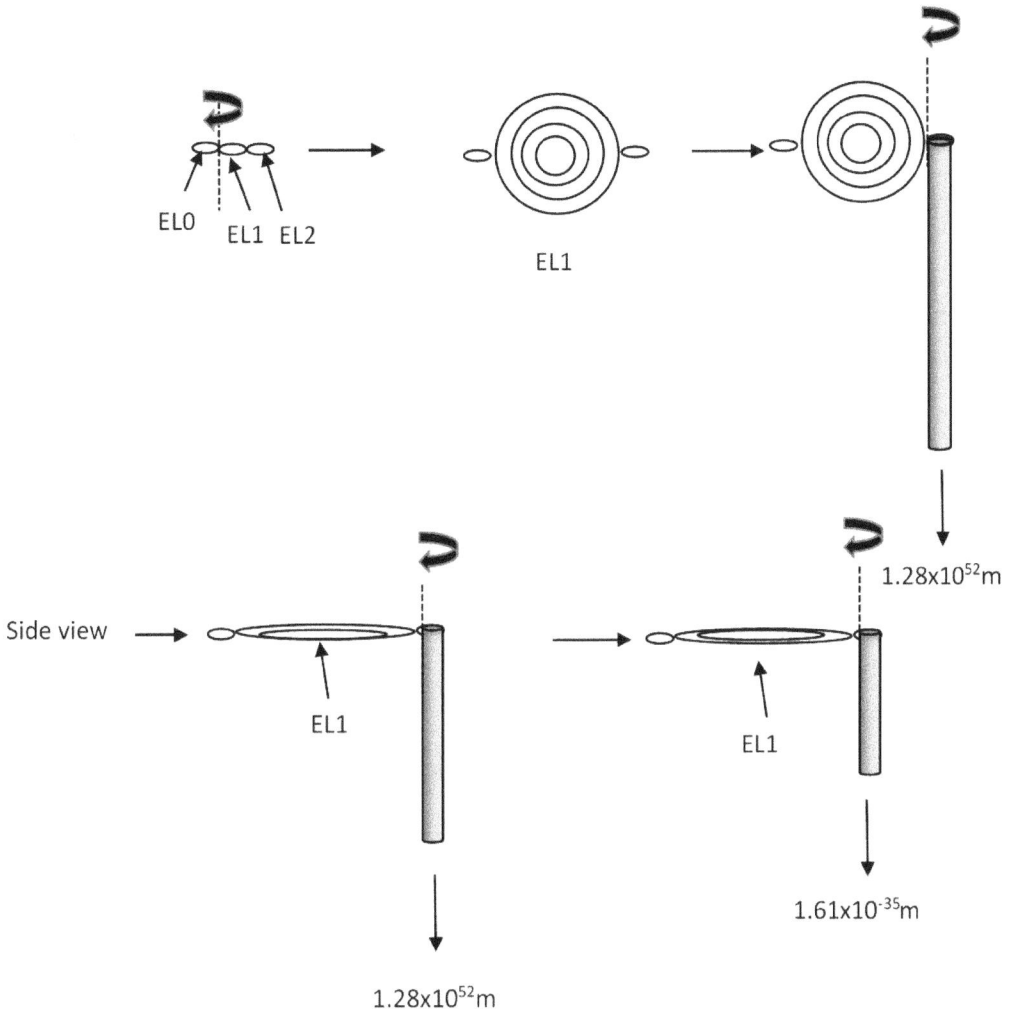

Fig. 8-7 - Transformation of a two-dimensional E-loop string to a two-dimensional Planck length open string and compaction of its mass.

The process for the formation of the three-dimensional activated string is similar to the looped string as described earlier. A disk with a radius of 1.61×10^{-35} will have 5×10^{31} segmented E-loop gauges forming the base disk. The difference is that the base disk of an open string is not created by the kinetic energy of the E-loops, it is simply converting the potential energy of the E-loop strings of EL_1 to tension energy. As such, the Planck-size of the stretched E-loops of EL_1 remain massless. The mass of the open strings, similar to the looped strings, will be 2.18×10^{-8} kg which is stored in EL_2 as an E-portal at one end of the string. Similar to the looped string, the torque created by the orbiting mass will flip the disk vertically causing it to rotate horizontally and spin vertically, creating a three-dimensional sphere, Fig. 8-8.

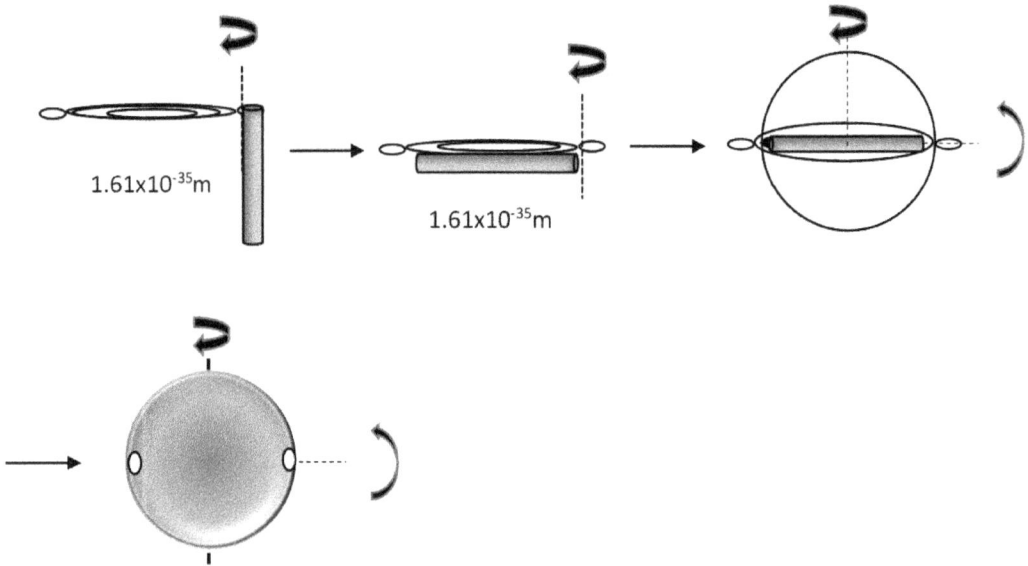

Fig. 8-8- Transformation of a two-dimensional Planck length open string to a three-dimensional open string. Note that open string has two E-poles on its sphere.

There are some differences in the formation of the three-dimensional structure that is generated between the activated looped and open string:

1. The expansion of the Planck disk in an open string is the result of application of equal forces on the middle E-loop EL_1 stretching it to the Planck dimension. The energy for the stretched E-loop is created by conversion of its potential energy of EL_1 into tension energy. Given that the middle E-loop, EL_1, starts with 0.28×10^{92} individual E-loops, creation of the Planck base disk will result in 8×10^{60} massless Planck disks.

2. Since there are two E-loops, EL_0, and EL_1 on both sides of the Planck disk, the resulting three-dimensional open string will have two E-poles on both sides of its sphere. This is very significant because it allows simultaneous attachment

of two-dimensional strings to its poles which will create opposing forces on both sides of the string and cause the three-dimensional string to break-off quickly. This means that any attachment to a three-dimensional open string will be short lived and transient not allowing sufficient time to pull and expand a three-dimensional open string similar to a three-dimensional looped string. As such, the dimension of a three-dimensional open string will remain constant at about Planck radius throughout the universe.

3. The spherical three-dimensional open string after decay will collapse into an elongated two-dimensional string with a width of:

4. Width = $1.6 \times 10^{-101} / 1.61 \times 10^{-35} = 1 \times 10^{-66}$ m

Which is 10^{31} times smaller than Planck length. Fig. 8-9.

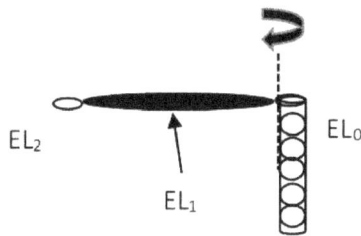

Fig. 8-9 - An elongated E-loop string representing a two dimensional open string.

In other words, a two-dimensional open string does not retain a circular geometry like a looped string after decay of its spherical three-dimensional structure. The reason for this is that the E-Loop, EL_1, will always be under unequal pull from either side. Unequal forces on the E-pole of the two-dimensional strings means that EL_1 is not stretched equally to maintain its circular symmetry. As a result, in a universe with so many attachments of difference forces, the two-dimensional structure of an open string will collapse into a very long and thin two-dimensional string with its E-portal mass at one end.

In classical string theories, an open string is depicted as a one dimensional, line like object. A true one-dimensional open string that resembles a line cannot exist in nature because it does not contain or create "space".

An open string is extremely narrow (1.61×10^{-35}m x 1×10^{-66}m) making it appear one-dimensional, but in reality, like an E-loop, it has a width of 1×10^{-66}m and a thickness of 8.9×10^{-160}m. In essence, a two-dimensional open string is a looped string disguised as a long and narrow one-dimensional string.

5. Because a two-dimensional open string will collapse into an elongated E-loop, it has no surface, unlike the two-dimensional Planck-size looped string which will maintain its surface. As such, multiple layers of EL_1 can exist as a two-dimensional open string. When the energy from the E-portal, EL_2, is used, the multiple layers of a two-dimensional open string will then stretch into a much longer string that is quantized by Planck length, as shown in Fig. 8-10.

EL1

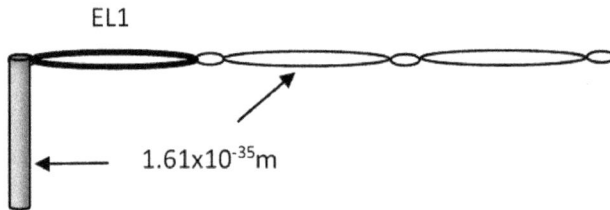

1.61×10^{-35}m

Fig. 8-10- Propagation of two dimensional open string. The bold string represents the 8×10^{60} pre-existing layers of Planck length strings.

In fact, a two-dimensional open string when stretched can readily expand to the length of the universe:

$8 \times 10^{60} \times 1.61 \times 10^{-35} = 1.3 \times 10^{26}$m

Since the two-dimensional open strings use the rotational energy of pulsating strings for their orbital rotation, the stored energy of EL_2 becomes available as kinetic energy for the massless strings of EL_1 to propagate. This is a significant propagation advantage as compared to two-dimensional looped strings. A two-dimensional looped string can only propagate when its E-portal is attached to the E-loop of a two-dimensional open string which in turn will spin-off two-dimensional Planck looped strings. Whereas a two-dimensional open string has 8×10^{60}, pre-stretched Planck length E-loops in EL_1 that can use the stored energy of the E-portal, EL_2 as kinetic energy to stretch up to the radius of the universe, Fig. 8-10.

Let us now look at the closed energy loop for creation of an activated three-dimensional open string. The E-loops of EL_0, using their potential energy, create the orbital rotation for E- loops EL_1 and EL_2. The potential energy of EL_1 is used to stretch 2×10^{30} E-loop gauges that create the Planck surface of EL_1. The orbital rotation of the E-loop EL_2, results in a kinetic energy that is stored in EL_2. In essence, the change in the energy of EL_0 is conserved in EL_2 and becomes the mass of the two-dimensional Planck string, 2.18×10^{-8}kg.

In a three-dimensional open string, the stored energy (mass) of the E-portal EL_2 utilizes its kinetic energy with a velocity of 3.5×10^{16}m/s to stretch as the energy of one Planck disk which then collapses back to an E-loop and provide its energy to the E-loop EL_0 as the source of energy for its orbital rotation. This orbital rotation, similar to the looped string, will have two components, mass and momentum comprising the energy momentum relation for the activated open strings.

Propagation of two-dimensional looped and open strings

Activated three-dimensional looped and open strings with opposite rotation annihilate each other and cause the rotation perpendicular to its plane of mass to cease to exist. The three-dimensional string is then left with its two-dimensional Planck disk still orbiting horizontally to sustain its mass inside the trapped vacuum left behind by the three-dimensional structure.

The two-dimensional looped and open strings utilize the orbital energy created by the pulsating extra dimension strings. As such, the momentum energy components for the two-dimensional string come at no price and the energy stored in its E-portal will be used to propagate new two-dimensional Planck-size strings:

$$E^2 = (mc^2)^2 + (P_z c)^2$$

As we saw earlier, the two-dimensional looped and open strings use the available momentum energy, which is constant at 1.96×10^9J, as its kinetic energy to propel itself. The remaining energy will then be used to create two-dimensional space mass:

$$E = mc^2$$

The amount of energy used to propagate and create two-dimensional strings depends on the energy content of the three-dimensional strings before decay and the amount of energy remaining in its E-portal storage. However, a significant portion of the propagation energy will be consumed and transferred to orbital extra dimension cylinders as wave energy. This will be discussed further in the upcoming sections and chapters.

Propagation of two-dimensional looped strings

In describing the propagation mechanism of the two-dimensional looped strings, I will keep our focus on the decay of activated strings attached to orbiting extra dimension cylinders and inside the spacetime cavity. Recall that outside the spacetime cavity where looped strings are significantly larger, the decay mechanism leads to creation of about 100 Higgs Boson which is essentially 100 spacetime cavities. Propagation of non-orbiting two-dimensional strings will be discussed in the next section under "open string regeneration".

When two oppositely rotating activated looped strings decay, the vertical rotation motion of the Planck disk ceases to exist. The two disks of the parent strings rotating horizontally join at the E-portals of its corresponding string a shown in Fig. 8-11. This attachment is due to the vacuum energy of the two E-portals creating an attraction between the two E-portals.

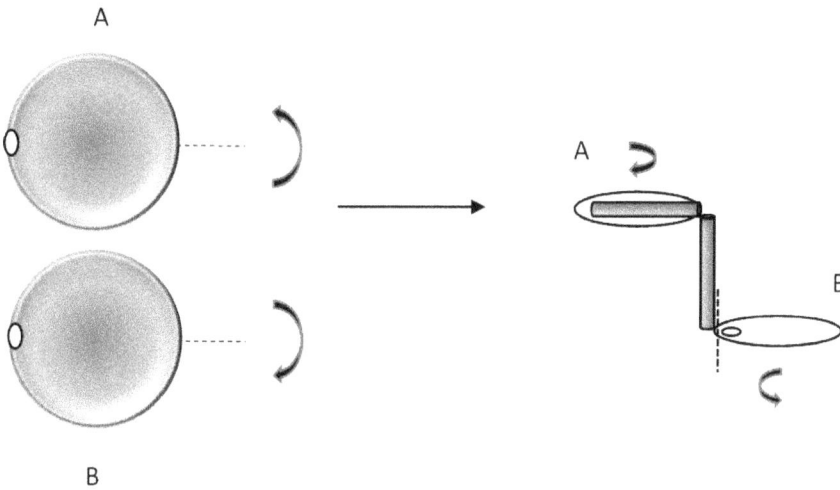

Fig. 8-11- Decay mechanism of two activated strings into two entangled 2D strings with opposite rotations. A) Forming the low-vac extra dimension cylinder, B) high-vac extra dimension cylinder.

Without the restriction of the vertical rotation of the two-dimensional string which creates the three-dimensional string, the E-portals are free to inflate and expand to their maximum allowable field radius. The E-portal located in the long column of the high-vac extra dimension cylinder, inflates un-hindered and propagates two-dimensional looped strings along its path using the "g" force created by the orbital rotation of the Planck disk. The mechanism of creation of the two-dimensional Planck-sized looped string is the same as described earlier in this chapter. Three-dimensional looped strings cannot be regenerated due to the interference of other strings. Two-dimensional strings inside the high-vac extra dimension cylinder capitalize on the orbital energy of the cylinders for their kinetic energy as calculated in Chapter 3, allowing their mass to be used for propagation inside the cylinder. As shown in Chapter 3, the speed of propagation of the strings inside the high-vac extra dimension cylinder quickly exceeds the speed of light because there is no restriction in the expansion of the E-portal and the two-dimensional strings inside the long column of the vacuum of the high-vac cylinder. A schematic representation of this model is presented in Fig. 8-12.

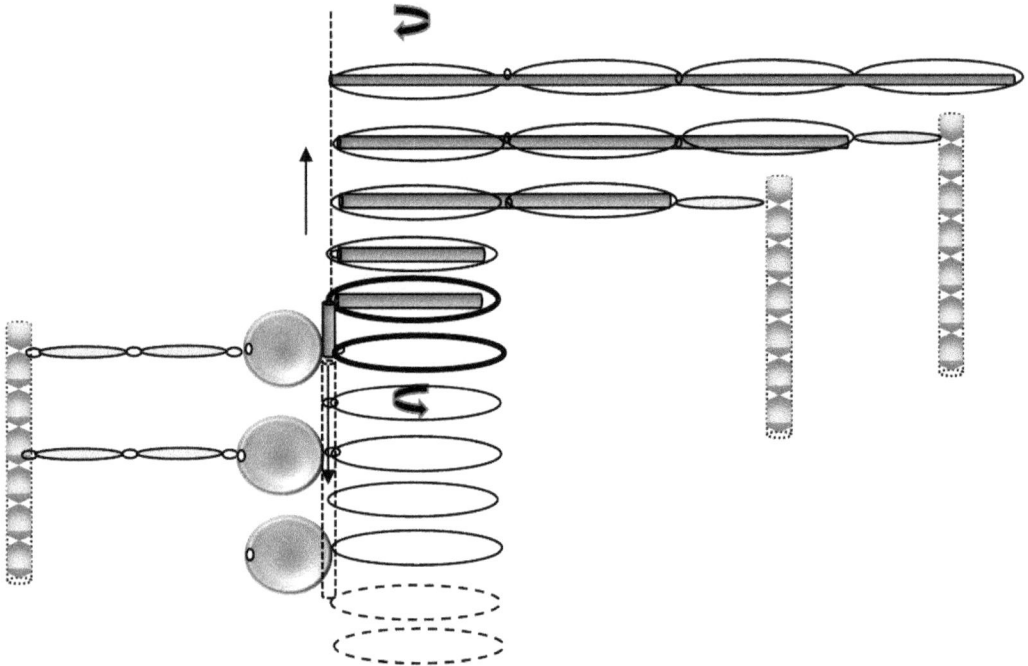

Fig. 8-12- Propagation mechanism of the two entangled two-dimensional looped strings. Pulsating strings provide the orbital rotation of the two-dimensional open strings which attach to the E-portal of the looped string and spin-off two dimensional looped strings. Bold strings represent the two original two-dimensional strings entangled.

The E-portal of the second string will also propagate perpendicular to the path of the high-vac extra dimension cylinders. Since the E-portal of the second string does not travel inside the long column of the vacuum, it is restricted by interference with other strings in infinite momentum frame, as such its propagation speed is restricted to the speed of light. The radius of propagation of this E-portal is calculated from equation

(3-19), Chapter 3. The E-loop at the end of the two-dimensional open strings then attaches to this E-portal and spin-off two-dimensional strings as described earlier, Fig. 8-12. I termed this the low-vac extra dimension cylinder because it propagates randomly in infinite momentum frame outside the organized cylindrical vacuum of the high-vac cylinder. Note that creation of the two-dimensional Planck strings in one case is dependent on the orbital rotation of the Planck disk, and in the other, attachment of the E-portal to the E-loop of an orbiting two-dimensional string.

Referring to Fig. 8-12, the two-dimensional strings in a high-vac extra dimension cylinder propagate in one direction inside the vacuum column of the two-dimensional strings, whereas, in a low-vac cylinder, the E-portal first becomes quantized by Planck length, then propagates in random columns as the environment may allow.

Both high and low-vac strings revert to their Planck dimension after annihilation regardless of its position inside the spacetime cavity and the dimension of its parent three-dimensional looped string. Two-dimensional looped strings are regenerated from their E-portal space as described in Chapter 8, which can only propagate to Planck dimension.

As you can see above, the propagation of high-vac extra dimension cylinders is dominated by the inflation speed of the E-portal which is significantly higher than the speed limitation in Hilbert space. This means that theoretically, the E-portal can achieve a speed of $1x10^{47}$m/s and traverse the universe in about $2.5x10^{-21}$s:

$$t= 2.5x10^{26}/1x10^{47} = 2.5x10^{-21}s$$

A substantial portion of the energy of the decayed string that creates the low-vac string is consumed as the kinetic energy of the orbital rotation of the extra dimension cylinder and the energy of the wave created by the high-vac extra dimension cylinder. The wave energy is consumed as the E-portal pushes against the field of activated open and looped strings while inflating and transferring that energy to the high-vac extra dimension cylinder (the vertical loops in Fig. 8-12). This energy in classical physics is the Hamiltonian of the wave function (see Chapter 11).

Recall that the mass of a single low-vac string located at the center of mass of the particles determines the baryonic mass of fundamental particles. This energy is far below the actual total energy of the decayed string. Recall that the mass of a proton is about 2×10^{-27}kg while the mass of its parent string is in the order of 10^{-12}kg (Chapter 4). The difference in the two masses (energy) becomes the energy of the wave created by the extra dimension cylinder.

As you have seen above, the E-portals of the high-vac and low-vac extra dimensional cylinders are co-joined, even though each string propagated rotates in an opposite direction from each other. This is quantum entanglement.

As you will see later, two particles that are entangled can be separated from each other, but the spin of one will be always the opposite of the other. The distance this information may travel could theoretically be the length of the universe because it will only take 2.5×10^{-21}s for this transmission. This is if there is no interference causing the path of the E-portal to be discontinued by other extra dimension objects in the universe. This is just simply to demonstrate that for two entangled strings, distance and time are not a factor in the way they communicate.

Propagation of two-dimensional open strings

Two-dimensional open strings propagate differently from the two-dimensional looped strings. The first major difference is that open strings cannot propagate via E-portal expansion. This is because generation of an open string requires three E-loop strings with the mass located at the end where the energy is stored in the E-portal. Attachment of the E-loop string of two existing two-dimensional open strings to the E-portal of a decayed string on both sides will place the E-portal with the energy in the middle of the two E-loops which is unacceptable. However, open strings make up for this limitation because the stretched E-loop EL_1 which gives its Planck length, has 8×10^{60} pre-stretched, massless open strings waiting to be propagated. As shown in Fig. 8-13, when two three-dimensional open strings annihilate each other, the plane of the two-dimensional strings become adjoined at E-portal, EL_0.

Even though the E-portal of the string B is free to inflate similar to a looped string, the open string will attach to other strings which will pull and stretch the open strings stored in EL_1, directing the flow of the stored mass inside the E-portal towards the propagated strings and giving it its mass.

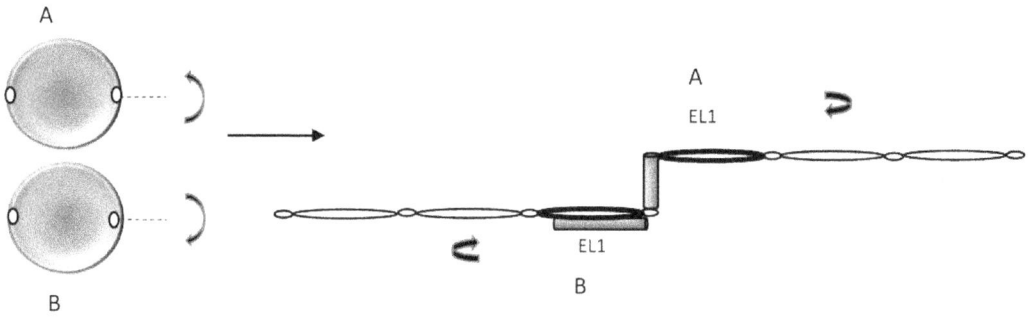

Fig. 8-13- Decay mechanism of two activated open strings into two entangled 2D open strings with opposite rotations. Bold strings represent multilayered two-dimensional open strings.

The E-portals of the two decayed strings will propagate delivering the wave energy and the orbital energy for the orbiting extra dimension cylinders in the same way as the low-vac extra dimension cylinders.

Another important difference in propagation of open strings is that inside the spacetime cavity, a preponderance of high-vac open strings of dimension 4.4×10^{-33}m are generated. As discussed earlier, stable two-dimensional open strings do not form inside the spacetime cavity until the depth of the field is 1.1×10^{-30}m which produces open strings of 4.4×10^{-33}m. The open strings that begin to form at this depth expand to their maximum field length. Since there are an overwhelming number of extra dimension cylinders formed inside the spacetime cavity, the open strings of the above length attach to the cylinders from both ends preventing them from propagating into longer strings, hence a preponderance of strings of 4.4×10^{-33}m are produced.

Recall that the two-dimensional open string of 4.4×10^{-33}m is a critical component in making a magnetic monopole, Tubular Monopole (charge), and photon. In the case of a photon, the open strings of the length 4.4×10^{-33}m attach to the same activated strings on both ends forming its basic structure.

Yet another feature of a two-dimensional open string, as mentioned before, is attachment to the E-loop of a pulsating looped string which not only is the source of the energy for the orbital rotation of all objects in the universe, but as you will see in the next section,

regeneration of activated open strings in our universe, Fig. 8-14.

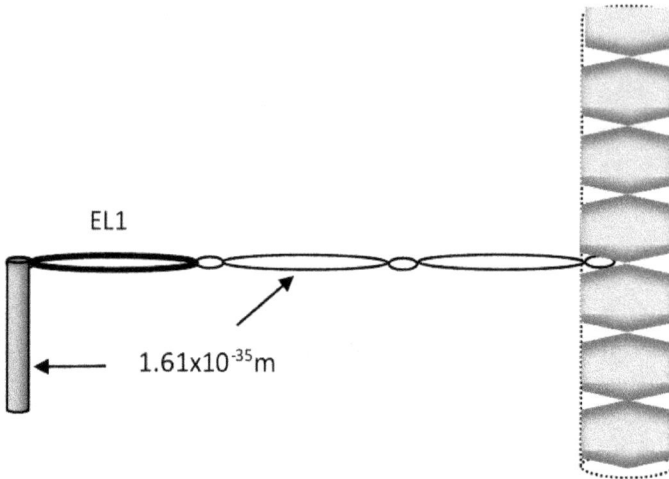

Fig. 8-14- Attachment of a two dimensional open string to a pulsating string. Creation of orbital rotation of extra dimension objects in restframe. Bold strings represent multilayered two-dimensional open strings.

Similar to the looped strings, two-dimensional open strings utilize a small fraction of their energy to propagate. The substantial portion of its energy is used as the displacement energy of the wave it creates.

Quantum Entanglement of particles

In the above sections, I described the process of quantum entanglement of two decaying strings. In this process, the E-portals of two decaying three-dimensional strings become adjoined. In the case of looped strings, the E-portal of one string becomes constrained by interference with other strings as it propagates, while the E-portal of the other is free to inflate to its maximum length and speed inside the vacuum of the orbiting extra dimension cylinder. Since the inflation of E-portals occurs at the speed of about $1x10^{47}$m/s, the propagation and formation of the strings is essentially instantaneous.

The two, two-dimensional looped strings that are entangled will have the same rotation as their parent three-dimensional strings. Therefore, if we measure the spin of one string, the spin of the other will be its opposite.

In practice, we cannot measure the spin of the two-dimensional strings because they are too small to be observable in our universe. However, we can measure the spin of their parent particles.

Recall from Chapter 3, all fundamental particles contain an extra dimension cylinder and two-dimensional open strings in their structure. Extra dimension cylinders house the disks of the entangled decayed strings that formed them. When two particles that are entangled are separated, they will contain two extra dimension cylinders that are connected by the E-portal of the two decayed strings. Meaning that information about one will always be measurable only by obtaining the information about the other. If we measure the spin of one particle which will be the same as the spin of the strings of its extra dimension cylinder, the spin of the entangled particle will always be opposite, Fig. 8-15. The entangled particles can be separated from each other and, barring any interference with the E-portal joining the two particles, the two particles can remain connected over very large distances. In theory, two entangled particles can be separated across the universe and if its E-portal remains intact, the two particles would communicate separated by the length of the universe in $2.5x10^{-21}$ seconds.

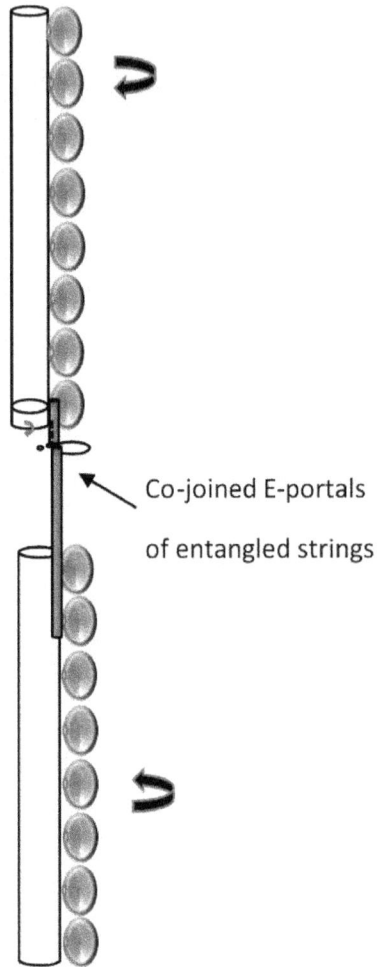

Fig. 8-15- Entanglement of two particles via entangled strings.

Quantum entanglement of strings has an enormous impact on the balance of the forces in the universe. In general, the direction of the orbital rotation of one entangled string creates the forces that are pointing upward and the other downward. The upward force contributes to the expansion of the universe and the downward force keeps the spacetime cavity of fundamental particles from disintegrating. The balance of these two forces establishes an equilibrium between the gravitational force and the expansion force each having a vector in opposite direction due to the entangled two-dimensional strings.

String Tension

In Chapter 1, I discussed a simple yet very important concept and formula:

$\rho = KR$

In which ρ is the energy density (or mass) of the string, R the radius of the field of string in restframe, and K is the string tension constant. Since the mass of the string in all cases represents a constant volume, the above equation is the same for a linear string mass or its density, albeit the value of K will change. Recall that I used this relationship to set up the fundamental equations that derived the energy field of open and looped strings in the universe.

This relation simply implies that strings have an intrinsic property in the restframe such that its energy density or mass is proportional to its length by a constant K which is the string tension. This means that in the restframe, if we pull and stretch a string, it will continue to generate new strings keeping its mass or energy/length constant. Similar phenomena in different forms are well accepted and common in classical physics and other field theories where the energy density of a field is linearly proportional to the length of the field, such as tubes of magnetic flux in super conductors.

Let us now see how this applies to different states of strings considering what we now know about the E-loops, two-dimensional Planck-sized looped and open strings, and activated strings in restframe.

E- loop strings

As described earlier, the E-loops are two-dimensional strings with no surface. As such they can be compacted in many layers yet remain as a two-dimensional object. In the previous sections, it was demonstrated that a mass of 0.15×10^9kg in E-loop space, which is equivalent to 2.18×10^{-8}kg in Planck space, can be compacted in a cylindrical volume having a length equal to its radius (2.3×10^{-51}m). The number of E-loops compacted in such a mass based on the mass of each individual E-loop is:

$$0.15 \times 10^9 / 5.6 \times 10^{-101} = 0.28 \times 10^{109}$$

This means we will have an E-loop with a mass of 0.15×10^9 kg which is comprised of 0.28×10^{109} individual layers of E-loops each having a mass of 5.6×10^{-101} kg.

The E-loops under the energy and torque created by the orbital rotation of a second attached E-loop inflate to 1.3×10^{58} m with a quantized length that equals the diameter of each E-loop.

Therefore, the mass per length of the strings remains constant at:

$$K = 5.6 \times 10^{-101} / 2 \times 2.3 \times 10^{-51} = 1.2 \times 10^{-50} \text{kg/m}$$

As we saw earlier, the mass per area of the E-loop string is constant at:

$$\rho = KA$$

$$K = 5.6 \times 10^{-101} / 1.6 \times 10^{-101} = 3.5 \text{ kg/m}^2$$

Therefore, $\rho = KR$ holds up for E-loop strings.

The mass per length for an E-loop reduces from about 1.2×10^{-50} kg/m to about 0.53×10^{-66} kg/m when it momentarily stretches to Planck length to create the Planck space using the kinetic energy of the orbital rotation and the orbital acceleration of 3.9×10^{67} m/s^2.

Two-dimensional Planck length looped and open strings

In constructing an activated looped and open string, we started with a disk of Planck radius having a mass of about 2.7×10^{-69} kg with an attached E-portal storing 2.18×10^{-8} kg as the core mass of the three-dimensional string. This mass is equivalent to 8×10^{60} Planck-size disks each having the same mass, 2.7×10^{-69} kg. When an activated string decays and reverts to a two-dimensional string, this mass will propagate throughout the universe creating the radius of the field of strings in restframe. The propagation of the strings is quantized in length and equal in Planck length. In the case of two-dimensional open strings, the strings are pre-stretched and layered as 8×10^{60} individual strings that are stretched each by its Planck length as the conditions demand.

The mass/length relationship for Planck length strings in restframe is:

$$K = 2.7 \times 10^{-69} / 1.61 \times 10^{-35} = 1.66 \times 10^{-34} \text{kg/m}$$

In fact, as stated earlier, if we stretch a two-dimensional string with a mass of 2.18×10^{-8}kg and 8×10^{60} layers, it will span across the radius of the universe:

$$1.61 \times 10^{-35} \text{m} \times 8 \times 10^{60} = 1.28 \times 10^{26} \text{m}$$

The mass per area of the two-dimensional Planck strings also remains constant:

$$K = 2.7 \times 10^{-69} / 8.14 \times 10^{-70} = 3.3$$

Which should remain the same as that of E-loops. Therefore, the $\rho = KA$ also holds for two-dimensional Planck strings.

Three-dimensional open and looped strings

Activated open strings maintain Planck dimension and mass as a quantum specie. When attached to extra dimension cylinders, or in restframe in general, their mass and length are quantized over any radius of restframe. The mass/length relationship for activated open strings or activated looped strings with a Planck radius is constant at:

$$2.18 \times 10^{-8} / 2 \times 1.61 \times 10^{-35} = 0.677 \times 10^{27} \text{ kg/m}$$

The dimension and mass of activated looped strings change with the radius of the energy field of the mass. However, outside the energy field of a mass, its mass/length relationship is constant at:

$$2 \times 10^{-23} \text{kg} / 2 \times 2.3 \times 10^{-20} \text{m} = 0.43 \times 10^{-3} \text{ kg/m}$$

As can be seen, all strings exhibit the same property regardless of its state and follow the same rule in restframe:

$\rho = KR$

However, this relationship has a limit based on the value of R. For E-loop strings, the maximum length for which this rule is applicable is $R = 1.3 \times 10^{58}$m, for two-dimensional Planck-size strings, $R = 2.3 \times 10^{26}$m, and for three-dimensional activated strings, $R = 1.15 \times 10^{26}$m.

There is one case where there is no limit to the value of R. This is the one-dimensional string I will discuss in Chapter 12. The one-dimensional string is comprised of what I have termed a zero-dimensional string with a dimension of 8.9×10^{-160}m. As you will see, this one-dimensional string forms the structure of bulk vacuum in the universe. The one-dimensional string can be stretched indefinitely creating zero-dimensional strings without any limitation to the value of R.

Regeneration of activated open strings

The graph of the field of activated open and looped strings in Chapter 1 indicates that after the Big Bang, the universe produced a significant number of three-dimensional strings we called the infinite momentum frame in a short time and continues to generate more to sustain the life of the universe for about 100B years. The constant mass /length property of the strings ($\rho =$KR) in restframe as described above indeed translates into generation of new three-dimensional strings in the universe. One may view it this way, as the strings in the restframe stretch and create the fabric of spacetime for the expansion of the three-dimensional universe, the universe has managed to fill the empty space with activated strings. This is a critical step because without regeneration of activated strings in the universe, the supply of existing three-dimensional strings will be consumed by the existing matter and blackholes in a short time. In this section, I will highlight a few ways activated strings are produced in the universe beyond the Big Bang.

Let us first look at pulsating extra dimension cylinders which represent two-dimensional looped strings in restframe2. Recall that pulsating extra dimension cylinders are created by decaying activated looped strings that are not orbiting as extra dimension cylinders. Under the force of the activated looped strings the Planck disk collapses to its allowable minimum dimension which is the E-loop string.

Upon decay of two-oppositely rotating activated looped stings, two extra dimension objects are created and propagated as described before. One is a high-vac pulsating string, the other a low-vac pulsating string. The propagation of the E-portal of the high and low-vac strings are restricted by the speed of light and will only pulsate one Planck length at a time moving forward as a "plug". The E-portal of the low-vac string propagates in a direction perpendicular to the high-vac string into a media where the availability of free bulk vacuum is scarce and random.

In a pulsating string, the dimension of the looped string shrinks down from Planck radius to 2.3×10^{-51}m or becomes an E-loop string and then expands back to its Planck radius with a frequency of:

$$f_1 = c/r = 3 \times 10^8 / 1.61 \times 10^{-35} = 1.86 \times 10^{43}$$

Under the conditions I described in the previous sections, when the two-dimensional looped string collapses into an E- loop, its energy will revert to 1.37×10^{25}J stored in an E-portal and converted to kinetic energy of the E-loop, with a "g" value of 3.9×10^{67}m/s and a spinning velocity of 3×10^{8}m/s.

However, in an existing universe, there will be an abundance of pre-existing two-dimensional strings due to the decay of activated open strings. When two two-dimensional open strings are attached to both sides of the collapsed E-loop, they will apply a uniform pull from both sides which will stretch the collapsed E-loop back to its Planck dimension. In this process, 5×10^{31} massless E-loops are created that will construct the new Planck disk. In the meantime, the attachment of a two-dimensional open string with a base mass of 2.7×10^{-69}kg to the rotational energy of the collapsed E-loop creates the conditions I described in the previous section for generation of an activated open string, Fig. 8-16. At the same time that a collapsed E-loop is being stretched back into a Planck-size string, a portion of its energy stored in its E-portal will be used to construct a new activated open string. On the next cycle, the process is repeated, and a new activated open string is generated with each pulse of the string. Fig. 8-16 presented below is a simplified version of the model that leads into generation of an activated open string.

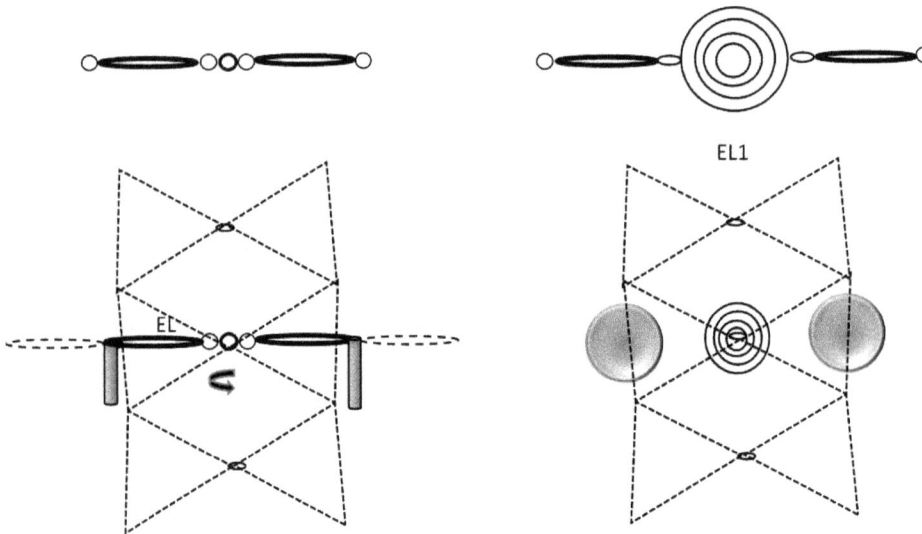

Fig. 8-16- Conceptual representation of pulsation mechanism and regeneration of an activated open string via attachment of two-dimensional open strings to a pulsating string.

The formation of a three-dimensional activated open string also requires a volume of vacuum that is equal to the volume of the activated open string. As mentioned earlier, high-vac pulsating strings contain a vacuum volume at least equal to Planck volume. This vacuum is the remnant of the decayed activated looped string. More importantly, since creation of this vacuum is instantaneous as a result of the collapse of the string into an E-loop, the space will be left without the interference of other objects allowing the formation of a three-dimensional sphere. Low-vac pulsating strings propagate in a media where bulk vacuum availability is scarce and random. As such, it will likely not regenerate as many activated open strings as the high-vac pulsating strings, if any.

Therefore, as the resframe2 stretches out in the form of a pulsating extra dimension cylinder moving forward one Planck length at a time, following the $\rho = KR$ rule, it creates new activated open strings. In this case, two activated open strings per Planck length. This confirms that the activated three-dimensional open strings also follow the $\rho = KR$ rule in restframe as discussed earlier.

Let us now look at the pulsation energy of a looped string. To do this we must first find the geometric average frequency of the looped string pulsating between a dimension of 1.61×10^{-35}m and 2.3×10^{-51}m:

$$f_3 = (f_1 f_2)^{1/2}$$

$$f_1 = c/2\pi r = 0.79 \times 10^{43} \text{ s}^{-1}$$

$$f_2 = c/2\pi r = 0.55 \times 10^{59} \text{ s}^{-1}$$

$$f_3 = 0.66 \times 10^{51} \text{ s}^{-1}$$

The energy of the pulsating string will be:

$$E = hf_3 = 6.626 \times 10^{-34} \times 0.66 \times 10^{51} = 4.37 \times 10^{17} \text{J}$$

This is the total orbital energy of a low-vac pulsating string. In the previous sections it was determined that an E-portal with a mass of 2.7×10^{-67} kg and energy of 1.3×10^{25}J will expand a distance of 2.5×10^{26}m. Correcting this based on the amount of energy available to a pulsating string:

$2.5 \times 10^{26}(4.3 \times 10^{17} / 1.37 \times 10^{25}) = 7.9 \times 10^{18} \text{m}$

Which is about the same as the R $= 3.9 \times 10^{18}$m, the radius of the low-vac pulsating extra dimension cylinder calculated in Chapter 3. A decayed activated looped string with an energy of 1.96×10^{9}J should theoretically pulsate a distance of 2.5×10^{26}m. This means that the vast majority of the energy of the low-vac pulsating string is consumed as the orbital energy of other extra dimension objects that attach to its E-portal.

Conversely, the radius of the high-vac string was calculated to be 0.66×10^{11}m which means that the E-portal will require an energy of:

$2.5 \times 10^{26}(E / 1.37 \times 10^{25}) = 0.66 \times 10^{11} \text{m}$

$E = 3.6 \times 10^{9}$J, in E-loop space.

When a two-dimensional string with an energy of 1.96×10^{9}J collapses into an E-portal, it will have an energy of 1.3×10^{23}J. As can be seen, only a small fraction of this energy is used for propagation into a distance of 0.66×10^{11}m. The significant portion of the energy of the high-vac pulsating string is used for regeneration of activated open strings.

Two-dimensional open strings that attach to the E-portal of a high-vac pulsating strings (Fig. 8-16), have at least consumed the energy of at least one Planck disk. Therefore, the E-portal of the high-vac pulsating string will have to deliver an energy equal to one Planck disk to regenerate the two-dimensional open string back to a three-dimensional string:

Energy of one Planck disk $= 2.7 \times 10^{-69} \times 9 \times 10^{16} = 2.43 \times 10^{-52}$J

Since two-dimensional strings that attach to the E-portal already contain the necessary E-loop (EL_1) in Planck space, this energy has to be delivered in E-loop space to be stored in the string's E-portal, EL_0. Therefore, the amount of energy above needs to be corrected for the E-loop space by multiplying by the conversion factor:

Energy required for conversion into a three-dimensional string $= 2.43 \times 10^{-52} \times 0.7 \times 10^{16}$

$E = 1.7 \times 10^{-36}$J

The total energy available from a decayed three-dimensional looped string that converts to a high-vac pulsating string is 1.96×10^9J. The total number of three-dimensional open strings that can be regenerated is then:

$$N = 1.96 \times 10^9 / 1.7 \times 10^{-36} = 1.15 \times 10^{45}$$

We can verify this by assuming that for every pulse the string travels a distance of 0.66×10^{11} two activated strings are produced:

$$2 \times 0.66 \times 10^{11} / 2 \times 1.61 \times 10^{-35} = 3.7 \times 10^{45}$$

Recall from Chapter 3 that since we have 3.8×10^{15} orbits ($2.5 \times 10^{26} / 0.66 \times 10^{11}$), and if we assume the same action is repeated, the total possible number of activated open strings generated will be about:

$$3.8 \times 10^{15} \times 3.7 \times 10^{45} = 14 \times 10^{60}$$

Attachment of two-dimensional open strings to the E-loop of a high-vac pulsating string not only regenerates two activated open strings per pulsation, but its orbital energy is used to create the orbital rotation of the restframe. This is because two-dimensional open strings, while attached to the E-loop at one end, are attached to a network of other strings and extra dimension objects on the other end. As such, the pulsating extra dimension cylinders are the axis and source of energy for creating the orbital rotation of the restframe in the universe as a whole.

Interestingly, the pulsation of high-vac extra dimension cylinders, with a length of 0.66×10^{11}m, produces enough activated open strings to equalize the consumption of the open strings contributing to the electromagnetic force of the universe:

$$M = 0.675 \times 10^{27} \text{ kg/m} \times 0.66 \times 10^{11}\text{m} \times 3.7 \times 10^{45} = 0.16 \times 10^{84} \text{ kg}$$

Regeneration of activated looped strings

Before I discuss the regeneration of activated looped strings, let us review one more time the reason activated looped strings expand and become larger while the activated open strings remain of the same dimension.

Activated open strings have two poles (E-loops) opposite each other on its three-dimensional sphere. Activated looped strings have only one pole (E-loop) on its three-dimensional sphere. As I mentioned before, the poles on the sphere surface are the point of attachment of other two-dimensional open strings due to the vacuum of the E-loop and the vacuum of the E-loop at the end of each two-dimensional open string which attracts each other. In an activated open string, both poles will be occupied by two-dimensional open strings of various lengths which are in turn attached to other extra dimension objects at the other end. This arrangement creates unequal and relatively large tension forces on each pole of the activated open string given how many subsequent attachments each two-dimensional open string may have. As a result, the attachments to activated open strings break off quickly and become transitory because of the tugging and pulling from both directions. In other words, the residence time of the attachment will be too short to allow any force to be exerted on the E-loop of the pole.

In an activated looped string, there is only one pole which allows the attachments to rotate with the sphere of the three-dimensional looped string, giving the attachment more residence time to exert a force on the E-loop of the pole.

Consider the case of two activated looped strings with opposite rotations attached to each other via a two-dimensional open string inside the spacetime cavity. The smallest two-dimensional string generated is 4.4×10^{-33}m at the depth of 1.1×10^{-30}m. When two activated looped strings with opposite rotation are attached to each other via this two-dimensional open string, they can exert a force on each other's E-loop causing the E-pole to stretch and expand, Fig. 8-17.

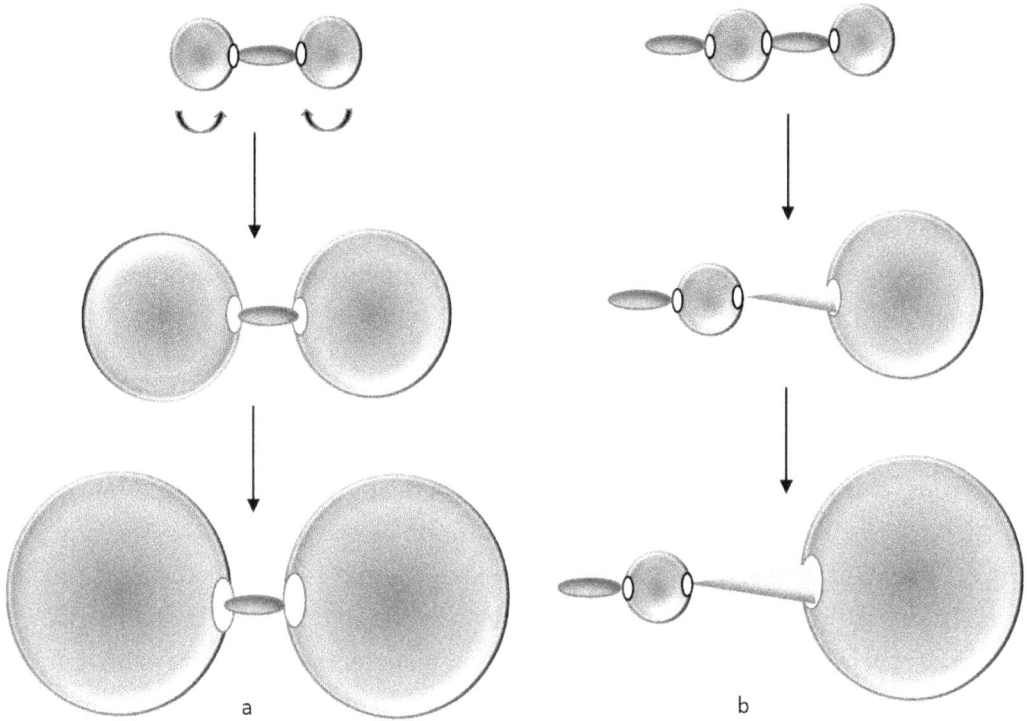

Fig. 8-17 – Sequential expansion of an activated looped string by the forces of attachments exerted on its pole, a) two looped strings with opposite rotations, b) stretched by other extra dimension objects.

When the E-loop string of the E-pole experiences equal forces on its circumference it will expand by creating more E-loops of the same surface area. In other words, more E-loops are generated and quantized by the same surface area ($1.6 \times 10^{-101} m^2$) to occupy the void generated. The newly generated E-loops, due to stretching of the E-pole, have no mass or kinetic energy. The source of the supply for the expansion of the E-loops of the pole and the two-dimensional Planck string at the core of the string is the mass of the string which is reduced from $2.18 \times 10^{-8} kg$ to $1.5 \times 10^{-23} kg$ when the looped string has fully expanded.

The increase in the E-loop diameter causes the gauges produced by it to expand correspondingly, increasing the radius of the E-portal, and the radius of the activated looped string correspondingly. This comes at the price of reducing the core mass of the activated looped string from Planck mass to match the energy required to expand the string dimensionally. As the two adjoining activated looped strings rise from the spacetime cavity of the matter into the infinite momentum frame, the dimensions of the E-loop will continue to increase and so does the radius of the activated looped string. Activated looped strings can attach to any other extra dimension object via a two-dimensional open string which in turn exerts a similar uniform force on the E-loop of the string causing it to expand.

I demonstrated in Chapter 2, the change of the radius of an activated looped string vs, the radius of the energy field of a mass inside and outside the spacetime cavity as equation (2- 3):

$r_a = 0.71x10^{-26} (Ra)^{1/4}$

The radius of the activated looped string, which at the nucleus of the spacetime cavity is $1.61x10^{-35}$m, will expand to $2.3x10^{-20}$m in infinite momentum frame. The radius of the E- loop at the pole of the activated string expands from $2.3x10^{-51}$m to about $1.6x10^{-36}$m. To give the reader a dimensional comparison, the size of the pole of an activated looped string in infinite momentum frame to its diameter is like the size of an atom to Earth. The pole of an expanded activated looped string in infinite momentum frame is about $1/10^{th}$ of the dimensions of activated open strings at Planck length ($1.61x10^{-35}$m).

Energy/mass of an expanded activated looped strings

Let us now see what happens to the energy of the expanded activated looped string.

The expansion of the activated looped string means that the gauges comprising the two-dimensional core of the string must also increase with the radius of the string. Since the core has a mass of $2.18x10^{-8}$kg, it will use the E-loops stored inside the E-portal of the core mass to create length, reducing its core mass. This conversion follows the Planck equation:

$m = h/2\pi rc$

When the activated looped string reaches a dimension of $2.3x10^{-20}$m, its mass will reduce from $2.18x10^{-8}$kg to $1.5x10^{-23}$kg substituting for r = $2.3x10^{-20}$m in the above equation.

We can also calculate the mass of the expanded activated looped string the same way I calculated the E-portal mass and inflation velocity of E-loop strings, i.e., by calculating the mass stored in its E-portal when the diameter of the string reaches 2.3×10^{-20}m. Recall that the kinetic energy of the E-loops stored inside the E-portal create the inflation velocity of V for the E-loops to stretch and expand:

$V^2 = 2rg$

$V^2 = 3.9\times10^{67} \times 2 \times 2.3\times10^{-20}$

$V^2 = 1.8\times10^{48}$

The energy of the string before expansion is 1.96×10^{9}J which in E-loop space is equivalent to 1.37×10^{25}J. We can calculate the mass of the activated looped string which is the mass stored in its E-portal located at one end of the string:

$E = 1/2mV^2$

$m = 2\times1.37\times10^{25}/1.8\times10^{48} = 1.5\times10^{-23}$kg

An instantaneous burst

As mentioned above, when an activated looped string expands, the entire E-portal dimension which is storing its mass expands accordingly. The radius of the E-pole expands to 1.6×10^{-36}m. This means that the surface area of the E-loops inside the E-portal will expand by a factor of:

$(1.6\times10^{-36}/2.3\times10^{-51})^2 = 0.48\times10^{30}$

As mentioned before, the expanded E-loops that generate a surface area which is 0.48×10^{30} times higher are massless with no kinetic energy, until the expanded E-Portal collapses back to its initial dimension of 2.3×10^{-51}m upon implosion of the looped string due to decay. However, we must first calculate the total number of E-loop strings that will end up inside the E-portal upon implosion of the string before multiplying it by the above number.

The total number of E-loops comprising the surface of the two-dimensional string of the expanded looped string is:

$$\pi(2.3 \times 10^{-20})^2 / 1.6 \times 10^{-101} = 1 \times 10^{62}$$

Similar to other activated strings, there will be 8×10^{60} of the above number stored in the E-portal of the expanded activated looped string before collapse:

$$1 \times 10^{62} \times 8 \times 10^{60} = 8 \times 10^{122}$$

This number corrected for the E-loop space using the Planck to E-loop space conversion factor of 0.7×10^{16} will be:

$$8 \times 10^{122} \times 0.7 \times 10^{16} = 5.6 \times 10^{138}$$

The total number of E-loop strings stored inside the E-portal of an activated looped string after decaying and collapsing into an E-portal will be:

$$5.6 \times 10^{138} \times 0.48 \times 10^{30} = 2.7 \times 10^{168}$$

When the strings collapse from their expanded E-loops to the dimension of 2.3×10^{-51}m, the above E-loop strings will gain a kinetic energy with a rotational speed of c, which can then create new mass.

If we divide the total number of E-loop strings above produced due to the decay of two expanded activated strings by the number of E-loops required to create a Planck-size activated looped string, we will obtain the total number of Planck-size activated looped strings regenerated:

$$2.7 \times 10^{168} / 4 \times 10^{92} = 6.7 \times 10^{75}$$

Let us now calculate approximately, the total number of activated strings that could occupy the volume of a spacetime cavity. First, the volume of a spacetime cavity as a cone with a base of $1.6 \times 1.8 \times 10^{-9}$m and a height of 1.8×10^{-9}m is about:

$$V = \pi(1.6 \times 1.8 \times 10^{-9})^2 \times 1.8 \times 10^{-9} / 3 = 15.6 \times 10^{-27} \text{m}^3$$

As calculated in earlier chapters, the average diameter of a looped string inside the spacetime cavity is about 2.7×10^{-32} m. We can estimate the total number of strings in a spacetime cavity based on the average volume of a looped and open string. The geometric average radius of strings inside the cavity will be:

$r = (2.7 \times 10^{-32} \times 1.61 \times 10^{-35})^{1/2} = 6.6 \times 10^{-34}$

Avg. volume of a string $= 4/3\pi \times (6.5 \times 10^{-34})^3 = 1.2 \times 10^{-99}$

No. of activated strings in a cavity $= 15.6 \times 10^{-27}/1.1 \times 10^{-99} = 1.3 \times 10^{73}$

Since we know the total number of Planck-size activated strings produced is 6.7×10^{75}, we can calculate the number of spacetime cavities produced. Recall that each spacetime cavity upon decay of an activated looped string represented a Higgs Boson.

Total number of spacetime cavities produced:

$6.7 \times 10^{75}/1.3 \times 10^{73} = 5.1 \times 10^2$

Fig. 8-18 displays the conceptual diagram for regeneration of activated looped strings and Higgs Bosons.

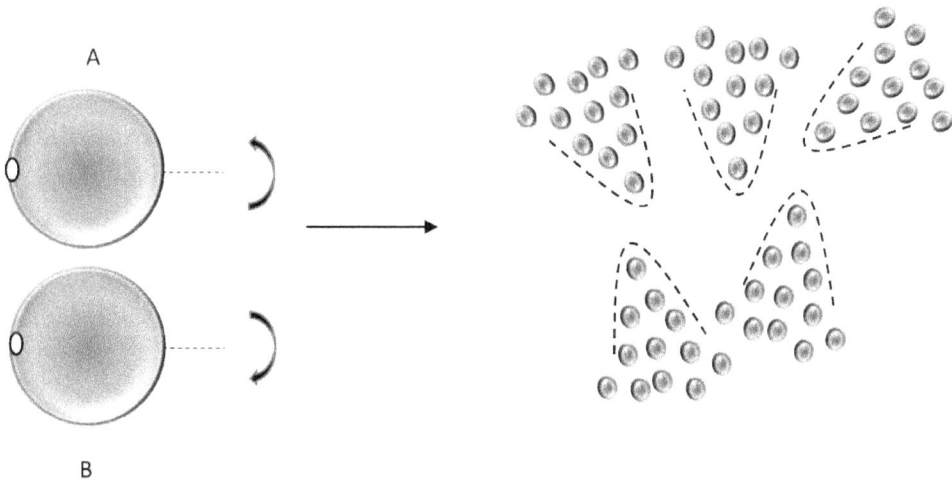

Fig. 8-18- Conceptual representation of decay of expanded activated looped strings and conversion to spacetime cavities, and generation of Higgs Boson particles.

This spacetime cavity will be substantially occupied with newly generated activated looped strings that will also occupy the 135 of 137 high-vac extra dimension cylinders inside the spacetime cavity. As discussed before, the particle associated with this short-lived spacetime cavity is a Higgs Boson having a mass of about 129 times the mass of a proton, or about 2×10^{-25}kg.

Since the mass of the expanded activated looped string before decay is 1.5×10^{-23}kg, the minimum number of Higgs Bosons formed will be:

$$1.5 \times 10^{-23} / 2 \times 10^{-25} = 75$$

The maximum number of Higgs Boson particles produced based on the number of spacetime cavities calculated above, at one particle per spacetime cavity, will be 510. Therefore, upon annihilation of two expanded activated looped strings in infinite momentum frame (outside the spacetime cavity), the number of Higgs Bosons produced per string decay will be about:

$$75 < \text{Higgs Bosons} < 510$$

After the decay of the activated strings and creation of the spacetime cavities, the strings inside the spacetime cavities will decay, entangle, and propagate similar to the model described in the previous sections.

The existence of a process that creates Higgs Boson spacetime cavities is critical for generation of new particles and matter. Wherever in the universe after the Big Bang that conditions allow, Higgs Boson spacetime cavities will be the seed for conversion of the cavities into the structure of atoms as described in the previous chapters.

CHAPTER 9

Strings, Entropy, and Temperature

LET US FIRST REVIEW A few items I have described in previous sections that are important to this topic.

First, in the generation of a two or three-dimensional Planck string, we begin with two E- loop strings each having a rotational speed of c, the speed of light. When the two E-loop strings join, one as the anchor creating an orbital rotation for the other, the potential energy of one is transferred to the other. This transfer creates the kinetic energy necessary to stretch the E-loop to Planck length. Recall that to create the Planck length, the E-loops inside the E-portal storing the mass reached a kinetic energy with an inflation velocity of 3.5×10^{16}m/s.

The anchoring E-loop, after transferring its energy to the orbiting E-loop, becomes a massless vacuum with no energy, essentially exiting the universe containing only information about its parent strings. The orbiting E-loop, after gaining the kinetic energy, stretches to create the Planck space and then collapses back to its original E-loop regaining its speed of light velocity.

In this process, we began with two E-loops of identical energy and lost one. In the process, we created Planck space and space mass. The energy of the lost E-loop string is used to stretch the second E-loop and create "space". How do we detect the change in energy of this space when the energy of one E-loop is transferred to create it?

The answer is that the strings have an Entropy which transforms this space energy into thermal energy we can measure by its temperature. We know that the classical relationship of energy and Entropy is:

$$T = dE/dS \qquad\qquad (9\text{-}1)$$

In the above process, for every two E-loop strings that create a Planck disk (space), an E-loop string is lost. The energy of an E-loop string is:

$$E = 5.4 \times 10^{-101} \times 9 \times 10^{16} = 48.6 \times 10^{-85} J$$

Since every activated string contains an equivalent 8×10^{60} Planck disks, the total energy used in creation of this space is:

$$dE = 48.6 \times 10^{-85} \times 8 \times 10^{60} = 3.9 \times 10^{-23} J$$

This is the change in the energy of the string to maintain a Planck space. Since the two-dimensional E-loop string has only two degrees of freedom (one for orbital rotation, and one for lateral movement), and the resultant three-dimensional looped string has 5, the change in degrees of freedom will be (5-2=3). Therefore, energy change per degrees of freedom will be:

$$dE = 3.9 \times 10^{-23} /3 = 1.3 \times 10^{-23} J/\text{degrees of freedom}$$

Let us define k_b as the entropy/degrees of freedom. Equation (9-1) then becomes:

$$T = dE/dS = dE/(\text{degrees of freedom} \times k_b)$$

If we define k_b as the amount of energy needed to raise the temperature 1 degree Kelvin:

$$1 = 1.3 \times 10^{-23}/k_b$$

$$k_b = 1.3 \times 10^{-23} J/K/\text{degree of freedom}$$

As you can see, k_b is the famous Boltzmann constant. In classical physics, this number is officially $1.38 \times 10^{-23} J/K$.

The Boltzmann constant, k_b, is the entropy/degrees of freedom of the most fundamental specie in the universe, an activated three-dimensional string. This means that the temperature of a single three-dimensional string in the universe is:

$$T = dE / dS = dE/ k_b = 3.9 \times 10^{-23}/1.38 \times 10^{-23} = 2.8 \text{ K}$$

Which is the same as the background temperature of the universe. The entropy of a system is the change in energy/degrees of freedom that contributes to the thermal energy. For a system comprising more than one activated string the entropy is then:

$$S = \text{no. of degrees of freedom} \times k_b$$

What this means is that the energy of the lost E-loop string, which creates the Planck space and its mass, manifests itself as the thermal energy of the string which we can measure. This means that creation of space = thermal energy.

Therefore, the process of thermal radiation in the universe, and what we feel and measure as temperature, is a side effect of expansion of space at string level.

Our three-dimensional universe is comprised of two different activated strings, open and looped. The dimension of activated open strings remains at Planck length throughout the universe. The dimension of activated looped strings increases from Planck length at the nucleus of the spacetime cavity to 2.3×10^{-20}m in infinite momentum frame.

When Planck-size activated looped strings are emitted from a body of mass, they will expand as they exit the spacetime cavity and create a larger space. The expansion of activated looped strings means expansion of the space itself at the sub-quantum level. This space expansion results in generation of the thermal energy described above.

As you will see in the next chapter, the largest bodies of mass in the universe are blackholes and this will be the mechanism of radiation from a blackhole.

Let us now calculate the thermal property of a three-dimensional looped string that is emitted from an object.

Recall that when a looped string is stretched from Planck length to its dimension in infinite momentum frame, 2.3x10^{-20}m, its lone E-pole is attached to a two-dimensional open string which is stretched uniformly causing its E-portal to expand to a dimension of 1.6x10^{-36}m, and the looped string itself to a radius of 2.3x10^{-20}m. The inflation velocity of the E-loop strings inside the E-portal which was 3.5x10^{16}m for Planck dimension will increase to 1.34x10^{24}m/s to enable the string to stretch to its new dimension:

$$V^2 = 2rg$$

$$V^2 = 3.9x10^{67} x 2 x 2.3x10^{-20}$$

$$V^2 = 1.8x10^{48}$$

$$V = 1.34x10^{24}$$

This means that the velocity of the E-loop strings increases by 0.38x10^8 to create the new space:

$$1.34x10^{24}/3.5x10^{16} = 0.38x10^8$$

Since temperature is proportional to the kinetic energy of the E-loop string, the stretching of the E-loop string from a Planck length to 2.3x10^{-20}m means that the temperature of the string will increase momentarily by a factor of $(0.38x10^8)^2$:

$$T = 2.8 x (0.38x10^8)^2 = 4x10^{15} K.$$

It takes the E-loop strings, 1.7x10^{-44} s to stretch to its expanded dimension:

$$t = 2.3x10^{-20}/1.34x10^{24} = 1.7x10^{-44}s$$

The string will observe a maximum temperature of 4x10^{15} K for 1.7x10^{-44}s. The temperature of the expanded string will quickly fall to 2.8 K when the space at macro-level expands to accommodate the new expanded string. The geometric average temperature of the string will be:

$$(4x10^{15} x2.8)^{1/2} = 1x10^8 K$$

The actual temperature observed or measured in the environment will depend on the number of looped strings with Planck dimension emitted, the frequency of the emission from a body of mass, and the amount of heat that dissipates with its surrounding strings.

The above example of expansion from a Planck length to 2.3×10^{-20}m in the universe only happens in blackholes or cosmic explosions. It is meant to demonstrate the thermal effect created by the expansion of space at the sub-quantum level.

Activated looped strings emitted from ordinary matter will have a dimension of about 4.6×10^{-29}m which means less expansion and significantly lower temperatures. Substituting in the above equations, we will find that the inflation velocity of the E-loop propagation inside the E-portal at this dimension will be:

$$V^2 = 3.9 \times 10^{67} \times 2 \times 4.6 \times 10^{-29}$$

$$V = 6 \times 10^{19}$$

The ratio of the temperature increase will be:

$$6 \times 10^{19} / 3.5 \times 10^{16} = 1.7 \times 10^3$$

Therefore, the maximum momentary temperature increase will be about:

$$T = 2.8 \times (1.7 \times 10^3)^2 = 8 \times 10^6 \text{ K,}$$

And the geometric average temperature of the string will be:

$$T = (2.8 \times 8 \times 10^6)^{1/2} = 4.76 \times 10^3 \text{K}$$

Again, the actual temperature measured will depend on the number of activated looped strings emitted and the frequency of the emission as the expanded temperature shown above is immediately moderated by the existing activated strings surrounding the emitted one having a temperature of only about 2.8K.

The important take away here is that the thermal energy we observe and measure as temperature is the result of the energy conversion of E-loop strings into space at string level. The temperature is proportional to the kinetic energy of the E-loop string stretching to create such space.

Since thermal energy is the result of the expansion of activated looped strings, the temperature of the atmosphere of stars will always be higher than the surface of the stars where the activated looped strings are emitted. This is because the expansion of activated strings takes place as the string moves away from the gravitational field of the stars which is higher on the surface. In other words, the gravitational field of the stars prevents the activated looped strings from expanding immediately causing the temperature to be lower on the surface. Recall that activated looped strings expand as a function of R_a, the energy field of the mass in infinite momentum frame.

This effect is not noticeable for smaller masses because the gravitational field of smaller masses is too small to affect the immediate expansion of activated looped strings upon emission.

As shown above, the temperature of an activated string is about 2.8K which is about the same as the background temperature of the universe. We can calculate this also by calculating the entropy of the current universe.

The entropy of the universe attributed to the activated looped string can be calculated as follows:

The radius of the observable universe is about 1.15×10^{26}m. If we envision the universe as one body of mass at its event horizon, the total number of activated looped strings that will fit on its surface will be:

Number of activated looped strings on the surface of the event horizon =

$$4\pi(1.15 \times 10^{26})^2 / \pi (2.3 \times 10^{-20})^2 = 1 \times 10^{92}$$

Where 2.3×10^{-20}m is the radius of the activated looped string. Since each activated string has 5 degrees of freedom:

$$dS = \text{degrees of freedom} \times k_b = 5 \times 1 \times 10^{92} \times k_b$$

The energy of the activated looped strings based on the mass of the universe (which is mostly dark matter mass) calculated earlier is:

$dE = 2x10^{53}x9x10^{16} = 1.8x10^{69} J$

The temperature of the universe based on the entropy of the activated looped strings will then be:

$T = dE/dS = 18x10^{69}/5x10^{92}x k_b = 2.6 K$

Let us now calculate the temperature of the universe based on the entropy of the activated open strings in the universe.

No. of activated open strings on the surface of the event horizon =

$4\pi(1.15x10^{26})^2/\pi (1.61x10^{-35})^2 = 2 x10^{122}$

The entropy of the universe based on the activated open strings is then:

$dS = $ degrees of freedom $x k_b = 5x2x10^{122} x k_b$

The energy of the activated open strings based on its mass calculated earlier is:

$dE = 0.4x10^{84}x9x10^{16} = 3.6x10^{100} J$

The temperature of the universe based on the entropy of the open strings will then be:

$T = dE/dS = 3.6x10^{100}/1x10^{123}x k_b = 2.6 K$

As you can see, the temperature of the universe regardless of which method is used is about 2.6K which is about the same as the 2.8K calculated earlier as the temperature of an activated string.

CHAPTER 10
Blackholes

Blackholes, structure, gravity

IN CHAPTER 2, I RESCALED the graphs of the energy fields of activated open and looped strings in infinite momentum frame to describe the fields of the strings reflecting the space dimensional transformation in the universe. This is possible because the graphs of the field of energy for both activated looped and open strings were obtained independent of the mass, allowing us to explore the universe with its geometric dimension.

This leads to the discovery of a spacetime cavity from 1.61×10^{-35}m to about 1.8×10^{-9} (an order of $10^{26)}$. The spacetime cavity unveiled many features of the quantum world and the extra dimension objects that make up our universe. We also learned of a new space dimension, an E-loop string, with a radius of 2.3×10^{-51}m which leads us to yet another space transformation from 1.61×10^{-35}m to 2.3×10^{-51}m (a factor of 10^{16}). Again, since the graphs are independent of the mass, it will allow us to rescale them to discover the properties of the new dimension. Fig. 10-1 is a representation of the rescaling of spacetime by a factor of 10^{16} to accommodate the new dimension. This spacetime cavity represents the transformation of the Planck space into an E-loop space and vice versa. The differentiation between looped and open string disappears once we have reached the E- loop space because as you saw earlier, the E- loop is the fundamental building block of all strings and the graph of the open strings represents the space transformation for the E- loop space. As discussed in Chapter 8, equation (1-27) from Chapter 1 also represents the energy field of the E-loop space.

Fig - 10-1- Dimensional transformation of the E-loop string energy field to Planck space

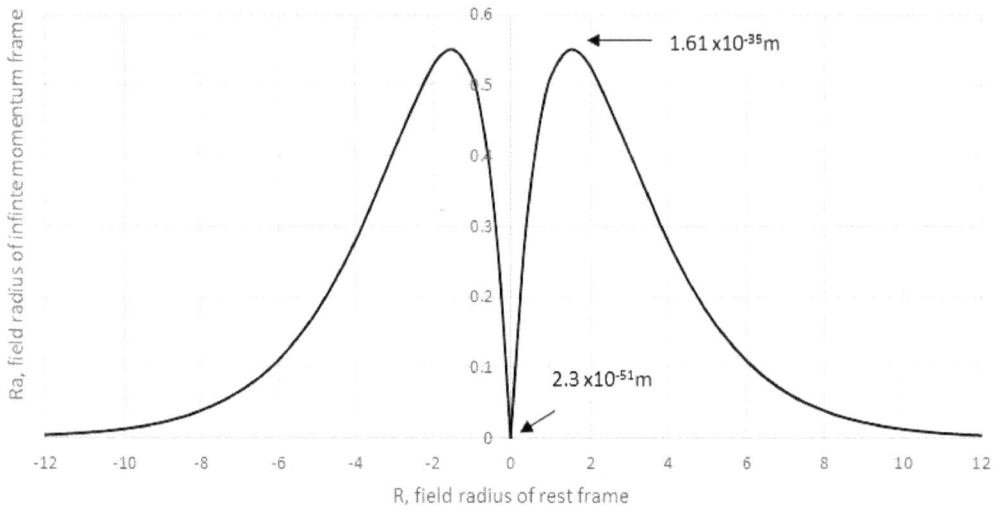

Fig. 10-2 represents the three stages of space dimensional transformation in one graph. As you can see in both Figures 10-1 and 10-2, the origin of the axis represents 2.3x10^{-51}m, the radius of the E-loop string, the fundamental building block of the universe.

Fig. 10-2 -Three stages of dimensional transformation of the space

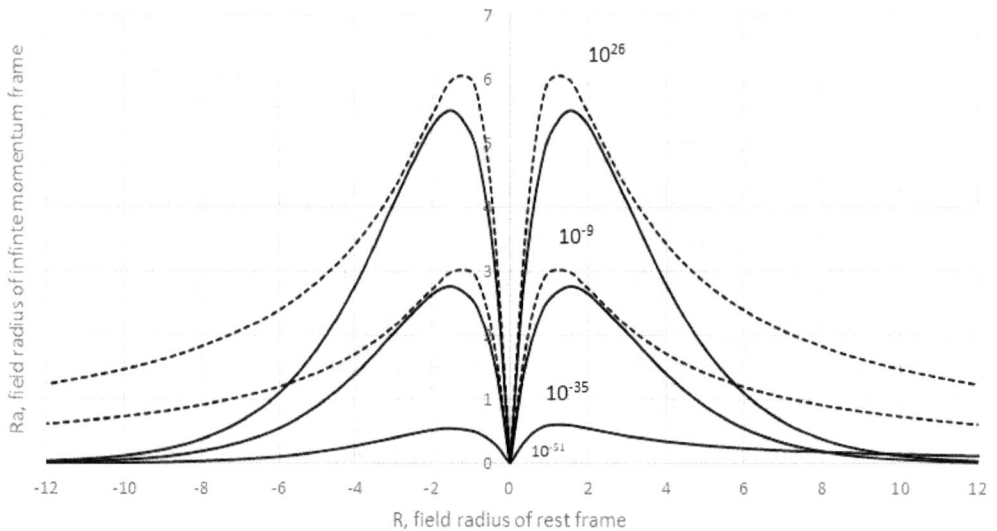

The magnitude of the radius of a mass will determine where the mass lies on the axis of R_a, the radius of the energy fields in infinite momentum frame. However, all objects and matter in our universe, large or small, will contain a combination of the three spatial dimensions down to 2.3×10^{-51}m.

As seen in Fig. 10-2, the field of energy with a Planck dimension which was previously limited to the nucleus of the spacetime cavity now expands and covers the entire radius of the field of restframe. This means that in the new space transformation, there is a much larger field with Planck dimension than just the nucleus of fundamental particles before we reach the E-loop dimension of 2.3×10^{-51}m. This is the first stage of a blackhole. A stage where all matter will be transformed into the two fundamental three-dimensional strings, looped and open with all at Planck length. The range of this field is no longer limited to the Planck dimension as it was in the nucleus of spacetime cavity, and it extends into infinite momentum frame with a radius that in classical physics is known as the Schwarzschild radius.

Since stable matter with a nucleus, such as an atom, has a stereo graphic spacetime, i.e., a mirrored spacetime cavity, a blackhole will also have a dual, north and south cavity as shown in Fig. 10-3.

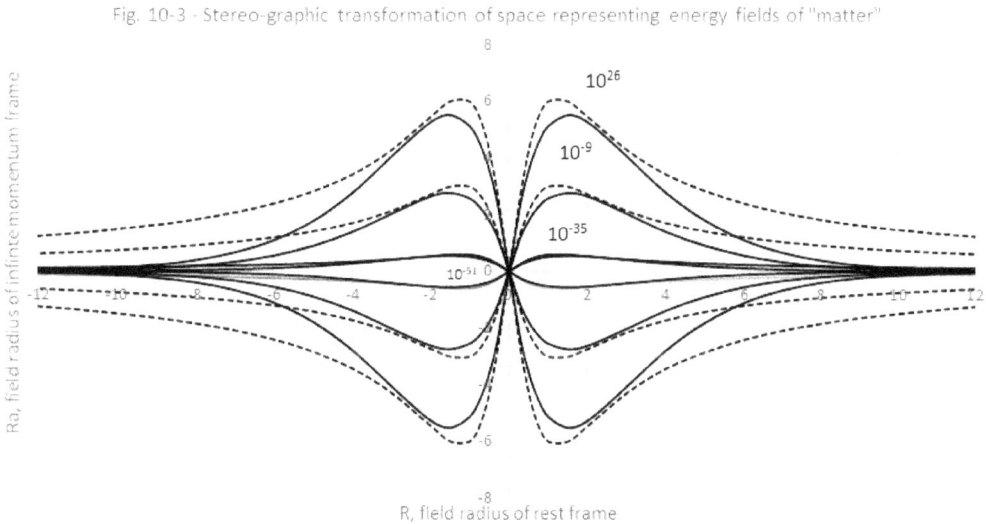

Fig. 10-3 - Stereo-graphic transformation of space representing energy fields of "matter"

Let us now see how such a large field with activated strings all in Planck dimensions might be created considering what we have learned so far.

As I described in previous chapters, two forms of extra dimension cylinders are created in the universe due to the decay of activated looped strings: orbiting and pulsating extra dimension cylinders. Both extra dimension objects are means of transporting the resulting two-dimensional strings out of the spacetime cavity when the activated strings decay. These extra dimension cylinders transport the two-dimensional strings to the outer edge of the universe but at different speeds. The speed of this transport is the same as the speed of light for the pulsating extra dimension cylinder and increases to the order of 10^{39}m/s for the orbiting extra dimension.

Up to now, we have discussed the process of decay of activated strings, gravity, the formation of extra dimension objects in the universe under normal operating conditions of universe where there is a balance between the forces of the universe, matter and its energy, and three and two-dimensional strings. Under these normal conditions, the two-dimensional strings formed as a product of the decay within the structure of fundamental particles that form the matter are transported out of the spacetime cavity of the atoms into the infinite momentum frame and eventually to the outer edge of the universe.

What happens when these portals for the transport of the two-dimensional strings out of the spacetime cavity of atoms are disrupted?

Let us look at a common way this disruption occurs in our universe. Consider an atom like iron which as discussed earlier has a strong magnetic property and atoms which can align themselves such that the spacetime cavities of north and south are pointing in the same direction. As I discussed in Chapter 4, the alignment of atoms results in the forces created by the angular momentum of north and south cavities to add up and become larger. This self-alignment of the atom was the feature that gave the iron material its magnetic property magnifying an otherwise weak magnetic field. Another force that adds up when the north and south cavities are aligned is the force created by the angular momentum of activated open strings creating the force of the "charge" around the orbiting extra dimension cylinders of Tubular Monopoles. Recall the balance of these forces are critical in maintaining the integrity of the orbiting extra dimension cylinders to which activated opens strings attach in a Tubular Monopole extra dimension.

We also know stars go through a process of fusion when they run out of fuel and its core begins to convert into various elements as a result of this fusion. When the fusion reaches a point where the elements produced are substantially iron (or similar atoms), the core of the star begins to feel a different force as a result of the alignment of the north and south spacetime cavities of these atoms. The massive amount of iron that is present causes the forces described above to become very large and create changes inside the spacetime cavity of the atoms.

The first change will be that the excessive forces exerted on the orbiting extra dimension cylinders will distort, and disrupt, the integrity of the orbiting extra dimension cylinders inside the spacetime cavity. Under excessive force of the torque created by the addition of the angular momentum of each individual spacetime cavity where its vectors are aligned, the Tubular Monopoles will become distorted and collapse. As you will see in Chapter 11, the 137 Tubular Monopoles convey the strong force inside their structure which will help balance the force inside the spacetime cavity and the electromagnetic force of the universe. When the Tubular Monopoles are lost, the electromagnetic force of the universe (4×10^{74}N), which is significantly higher than the maximum gravitational force inside the spacetime cavity (2.9×10^{44}N), will cause the spacetime cavity to collapse. At the same time the path of the high-speed transport of two-dimensional strings out of the nucleus will be disrupted.

Without the huge number of orbiting extra dimension cylinders which were the quick exit path of two-dimensional Planck-sized strings out of the spacetime cavity, the two-dimensional looped strings convert to pulsating strings. Creation of more pulsating strings means increased number of axis and energy for orbital rotation of strings inside a field containing the distorted spacetime cavities of fundamental particles of matter. As a result, the number of attachments of pulsating strings inducing orbital rotation in restframe2 increases substantially.

Disruption of the boundaries of spacetime cavities means that a larger field is produced that is a mixture of the strings that make up the spacetime cavity. Since the vacuum released due to the decay of three-dimensional strings does not have an organized pathway to exit, it will become dispersed and create a large field in which the activated strings are predominantly in restframe1 state. Recall that inside the spacetime cavity of ordinary matter this field was limited to the dimension of less than 1.4×10^{-20}m. In a blackhole, the field of restframe1 state extends to a significantly larger field. Let us designate this field with the radius R_s. In classical physics, this field is referred to as the Schwarzschild field. A feature of this larger field is that two-dimensional strings cannot propagate in the same way as they do in the spacetime cavity of ordinary matter as described earlier.

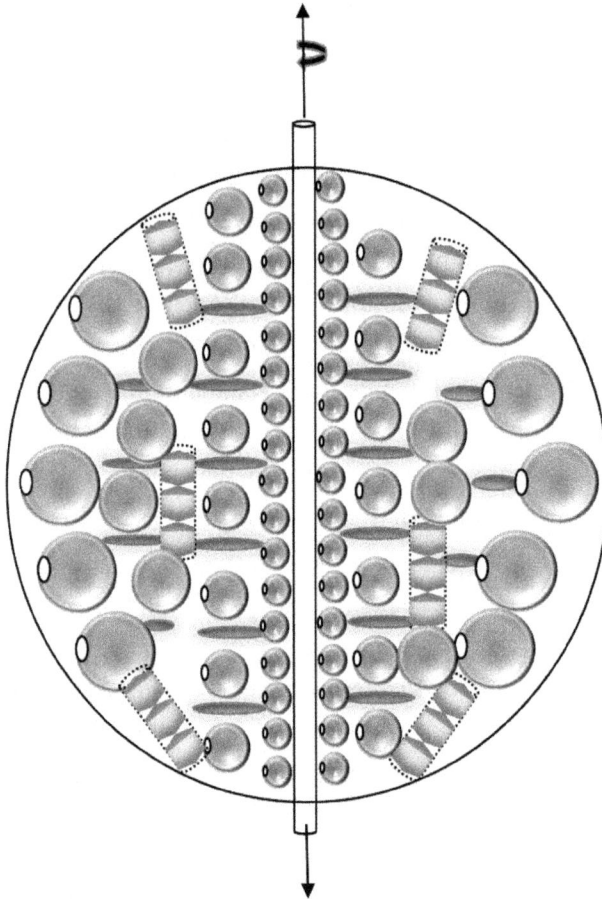

Fig. 10-4- Conceptual representation of the structure of strings and pulsating extra dimension cylinders inside the Schwarzschild field of the blackhole. a) three-dimensional looped strings shrink to Planck length at the singularity, b) there is one main orbiting extra dimension cylinder at the center of the field.

Since orbiting extra dimension cylinders no longer exit inside the blackhole as an organized pathway, the vacuum of these objects randomly distributes inside the field with radius R_s. One exception is the creation of a lone extra dimension cylinder at the center of the field. This is because the decay of the activated Planck-size strings is the highest at the center of the blackhole which leads to the formation of a stable extra dimension cylinder compacted with high energy density E-portals. As you will see shortly, this lone extra dimension cylinder with a radius of Planck length at the center of this field is critical in many ways for the existence of a blackhole, and it is where the singularity of a blackhole resides.

This new object with a mass "M" and a new radius of R_s (Schwarzschild radius) is a blackhole.

Let us now look at the gravitational field of the blackhole. Recall from Chapter 6, the gravitational field equation of the activated looped strings in restframe is:

$g_\perp = c^2/2R_a + c^2/2R$

$R_a = Rc^2 /(2Rg_m - c^2)$

Choosing natural units, c =1, the above equation becomes:

$R_a = R /(2Rg_m - 1)$

This equation also holds for the gravitation field of the blackhole with a few exceptions.

A blackhole goes through two stages of space transformation. In the first stage, activated looped strings shrink from a dimension of 2.3×10^{-20}m in infinite momentum frame down to 1.61×10^{-35}m just as it does inside ordinary matter. As such, the blackhole behaves like an ordinary macro-mass.

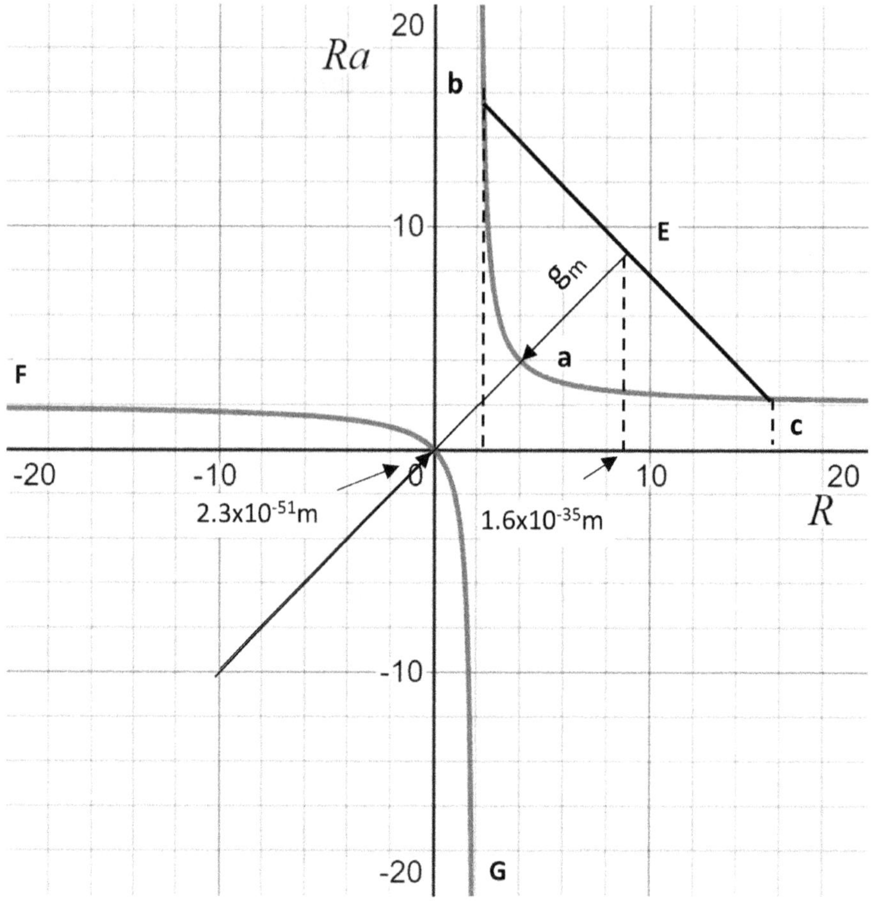

Fig. 10-5A – Spacetime curvature created by gravitational force of a blackhole with a mass "M", radius R_s. Upper right quadrant represents gravitational field of Schwarzschild radius as a macro-mass, lower left quadrant represents gravitational field of Planck size extra dimension cylinder.

Fig. 10-5A demonstrates that the gravitational field of a blackhole as a macro-mass is essentially identical to that of ordinary matter as demonstrated in Chapter 6 with the rescaling of the origin to Planck radius.

The graph in upper right quadrant represents the blackhole as a macro-mass with a field radius of R_s and curvature of spacetime corresponding to coordinates A, B, C, F, and G as described in Chapter 6.

As an ordinary macro-mass, the gravitational acceleration corresponding to points A, B, and C in the restframe would be the same as what we saw in Chapter 6 for any macro-mass except that unlike ordinary matter where the "$g_\perp$" on the mass was:

$$g_\perp = c^2/R_m$$

On the surface of the Schwarzschild sphere which is in restframe1 state, the orbital acceleration is predominant, and the equation:

$$g_\perp = c^2/2R_a + c^2/2R$$

is reduced to:

$$g_\perp = c^2/2R$$

or

$$g_\perp = c^2/2R_s$$

Recall from Chapter 6, the gravitational acceleration corresponding to Point C not on the Schwarzschild sphere would then be:

$$g_\perp = 0.65\, c^2/2R_s$$

Point B not on the Schwarzschild sphere:

$$g_{rf} = 5.2c^2/2R_s$$

In the second stage of space transformation, we must rescale the origin of the graph of Fig. 10-5A from 1.16×10^{-35}m to 2.3×10^{-51}m, a factor of 0.7×10^{16}, to account for the dimensional transformation of spacetime as the Planck-size strings are converted to E-loop strings. The origin on the graph, which is now 2.3×10^{-51}m, is the dimension and location of the E-portals which as you will see later becomes the singularity, Fig. 10-5B.

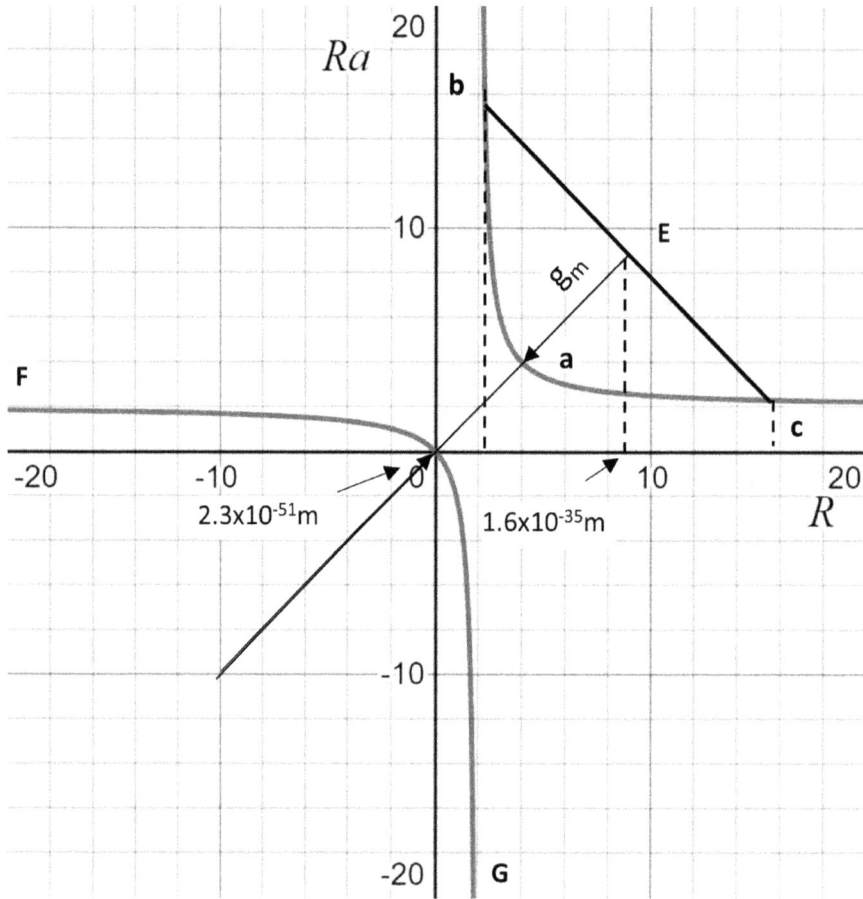

Fig. 10-5B – Spacetime curvature created by gravitational force of a blackhole with a mass "M", radius R_s. Upper right quadrant represents gravitational field of Planck space extra dimension cylinder, lower left quadrant represents gravitational field of E-loop space and singularity.

Since I have referred to Planck scale as the quantum scale, let us refer to this stage of transformation as the sub-quantum stage which contains strings in both Planck dimension and E-loop dimension.

The upper right quadrant in Fig. 10-5B represents the field of the extra dimension cylinder where all strings attached to the cylinder are of Planck length. R_s, which is the same as the Schwarzschild radius of the macro-mass, is the length of the extra dimension cylinder which contains or is occupied by the Planck-size strings.

The gravitation acceleration of the strings attached to the extra dimension cylinder with a Planck radius is:

$$g_\perp = c^2/2r = c^2/\ 2x1.61x10^{-35} = 2.8x10^{51}\ m/s^2$$

The gravitational acceleration corresponding to coordinate c, is:

$$g_\perp = 0.65\ c^2/r = 0.65x2.8x10^{51} = 1.8x10^{51}\ m/s^2$$

The gravitational acceleration corresponding to Point b is:

$$g_\perp = 5.2\ c^2/2r = 5.2\ x2.8x10^{51} = 14.5\ x10^{51}\ m/s^2$$

Without the spacetime cavities of the individual atoms, the activated looped strings of different dimensions that normally exist in the order described for fundamental particles, are present within the Schwarzschild radius in a disordered fashion but are gradually shrinking as they approach the extra dimension cylinder through the center of the blackhole. At the center, where the extra dimension cylinder with the Planck radius lies, all three-dimensional strings, looped and open are at Planck length, Fig. 10-6.

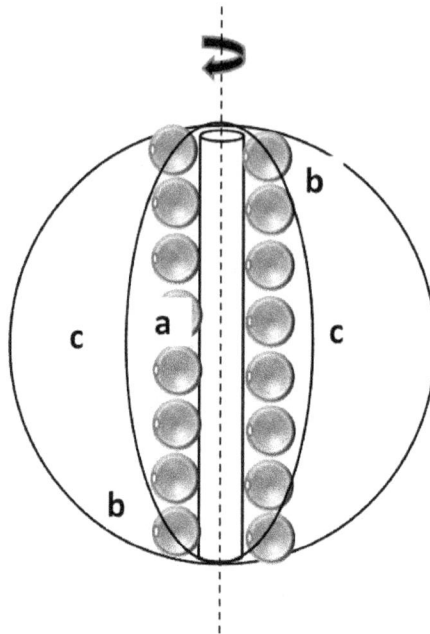

Fig. 10-6- Conceptual representation of the core mass of the blackhole inside Schwarzschild field. Planck size activated open and looped strings attached to an orbiting extra dimension cylinder. Location of points "b" and "c" represent the highest and lowest gravitational acceleration points of the blackhole as compared to point "a" on the body of the mass.

The restframe dimension of coordinate E in Fig. 10-5B with respect to the center line is 1.61×10^{-35}m, the radius of the extra dimension cylinder. The restframe dimension of coordinate "c" will be 3.2×10^{-35}m, and the restframe dimension of coordinate "b" will be about 0.4×10^{-35}m. This suggests that the extra dimension cylinder has a small orbital rotation around its center axis creating a spacetime curvature of the coordinates b and c as shown in Figs. 10-6 and 10-8B.

Referring to Fig. 10-5B, the field presented by the mass "M" and radius R_s experiences the maximum gravitational acceleration of 14.5×10^{51}m/s² on the restframe as calculated above. The dimension of activated looped strings inside the Schwarzschild radius vary with the radius R_s and as such the force created by them changes as a function of R_s:

$\rho g_1 = 0.96k/R_s^2$ (see Chapter 6)

The force created by the activated looped strings must equal the force of the mass "M" on its restframe. We can calculate the radius of the field R_s exposed to the above gravitational field by equating the two forces:

$$V(\rho g_1) = Mg_2 \hspace{4cm} (10\text{-}1)$$

$$V = 4/3\pi R_s^{\,3}$$

Substituting for ρg_1 from the above equation and for g_2 the maximum gravitational acceleration exerted by point "b":

$$4/3\pi R_s^{\,3}\,(0.96k/R_s^{\,2}) = M\,x14.5\,x10^{51}$$

$$R_s = 1.49x10^{-27}M \hspace{4cm} (10\text{-}2)$$

We can also obtain the Schwarzschild radius by setting the gravitational acceleration of the blackhole as a macro-mass in infinite momentum frame equal to that of its gravitational acceleration in restframe as shown earlier:

$$g_m = GM/R_s^{\,2} = g_\perp = c^2/2R_s$$

Indeed, R_s derived above is the same as Schwarzschild radius in classical physics:

$$R_s = (2G/c^2)M$$

The constant $1.49x10^{-27}$ calculated above in equation (10-2) is the same as $2G/c^2$ which is about $1.48x10^{-27}$.

Let us now take a closer look at the structure of the strings inside the field R_s of a blackhole. As I mentioned, the spacetime cavity of all individual atoms forming the matter "M" inside the blackhole collapse and only one orbiting extra dimension cylinder exists at the center of the blackhole. This extra dimension cylinder is one of $2x10^{30}$ high-vac extra dimension cylinders calculated in Chapter 3 that will reach the outer edge of the universe with a length of about $2.5x10^{26}$m. A portion of the length of this extra dimension cylinder inside the blackhole will be equal to $2R_s$ to which activated looped and open strings will attach. The radius of this extra dimension cylinder is

1.61×10^{-35}m, see Figs. 10-4, 10-6.

The mass of this object will follow the $\rho = KR$ rule I have discussed throughout this book. In which K is the linear mass of each Planck-size activated string or,

$$K = 2.18 \times 10^{-8}/2 \times 1.61 \times 10^{-35} = 0.67 \times 10^{27} \text{kg/m}$$

This means that the core mass of a blackhole is the collective mass of all activated looped strings with Planck length (2.18×10^{-8}kg) attached to the extra dimension cylinder at the center of the blackhole with a radius of R_s. Since only half of these strings are activated looped strings (the other half are activated open strings), the total mass will be:

$$M = 1/2(2KR_s) = KR_s$$

The Schwarzschild radius for any mass "M" can then be calculated as:

$$R_s = M/K$$

Substituting for $K = 0.67 \times 10^{27}$kg/m

$$R_s = 1.49 \times 10^{-27}M \qquad\qquad\qquad (10\text{-}3)$$

Which is the same as the radius calculated by balancing the forces of activated looped strings in infinite momentum frame and restframe (10-2).

Recall that for baryonic mass of the ordinary matter, I defined mass as the change of energy of the activated looped string that exited our universe (Hilbert space). This change of energy was the mass of a single two-dimensional looped string at the center of the mass of the particle.

In a blackhole, the entire mass of a three-dimensional looped string exits our three-dimensional universe when it attaches to the sole extra dimension cylinder with the length of $2R_s$ and then decays. The reason for this is that the two-dimensional strings cannot propagate inside the Schwarzschild radius of the blackhole similar to the decay in ordinary matter. Therefore, the mass of a blackhole will become a linear addition of the mass of activated strings that decay and disappear from our three-dimensional universe via the extra dimension cylinder at the center of the blackhole, Fig. 10-6.

The graph in the lower left quadrant of Fig. 10-5B represents the field inside the extra dimension cylinder, at the center of the blackhole, and the transition between the Planck space, 1.61×10^{-35}m and E-loop space, 2.3×10^{-51}m.

Unlike the extra dimension cylinders described inside the spacetime cavity of ordinary matter where they are segregated by the spin of the strings forming them, the two-dimensional strings of Planck length with both right- and left-hand spins are randomly mixed. As such, they can no longer propagate inside the lone extra dimension cylinder of the blackhole, and with opposite rotations, they annihilate each other and are reduced to E-portals inside the extra dimension cylinder. The E-portals then inflate instantaneously to the maximum length given that each E-portal will begin with an energy or mass of 2.18×10^{-8}kg in Planck space.

Therefore, all E-loop strings inside the extra dimension cylinder are subjected to the gravitational acceleration of 3.9×10^{67}m/s²:

$g = c^2/r = 3.9 \times 10^{67}$m/s^2

in which $r = 2.3 \times 10^{-51}$m

The energy of an activated string, which is 1.96×10^9J in Planck space, remains the same as a two-dimensional Planck string which then becomes 1.37×10^{25}J in E-loop space when the strings are all converted to E-loop strings.

The E-loop strings formed inside the sole extra dimension cylinder of the blackhole create what is referred to as the singularity in classical physics.

The E-loops will exit our universe via the extra dimension cylinder (singularity) in two ways.

First, they create E-portals with a rotation corresponding to its parent two and three-dimensional Planck strings. Therefore, the extra dimension cylinder becomes packed with E-portals with a radius of 2.3×10^{-51}m propagating in opposite directions.

The E-portals begin with an energy of 1.37×10^{25}J in E-loop space and propagate a distance of 1.3×10^{58}m at an inflation speed of 0.71×10^{63}m/s:

$1.37 \times 10^{25} = \frac{1}{2}(5.4 \times 10^{-101})V^2$

$V = 0.71 \times 10^{63}$m/s

$R = V^2/g = (0.7 \times 10^{63})^2/(3.9 \times 10^{67}) = 1.3 \times 10^{58}$m

In which the volume of the E-portal is:

$1.3 \times 10^{58} \times 1.6 \times 10^{-101} = 2 \times 10^{-43}$m³

The orbiting extra dimension cylinder at the center of the blackhole can sustain a fixed number of E-portals (Fig. 10-7). If we take the dimension of the extra dimension cylinder at the highest point of acceleration (b) to be about 1/4 of Planck length, the number of E-portals will be:

$1/16 \ x\pi(1.61 \times 10^{-35})^2/ \ 1.6 \times 10^{-101} = 3 \times 10^{30}$

In which 1.6×10^{-101}m² is the surface area of an E-portal.

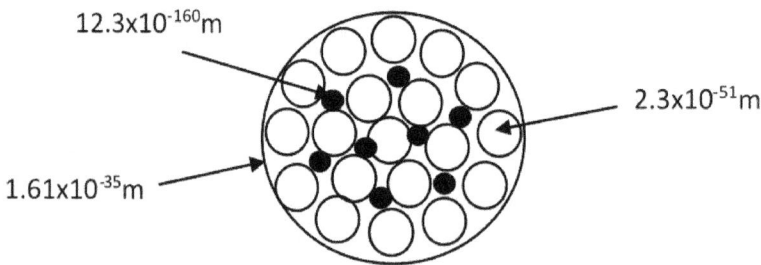

Fig. 10 –7 – Cross sectional view of the extra dimension cylinder (singularity) comprised of E-portals and one dimensional strings.

This means that there is sufficient free vacuum inside the extra dimension cylinder at the center of the blackhole to sustain 5×10^{31} E-portals. Each E-portal carries a mass of 2.18×10^{-8}kg out of the blackhole as the three-dimensional strings decay and convert to E-loop strings. Therefore, the total mass that the extra dimension cylinder at the center of the blackhole will sustain as E-portals is:

$$3 \times 10^{30} \times 2.18 \times 10^{-8}\text{kg} = 6.5 \times 10^{22}\text{kg}$$

Since the above mass calculated is independent of the Schwarzschild radius and the mass of the blackhole itself and exists in all blackholes equally, this suggests that the minimum mass for a stable and sustainable blackhole is about 6×10^{22} kg. In other words, blackholes with smaller masses will dissipate quickly because they cannot sustain a Planck-size extra dimension cylinder as the housing for its singularity. A singularity that is not packed with E-portals will collapse under the force of the blackhole's gravity. To date, the smallest blackhole discovered in the universe is about 3 solar masses or about 6×10^{30}kg which is significantly above the minimum threshold calculated above. Blackholes with smaller masses may form but may be unstable and transient because their mass will be unable to sustain a stabilizing extra dimension cylinder at the center of the blackhole as calculated above. The extra dimension cylinder at the center of a blackhole is the anchoring axis of its orbital rotation.

What happens to the remaining mass of a blackhole when the extra dimension cylinder at its center is fully compacted with E-portals? The E-portals can only sustain a mass of about 6.5×10^{22} kg. How about the remaining mass of larger blackholes, those with a mass greater than 6.5×10^{22}kg?

The remaining E-loop strings inside the singularity of a blackhole are stretched to their maximum length using the energy of the E-portal. In this case, the torque created by the orbital rotation of two attached E-portals will stretch the E-loop strings to their maximum length:

$$F \times R = E = 2 \times 1.37 \times 10^{25} \text{J}$$

$$F = mg = 5.4 \times 10^{-101} \times 3.9 \times 10^{67}$$

$$R = 1.3 \times 10^{58}\text{m}$$

Since the surface area of the E-loop string with the above length remains the same, the width of the stretched E-loop string will become:

$$1.6 \times 10^{-101} / 1.3 \times 10^{58} = 12.3 \times 10^{-160} m$$

Which is about the same as its thickness calculated earlier, at $8.9 \times 10^{-160} m$.

In essence, the two-dimensional E-loop string becomes what one may regard as a one-dimensional object with a width of $12.3 \times 10^{-160} m$, a thickness of $8.9 \times 10^{-160} m$ and a length of $1.3 \times 10^{58} m$. This process in classical physics is regarded as spagettification. The extra dimension cylinder at the center of the blackhole will then carry 5×10^{31} E-portals and the remainder of the mass of a blackhole is converted to one dimensional strings that fit comfortably between the void volume of the space between the E-portals, Fig. 10-7.

As you will see later, this one-dimensional object is the fundamental structure of bulk (free) vacuum in space, or the universe itself. The energy density of this one-dimensional object remains the same as that of E-portals or $0.59 \times 10^{144} kg/m^3$ in Planck space because the volume and mass of the E-loop does not change as it stretches. It will also be the lightest object in the universe with a mass per length of:

$$K = 5.6 \times 10^{-101} / 1.3 \times 10^{58} = 43 \times 10^{-160} kg/m$$

As compared to an activated string with:

$$K = 2.18 \times 10^{-8} kg / 2 \times 1.61 \times 10^{-35} = 0.67 \times 10^{27} kg/m$$

Activated three-dimensional looped strings decaying anywhere away from the extra dimension cylinder and inside the Schwarzschild radius will also exit the blackhole as E-portals. The loss of this energy establishes the mass of the sphere of the blackhole which is smaller than the mass at the center of the blackhole. The reason for this is that the rate of decay of activated strings away from the vacuum of the extra dimension cylinder at the center of the blackhole is significantly lower resulting in the core mass at the center being the dominant mass we observe.

Referring to Fig. 10-5A, the boundaries of the gravitational field of the strings in the lower left quadrant of the graph extend by $2R_s$ from the center of the blackhole. This means that similar to the quantum field of ordinary matter, a blackhole as a macro-mass creates a large spacetime cavity with a field radius of $2R_s$ as shown in Fig. 10-8A with blue lines. Let us call this the "Schwarzschild spacetime cavity". Note that the field with Schwarzschild radius falls at the same location inside the Schwarzschild spacetime cavity as in the spacetime cavity of ordinary matter where the strings exist predominantly in restframe1 state due to overwhelming production of bulk vacuum, at $R \approx 0.56$. The coordinates A, B, C, E, F, G shown in Fig. 10-5A and a, b, c in Fig. 10-5B are also shown in the graph of the energy field of the strings in Figs. 10-8A and 10-8B.

Fig. 10-8A - Symmetric curvature of spacetime of Schwarzschild sphere and spacetime cavity

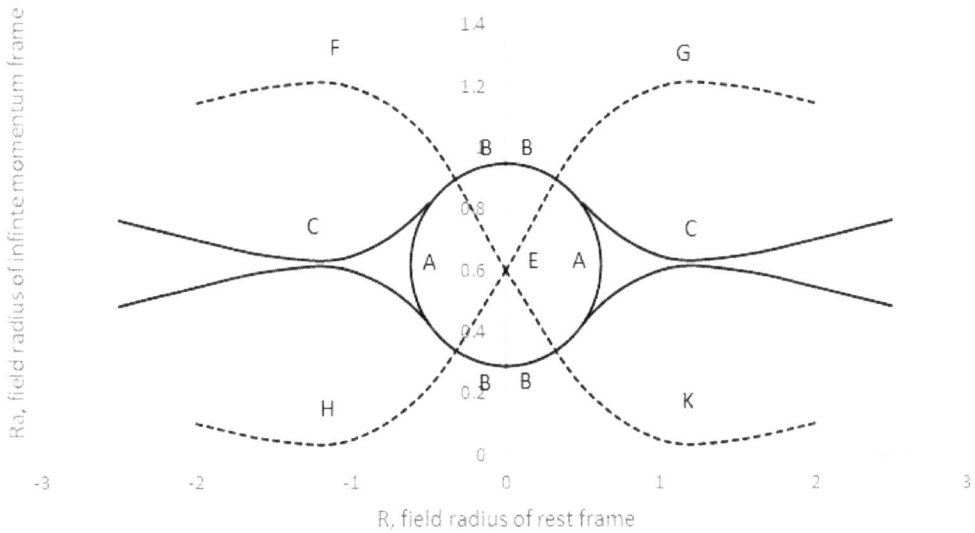

Fig. 10-8B - Symmetric curvature of spacetime of Schwarzschild sphere as a macro-mass and quantum mass

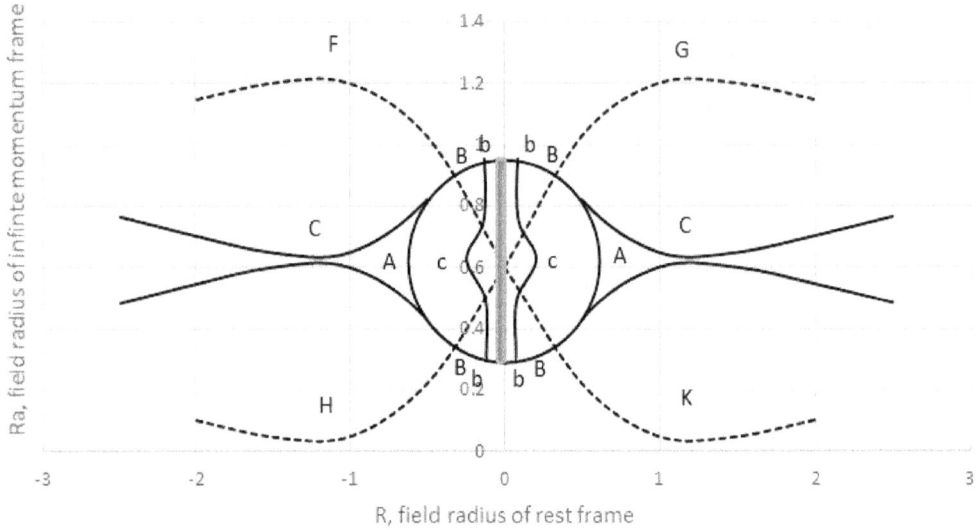

It is important to note that the direction of the rotation of the E-loop of the E-portals will be the same as its parent three-dimensional string. Therefore, decay of three-dimensional strings will result in creation of E-portals that expand in opposite directions. For this reason, a blackhole will also have a north and south cavity in which the direction of the rotation of the E-portals of its singularity will be opposite each other. Therefore, the singularity of a blackhole is not unidirectional, Fig. 10-9.

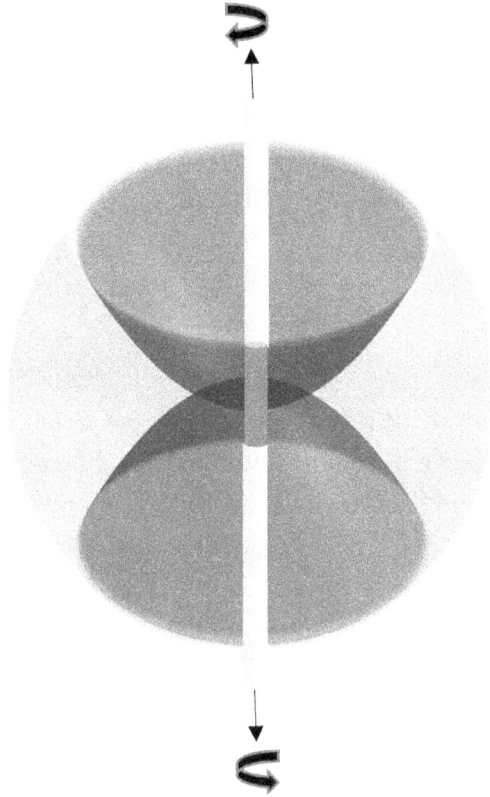

Fig. 10 -9 – Conceptual representation of Schwarzschild sphere, the field of activated looped strings (Schwarzschild spacetime cavity), and its relation to the singularity.

Space itself, regardless of whether it is inside the Schwarzschild radius or in open infinite momentum frame, is dominated by the dimension of activated looped strings which are 10^{15} times larger than activated open strings. In essence, in our three-dimensional universe we live on the surface of these strings and the rest of the strings and extra dimension objects are the fillers of space.

Reduction in size of the three-dimensional looped strings will be the first experience of "matter" under the gravitational force of the blackhole. This means that spacetime itself will shrink. The reduction in the size of activated looped strings and the shrink-age of space itself begins at a field radius of $2R_s$ where the Schwarzschild spacetime cavity begins. The dimensions of activated looped strings will experience a reduction following the equation:

$$r_a = 0.71x10^{-26} \ (Ra)^{1/4}$$

Activated looped strings outside the Schwarzschild spacetime cavity will have a dimension of $2.3x10^{-20}$m in infinite momentum frame. This dimension will shrink down to $1.61x10^{-35}$m at the center of the blackhole when it reaches the extra dimension cylinder or the singularity.

The shrinkage of spacetime will create a time dilation effect observed by an outside observer. In other words, the time that it will take for an object to travel at the speed of light between the same two points which appear to have a fixed distance will be significantly shorter inside the Schwarzschild spacetime cavity than it is outside this field. As such, motions inside this radius will appear slower or motionless altogether to an outside observer. A detailed description of this effect and space and time is given in Chapter 11.

As mentioned earlier, this effect also exists for ordinary macro-masses, but since $2R_m$ is usually significantly larger than $2R_s$, we only see a small shrinkage of the activated looped strings and time dilation effects compared to the mass of a blackhole.

Activated looped strings shrink in dimension which increases their mass and energy density as they move towards the singularity of the blackhole. Planck-size activated looped strings that exist in the spacetime cavity of matter that falls into a blackhole, or random Planck-size strings inside the blackhole, will be directed towards the surface of the Schwarzschild sphere, under the centrifugal force of the orbital rotation inside the blackhole. Therefore, while there is a flow of larger and lighter activated looped strings towards the center, there is a flow of smaller and heavier Planck-size looped strings towards the surface. Activated Planck-size looped strings at the surface are then emitted to its surroundings which as you will see in the next section will establish the entropy of the blackhole.

Entropy and Temperature of a Blackhole

To calculate the temperature of the blackhole, we must first understand and calculate the entropy of a blackhole. The temperature and entropy of a blackhole are related by the thermodynamic equation shown below:

$$dS = dE/T \qquad\qquad\qquad (10\text{-}4)$$

In which dE is the change in energy of the blackhole.

Where does the entropy of a blackhole come from?

Recall that the Schwarzschild sphere of a blackhole is the region where spacetime cavities of all the atoms of ordinary matter that comprise the blackhole are disintegrated into a mixture of three and two-dimensional strings of various dimensions. This leads to creation of a Schwarzschild spacetime cavity, where the dimension of activated looped strings is determined by its location inside the cavity as a function of (R_a). However, Planck-sized activated looped strings that are remnants of the original matter and those pushed to the perimeter of the sphere due to the heavier mass of the string will be emitted from the surface of the Schwarzschild sphere back into the infinite momentum frame beyond the event horizon of the blackhole.

Planck-size activated looped strings that are inside the Schwarzschild sphere are at restframe with 2 degrees of freedom. When they are emitted into infinite momentum frame, they will have 5 degrees of freedom. This comes at the expense of reducing its energy density from $0.2 \times 10^{97} \text{kg/m}^3$ to $0.2 \times 10^{36} \text{kg/m}^3$ when it expands from Planck dimension to $2.3 \times 10^{-20} \text{m}$. This change in energy is observed as the radiation energy of the blackhole. Recall that each string is associated with an entropy and a temperature associated with this entropy as described in Chapter 9. We can calculate this temperature in the same way I demonstrated earlier for a single string.

At the event horizon of the blackhole there is a significant inflow of activated looped strings much larger in dimension than that of Planck dimension which establishes the gravitational force of the blackhole. At the same time there is a small outflow of Planck-size activated looped and open strings that are emitted from the surface of the blackhole as radiation.

The entropy of the blackhole is dependent on the degrees of freedom of the constituents that are emitted from the blackhole. In this case, the activated looped strings undergo an energy and dimensional change upon emission from the Schwarzschild sphere. Activated open strings do not contribute to this radiation because their dimension remains unchanged upon emission.

The entropy of a blackhole is therefore the number of degrees of freedom associated with the Planck-size activated looped strings that form the individual constituents of the blackhole multiplied by the Boltzmann constant, k_b. As you saw earlier, the Boltzmann constant itself is the entropy/degrees of freedom of the most fundamental specie in our universe, a single three-dimensional string.

S = degrees of freedom x k_b

To do this, we must first find the total number of activated looped strings that can be emitted from a blackhole with a Schwarzschild radius of R_s.

We know that the energy density of activated looped strings vary by R_a. Therefore, from Chapter 1, the energy density of the looped strings inside the Schwarzschild radius is:

$\rho = 1.2k/c^2(1/R_s)$

$V = 4\pi/3(R_s)^3$

The total mass of activated looped strings inside the Schwarzschild radius will then be:

$$m = \rho V = 5k/c^2 R_s^2 = 1.33 \times 10^{62} R_s^2 \qquad \text{10-5}$$

The total number of Planck-size activated looped strings:

$$\text{Number} = 1.33 \times 10^{62} R_s^2 / 2.18 \times 10^{-8} = 0.61 \times 10^{70} R_s^2$$

Recall that activated strings have 5 degrees of freedom in infinite momentum frame, and at restframe will lose their translational freedom (3 for x, y, and z). Therefore, the number of degrees of freedom will be the total number of activated looped strings multiplied by their degree of freedom:

Degrees of freedom = 2 x number of Planck-size strings = $2 \times 0.61 \times 10^{70} R_s^2 = 1.22 \times 10^{70} R_s^2$

The entropy of the blackhole will then be the above multiplied by the entropy of a single string which is k_b, the Boltzmann constant:

$$S = k_b \times 1.22 \times 10^{70} R_s^2$$

If we substitute for $R_s = 1.48 \times 10^{-27} M$:

$$S = 2.67 \times 10^{16} k_b M^2 \qquad (10\text{-}6)$$

This is the entropy of the blackhole which is the same as the famous (Bekenstein, Hawking, Strominger & Vafa) blackhole entropy:

$$S = (4\pi k_b G/ch) M^2 = 2.65 \times 10^{16} k_b M^2 \qquad (10\text{-}7)$$

We can now calculate the temperature of the blackhole by substituting for dE and dS in equation (10-4):

$$T = dE/dS = d(Mc^2)/d(1.22 \times 10^{70} k_b R_s^2) \qquad (10\text{-}8)$$

Or

Substituting for $R_s = 1.48 \times 10^{-27} M$:

$$T = dE/dS = d(Mc^2)/d(2.67 \times 10^{16} k_b M^2) = 1.68/k_b M \qquad (10\text{-}9)$$

Which is the same as Hawking's temperature:

$$T_H = \hbar c^3/8\pi G k_b M = 1.7/k_b M \qquad (10\text{-}10)$$

Notice that from equation (10-8), temperature is the change of energy with respect to space (R^2). Earlier, I defined mass as the effect we observe due to the change of energy with time. Temperature is the effect we observe due to the change of energy with space, or the effect we observe when activated looped strings expand or shrink.

The energy associated with the above temperature, relates to the mass of activated looped strings emitted from the surface of the event horizon. Activated strings inside the Schwarzschild field are at restframe with two degrees of freedom corresponding to its spin and orbital rotation. Emitted Planck-size activated looped strings will gain all 5 degrees of freedom in infinite momentum frame after emission. The net change in degrees of freedom will be 3. The temperature of the blackhole calculated above is in equilibrium with the strings emitted. We can calculate the energy needed to raise the temperature of the string to that of the blackhole:

$$T = E /3k_b$$

Substituting for T from the above equation (10-9):

$$E = 5.1/M = mc^2 \qquad\qquad (10\text{-}11)$$

The mass of the looped strings that radiate from the blackhole will be:

$$m = E/c^2 = 5.1/Mc^2 \qquad\qquad (10\text{-}12)$$

If we assume the rate of the string turnover is about Planck time, the rate of looped string radiation will be:

$$Q = m.f = (5.1/Mc^2)0.29\text{x}10^{43} = 1.48\text{x}10^{27}/M \ \ kg/s$$

Where "f" is Planck frequency.

We can calculate the time it takes for all three-dimensional looped strings to be radiated:

From equation (10-5),

$$m = 1.33\text{x}10^{62}R_s{}^2 = 1.33\text{x}10^{62}(1.48\text{x}10^{-27}M)^2 = 2.91\text{x}10^8M^2$$

$$t = m/Q = 2.91\text{x}10^8M^2 / (1.48\text{x}10^{27}/M) = 1.96\text{x}10^{-19}M^3 \ s$$
$$t = 1.96\text{x}10^{-19}M^3/3.1\text{x}10^7 = 0.63\text{x}10^{-26}M^3 \ years$$

$$t = 0.63\text{x}10^{-26}M^3 \ years \qquad\qquad (10\text{-}13)$$

For example, if we substitute for M = 2x10^{30} kg which is equal to a solar mass, it will take

t = 0.5 x10^{65} years for the blackhole to disappear.

Magnetic monopoles

Recall the structure of a magnetic monopole from Chapter 3. It entails an activated open string attached to an activated looped string via a two-dimensional open string. In infinite momentum frame and outside the spacetime cavity of an atom, the activated looped string expands causing its pole to also expand from the radius of an E-loop, 2.3x10^{-51}m, to about 1.6x10^{-36}m. As you saw, the expansion of an E-pole adds a new gravitational acceleration component to the magnetic monopole. When the E-pole to which the open string attaches expands, the two-dimensional string will create a new acceleration that is based on the diameter of the E-loop of the E-pole.

The gravitational acceleration of the monopole with the radius of the E-pole fully expanded is:

$g_{imf} = c^2/2r = c^2/1.6 \times 10^{-36} = 2.8 \times 10^{52} m/s^2$

As mentioned in the previous section, a Blackhole emits Planck-size activated looped strings from its surface. These strings, once outside the Schwarzschild field, will attach to two-dimensional open strings and begin to expand.

There will be a zone just outside the Schwarzschild radius where the radius of the expanding E-pole is even smaller than 1.6x10^{-36}m before the activated looped string reaches the maximum expansion. This means that the acceleration and force created by the monopole, F_m (Chapter3), will be even higher than that of the monopole in infinite momentum frame.

As a result, the magnetic monopoles created immediately surrounding the field of Schwarzschild sphere will create a significantly higher force than the blackhole itself. In fact, the magnetic field created by these monopoles is so strong that it can control and prevent how matter is fed into the blackhole rendering some to become inactive.

Magnetic monopoles once outside the blackhole's Schwarzschild field can attach to extra dimension cylinders and form a Tubular Monopole which is the basic structure of a charge. Hence giving the blackhole its magnetic and electrical property.

Blackholes and the universe

In Chapter 3, I demonstrated that there are 2×10^{30} extra dimension cylinders that have a length of 1.5×10^{26}m inside the field radius of infinite momentum of the universe. In Chapter 5, I showed that these same extra dimension cylinders in combination with the expanded activated looped strings form the restframe1 structure of what we know as "dark matter".

Extra dimension cylinders created by ordinary matter is a vacuum containing two-dimensional Planck strings propagating across the universe. These extra dimension cylinders can break up, reform, reshape and reconnect as the environment of the space requires. However, the extra dimension cylinder created at the center of a blackhole (singularity) is packed with E-portals which are stable under any condition in our universe. Given the speed of inflation and the energy of each E-portal, these extra dimension cylinders are the best candidates for extending the length of the universe without disruption. If so, we can then link the total mass of "dark matter" and the universe to the number of blackholes in the universe. Since the universe has a fixed number of such extra dimension cylinders of the kind (2×10^{30}), the total mass of dark matter was calculated to be:

$M = 0.435 \times 10^{-3}$ kg/m x 2.3×10^{26}m x $2 \times 10^{30} = 2 \times 10^{53}$kg

Where the mass per length of activated looped strings is:

2×10^{-23}kg/$2 \times 2.3 \times 10^{-20}$m $= 0.435 \times 10^{-3}$ kg/m

In which 2×10^{-23} kg is the mass of each activated looped string in infinite momentum frame.

As such, one may view blackholes as key players in establishing the structural mass of the universe which is dark matter. Recall that Higgs Bosons, which are a by-product of the generation of activated looped strings, are produced as the looped strings of dark matter structure annihilate each other. Therefore, blackholes which create the extra dimension cylinders (singularity) that attract looped strings and create dark matter are responsible for generation of three-dimensional looped strings and maintaining a balance of energy in the universe. To the extent that our universe may survive for 120B years.

Information and blackholes

An E-loop string is the most fundamental building block of activated strings. As such it carries the information about its parent string, such as its spin and momentum. When activated strings decay, its information remains with the two-dimensional strings which eventually convert to E-loop strings. E-loop strings exit our universe via E-portals as a two-dimensional string forming the outer edge of the universe in restframe. Therefore, the eventual destination of information about strings that made up our universe lies outside the reach of our universe.

In a blackhole, there are three ways information is disseminated. 1) The segment of the strings that are converted to E-loop strings inside the singularity, or anywhere inside the Schwarzschild sphere, are transported to the outer edge of the universe as mentioned above. This information is not retrievable but not lost.

2) The segment of activated looped strings that are emitted back into our three-dimensional universe as blackhole radiation. Planck-size activated looped strings that are emitted from the blackhole carry the information about their parent matter that became part of the blackhole's mass. This information will remain in our three-dimensional universe and can be viewed as retrievable since activated looped strings are a component of any object with a mass and the space itself.

3) The segment of the strings converted to one-dimensional strings with Z-strings as its constituents. One may argue that this information is lost from a universe that E-loop strings are its fundamental constituents which includes our three-dimensional universe. But one can also argue that Z-strings retain that information as the background vacuum of the" forever universe". However, regardless of how one might interpret the fate of this information, it is not retrievable.

In our universe, information is carried by the E-loop strings upon decay of activated and two-dimensional strings which then resides at the circumference of the universe.

Final Space transformation

As shown above, a blackhole has three stages of quantum transformation, one is from macro scale ($>1.8 \times 10^{-9}$m) to Planck scale, the other from Planck scale to the E-loop scale, and then from E-loop scale to what may be regarded as a zero-dimensional cross section with a dimension of 12.3×10^{-160}m. This is the final transformation of space, Fig. 10-10.

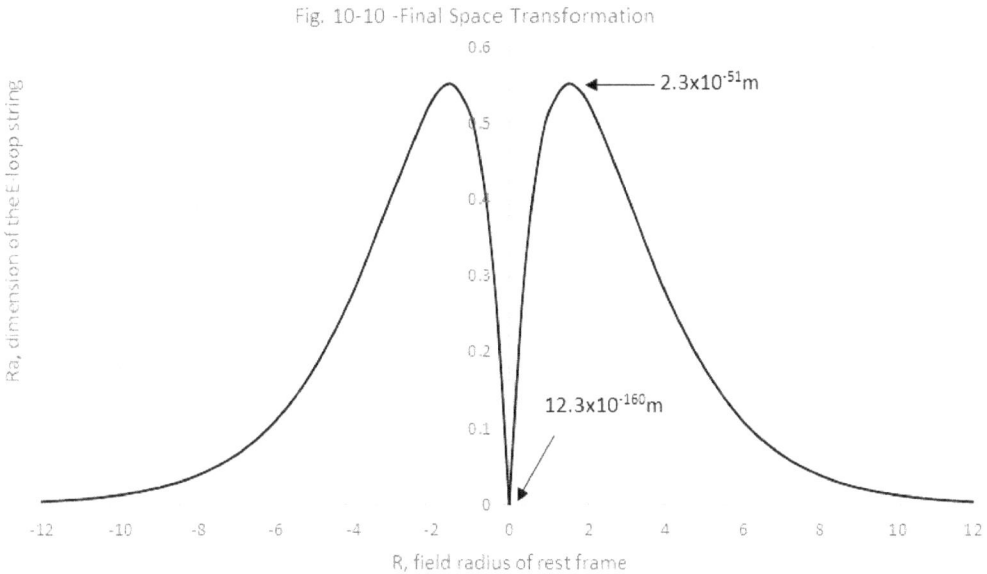

Fig. 10-10 -Final Space Transformation

Zero-dimensional vacuum strings or "Z-strings" are the smallest vacuum specie in the universe with a dimension of about 12.3×10^{-160}m. As you will see in Chapter 12, this string is the fundamental building block of "bulk vacuum" in the universe.

Therefore, one may view a blackhole as a place where ordinary matter is transformed to Plank space, then E-loop space, and finally to the one-dimensional bulk vacuum itself where the Big Bang process begins. Dust to dust, ashes to ashes.

CHAPTER 11
Strings and Time

Time

SO FAR, I HAVE DESCRIBED our universe as a space filled with activated three-dimensional looped and open strings as its major components and two-dimensional objects encased by a vacuum which in combination with the three-dimensional strings form extra dimension objects.

The number of extra dimension objects that form is too large to enumerate. However, I described a few that are critical to the formation of sub-atomic particles of matter, such as charge, photons, magnetic monopoles, etc. These are the extra dimension objects that play a significant role in the creation of our universe.

Therefore, collectively, the activated three-dimensional strings and extra dimension objects form the fabric of spacetime in our universe. We learned that the main constituents of space, and its attachments, all move at the speed of light, c, 3×10^8m/s. Therefore, we must think of our space as a dynamic system which transports its content at the speed of light as compared to the objects moving in a static universe at the speed of light. For example, a photon moving at the speed of light in what appears to be a vacuum, in reality is a photon transported by its media (the space) at the speed of light.

With this background, let us now define time as a measure of change in relative position of moving objects or spaces with respect to each other or a point of reference. If the objects or spaces are not moving, there is no change in their relative position, thus their distance or geometric location is sufficient in defining their relative positions using the x, y, z coordinates. However, in the universe described above, everything is moving, and another coordinate is needed to define the relative position of objects and spaces. This coordinate is "time". For example, we can define the relative position of two cities by their distance, but since both cities are moving with respect to the rotation of earth or a reference point such as the sun, we can define the change in their relative positions by their respective positions on the earth which will become the time of day. Or we may define the start and end of an action by another coordinate when the position or location of the act (x, y, z) has been determined, such as how long it takes to get something done. Thus, time becomes another coordinate of space in describing an event similar to the familiar x, y, and z coordinates.

If time is another dimension that defines the coordinate of moving objects with respect to each other or a point of reference, is there a geometric dimension that describes it? What force causes the strings in restframe, and subsequently all objects with a mass which are in restframe, to move in order to create this coordinate?

Let us first start by defining the geometric dimension of time in the quantum world for species that move at the speed of light and space itself, then I will expand this to include objects with a mass.

As mentioned above, the two major components forming spacetime in our universe are the three-dimensional open and looped strings. We learned that the dimension of activated open strings does not change from its initial Planck length throughout the universe. However, the dimension of activated looped strings changes from Planck length (1.61×10^{-35}m) at the nucleus of the spacetime cavity to about 2.3×10^{-20}m outside the cavity in infinite momentum frame. In addition to the spacetime cavity of an atom or particle, this dimensional change occurs in the spacetime cavity of the quantum and gravitational field of large macro-masses as described earlier. The change at its maximum is substantial, a factor of about 10^{15}. The change of the dimension of activated looped strings in Chapter 2, equation (2-3) was given as:

$$r_a = 0.71 \times 10^{-26} \, (Ra)^{1/4}$$

In which R_a is the radius of the energy field of a mass and if the mass is the universe, the radius of the universe as the field of infinite momentum frame.

Naturalized dimensional change of activated looped strings as a function of R_a, the radius of the energy field of a mass, is shown in Fig.11-1 and for the quantum and gravitational field of a large mass in Fig. 11-2.

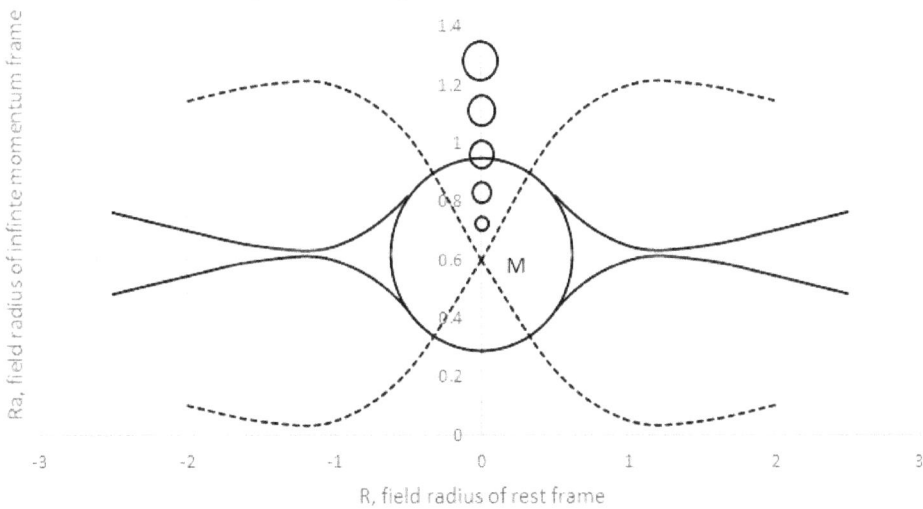

Fig. 11-1 -Radius of the activated looped string Vs Ra

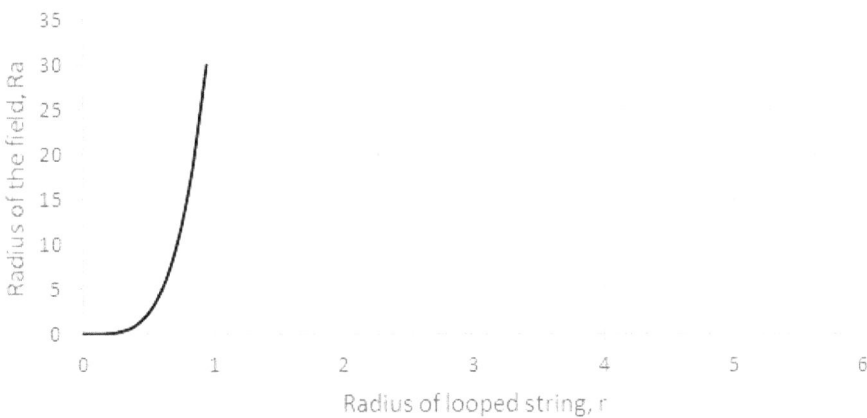

Fig. 11-2 - Dimensional change of activated looped strings inside the quantum and gravitaional field of mass M

The dimension of activated looped strings begins to shrink when it enters the quantum and gravitational field of large masses as well as inside the spacetime cavity of fundamental particles as shown above.

We must think of our universe as the wide-open space of infinite momentum frame. A space dominated by activated looped strings with a dimension of 2.3×10^{-20}m along with much smaller activated opens strings of 1.61×10^{-35}m filling the interstitial spaces between the activated looped strings, Fig. 11-4.

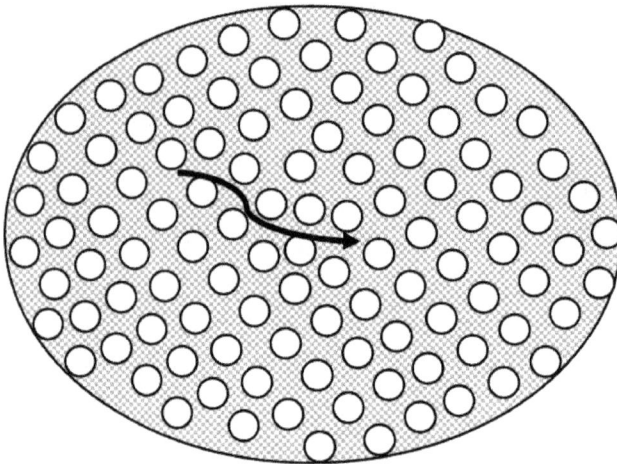

Fig. 11-3- Conceptual representation of the universe with activated looped strings, white circles, and activated open strings, small dots. Arrow shows the tortuous path of photons through activated open string space.

The shrinkage of the activated looped strings begins when the strings are subject to the quantum and gravitational field of large masses and becomes substantial as it enters the spacetime cavity of atoms and fundamental particles at dimensions below 1.8×10^{-9}m. The dimension of activated looped strings anywhere else in the universe not subject to the quantum and gravitational field of a mass or outside the spacetime cavity of an atom remains constant at 2.3×10^{-20}m.

Objects containing activated looped strings in their structure will experience a corresponding shrinkage in volume which is proportional to shrinkage of the smallest constituent comprising its structure. Shrinkage of the activated looped strings means that their mass will increase:

$$m = h/2\pi rc$$

Therefore, the mass of the objects entering a space zone affected by the quantum and gravitational field of a larger mass will increase correspondingly, as will the mass of the space itself.

Let us now look at the dimension of activated looped string and how its changes impact "time".

Referring to Fig. 11-4, activated open strings in the presence of much larger activated looped strings must move across the surface of the sphere of activated looped strings to move from point A to point B. The distance activated open strings or any specie such as a photon (since photons contains only activated open strings) travel will be the diameter of the looped string minus the diameter of the open string itself:

$$\ell = 2(r - 1.61 \times 10^{-35}) \tag{11-1}$$

In which r is the radius of the activated looped string, with activated open strings at Planck dimension. The time it takes activated open strings or a species such as a photon to travel (be displaced) any given distance of x is:

$$t = x/c \tag{11-2}$$

Since the dimension of the activated looped string in each space with a constant radius of the field of its mass (R_a) is the same, the distance given in equation (11-2) is quantized by ℓ which is constant. In other words, for any distance x, time in equation (11- 2) becomes the number of "quantum hops" made by the activated open strings:

$$N = x/2(r - 1.61 \times 10^{-35})$$

$$t = x/c = Nx2(r - 1.61 \times 10^{-35})/c$$

Since,

$2(r - 1.61 \times 10^{-35})/c = \text{constant} = k_t$, for any given space with a constant R_a;

$$t = k_t N \qquad\qquad (11\text{-}3)$$

In other words, "time" in any space is the number of "quantum hops" made by the activated open strings that are displaced a distance of x. Since photons contain only activated open strings in their structure, it will reside in the media of activated open strings and the time it takes for a photon to travel a distance of x equals the number of "quantum hops" it takes to get to its destination. The number of "quantum hops" will become the coordinate describing the position of a photon as a moving object with respect to a reference point.

Let us designate the distance ℓ in equation (11-1) as the "geometric dimension" of time.

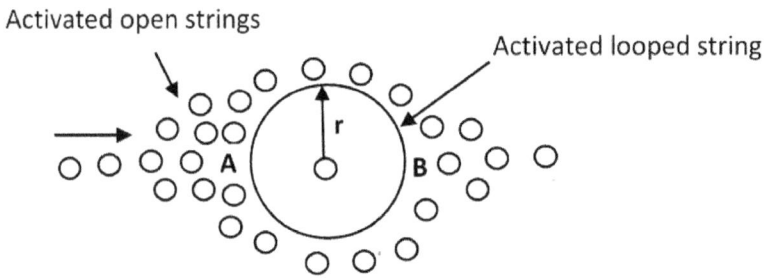

Fig. 11-4- Conceptual representation of flow path of activated open strings around an activated looped string in open space.

In most cases, except in and around the nucleus of the spacetime cavity, the radius of the activated looped string is significantly higher than Planck length, therefore, we can approximate ℓ as:

$$\ell = 2r, \qquad\qquad (11\text{-}4)$$

The geometric dimension of time equals the diameter of activated looped strings and the number of "quantum hops" defining "time" will be simply:

$$N = x/2r \qquad\qquad (11\text{-}5)$$

And the constant $k_t = 2r/c$ or

$k_t = 2 \times 2.3 \times 10^{-20} / 3 \times 10^8 = 1.53 \times 10^{-28}$ s/quantum hops

or,

$1/k_t = 0.65 \times 10^{28}$ quantum hops/s

In which $1/k_t$ may be considered the universe's internal clock.

The geometric dimension of time is a constant number that equals the diameter of the activated looped strings in infinite momentum frame but changes when subject to the quantum and gravitational field of the masses in the direction of R_a. One may envision this distance as the physical equivalent of the time interval between events in space as described by the Minkowski space and general relativity.

Furthermore, since the geometric dimension of time is interwoven into the three-dimensional strings that make up the space, one can say that "time" is imbedded in the geometric dimension of space itself.

We now know that the geometric dimension of time "ℓ" changes depending on the location of the activated looped string in the universe as a function of R_a. It will become smaller inside the quantum or gravitational field of a large mass and its dimension varies by the equation:

$$r_a = 0.71 \times 10^{-26} (Ra)^{1/4}$$

Therefore, time and distance will have a different value depending on the dimension of the activated looped string in that space.

Referring to Fig. 11-5, for the same distance *x*, "time" or the number of "quantum hops" increases or decreases for activated open strings, making it appear as if the internal clock for the object in that space has slowed or sped up.

Or, for the same "time" or the same number of "quantum hops" in a space where activated looped strings are much larger or smaller, activated open strings travel a longer or shorter distance as compared to a space where the dimension of activated looped strings remain constant (reference space). Note that the direction of the field where the dimension of activated looped strings remains constant is perpendicular to the direction of the field of its dimension change.

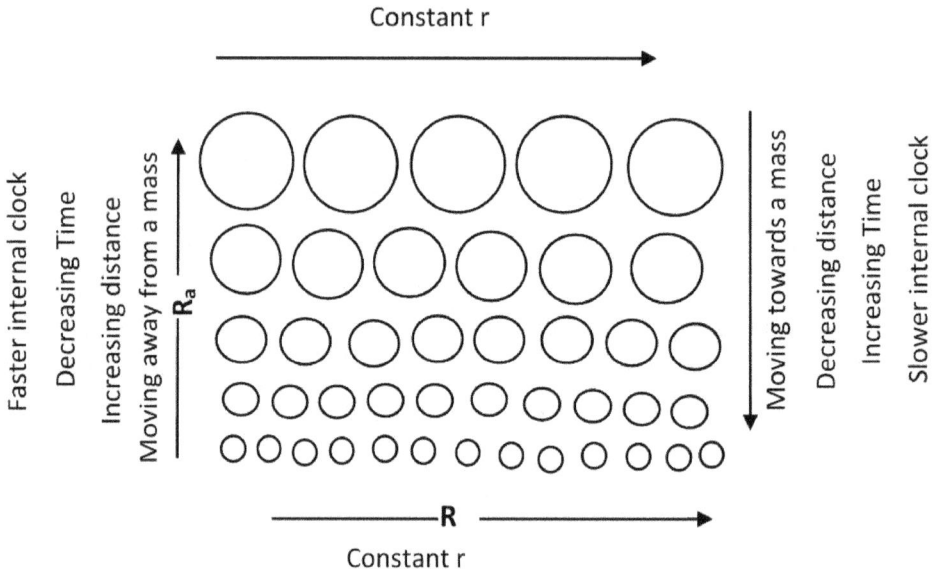

Fig. 11-5- Conceptual representation of dimensional change of activated looped strings subject to the energy field of masses and its effect on space and time.

In our universe, the dimension of the activated looped strings changes subject to the field of masses of stars or galaxies. Therefore ℓ, the geometric dimension of time, is never the same. As such, time changes depending on the location of the space in the universe.

Where the dimension of activated looped strings r, is the same as Planck length:

$$t = k_t N = N \ell /c = N \times 2(1.61 \times 10^{-35} - 1.61 \times 10^{-35})/c = 0$$

This means that time will not exist, i.e., the distance the string is displaced is zero, or the string does not move (no quantum hops).

Time and string resistance forces

Without movement of the strings in the universe, the concept of geometric dimension of time is meaningless. Strings in infinite momentum frame and restframe must experience a positive net force to cause them to move which would then necessitate the coordinate of "time" to locate their position. Referring to Fig. 11-4, the flow of strings across the surface of each other creates a force, I will refer to this as String Resistance (SR) Force which plays an important role in the movement and kinetics of strings in infinite momentum and restframe:

$$F_{sr} = \rho c^2 (2\pi r^2) \qquad\qquad (11\text{-}6)$$

In which ρ is the mass density of the string flowing across another string with a radius of r.

In the universe there are two places where time does not exist; at the singularity of a blackhole, and the nucleus of the spacetime cavity of fundamental particles where the dimension of activated looped string shrinks down to Planck length, rendering:

$$\ell = 2(r - 1.61 \times 10^{-35}) = 0$$

The difference between the two is that in a blackhole, the geometric dimension of time is zero along the entire length of the Schwarzschild radius, whereas it is limited to only Planck length at the nucleus of the fundamental particles such as protons and neutrons. As you will soon see, this is where the net force applied on the strings will also be zero.

As the dimension of the activated looped strings (r) shrinks inside the spacetime cavity or subject to quantum and gravitation fields of large masses, the magnitude of the geometric dimension of time ℓ becomes smaller, or the internal clock of the space slows down. This means that time or the number of quantum hops increases. And vice versa, where ℓ becomes larger, the internal clock of the space speeds up, meaning that the number of quantum hops (time) decreases.

Let us now address "time" for strings in restframe state which comprise objects with a mass. Objects with a mass also require a net positive force to initiate the movement which will establish the coordinate of time.

When activated looped and open strings are immobilized in restframe1 state, it creates a resistance to the flow of activated looped and open strings in infinite momentum frame, as shown in Fig. 11-6.

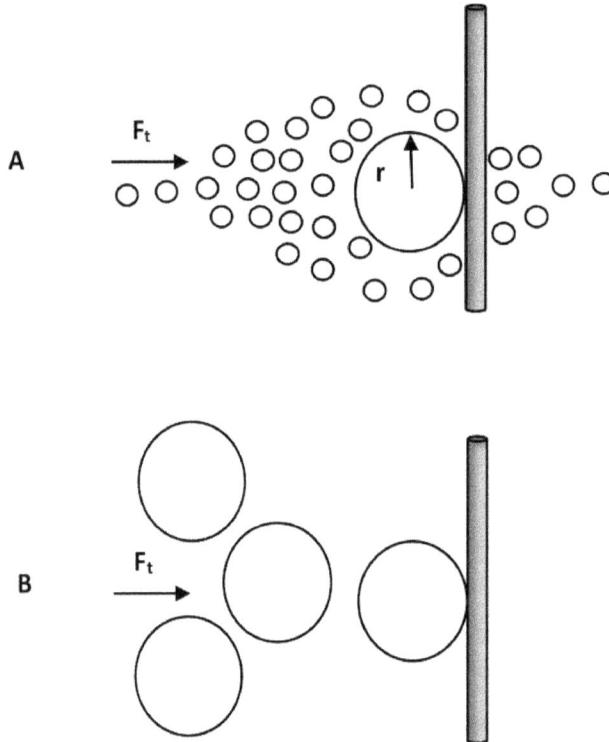

Fig. 11-6- Conceptual representation of activated looped string attached to an extra dimension cylinder at restframe. A) Activated open strings flowing past activated looped string create the string resistance force F_{srr}. B) activated looped strings flowing past activated looped string create the force F_{srr}.

Let us name this force String Restframe Resistance Force or SRR-force as an abbreviation. The magnitude of this SRR-force is the same as equation (11-6):

$$F_{srr} = \rho c^2 \, (2\pi r^2)$$

In which ρ is the energy density of the string free flowing in infinite momentum frame, and r is the radius of the string or a surface that creates the resistance to the flow.

Let us now calculate the magnitude of this force for the four different ways it is applied to masses outside the spacetime cavity of an atom or a mass. For this, I will start with the effect of this resistance force on the $2x10^{30}$ structural extra dimension cylinders that make up the mass of dark matter as shown in Chapter 5.

   a) Free flowing activated open strings exerting force on the surface of an activated looped string in restframe:

If we substitute for $\rho =0.2x10^{97}kg/m^3$, and r = $2.3x10^{-20}$m, the radius of the activated looped string:

The magnitude of the SRR- force is:

$$F_{srr1}= 0.2x10^{97}x9x10^{16}x2\pi(2.3x10^{-20})^2 =6 \ x10^{74} \ N$$

Which is slightly higher than the electromagnetic force in the universe:

$$F_e=10.3 \ x10^{-10}x0.28x10^{84} = 2.9 \ x10^{74}N$$

Recall, the gravitational force of the universe exerted by the activated looped strings was:

$$F_g=2.46x10^{-10}x2x10^{53} = 0.49 \ x10^{44} \ N$$

This means that the SRR-force F_{srr1}, is the dominating force acting on the extra dimension cylinders constituting the mass of the universe (dark matter mass).

Since the dimension of the activated strings along the length of the extra dimension cylinder is about the same in infinite momentum frame along the radius of the universe, this force is applied equally across the length of the extra dimension cylinder. Also, since the direction of all the resistance forces applied to the activated looped strings are parallel, the forces are not additive, and the force applied to one string is the same as the force applied to the entire extra dimension cylinder.

This force acts on a single activated looped string in restframe causing the mass attached to the string to move. Note that while the force, F_{srr1} is 6×10^{74}N applied to a structural extra dimension cylinder, its magnitude equals the magnitude of the electromagnetic force created by 2×10^{30} extra dimension cylinders that receive the same force. Thus, establishing an equilibrium between this force and the total electromagnetic force of the universe.

The movement (displacement) of the object with the mass creates the coordinate of "time" or the geometric dimension of time described earlier for that mass. Without this movement, the coordinate of "time" for masses in restframe will not exist.

The SRR-force applied to the extra dimension cylinders will cause it to move a distance of R_x:

$$F_{srr} \times R_x = mc^2 \qquad\qquad (11\text{-}7)$$

In which R_x is the distance the extra dimension cylinders move or are displaced as a result of the force exerted on activated looped strings attached to the extra dimension cylinder. If we substitute for F_{srr} from above and the mass of the universe:

$$6 \times 10^{74} \text{ N} \times R_x = 2 \times 10^{53} \times 9 \times 10^{16}$$

$$R_x = 3 \times 10^{-5} \text{m}$$

Recall from Chapter 3, the radius of the activated looped strings in restframe was 4.6×10^{-5}m for $R_a = 1.15 \times 10^{26}$m or the radius of the observable universe. In other words, SRR-force described above causes the extra dimension cylinders that define the structural mass of the universe to move or be displaced by a length equal to its restframe radius. The number of "quantum hops" that activated open strings make for this movement are:

$$N = 3 \times 10^{-5} / \ 2 \times 2.3 \times 10^{-20} = 0.65 \times 10^{15}$$

Which as mentioned earlier may be viewed as the internal clock of the universe. The time for the displacement is:

$$t = 3 \times 10^{-5} / 3 \times 10^{8} = 1 \times 10^{-13} \text{ s}$$

This means 0.65×10^{15} quantum hops equal 1×10^{-13}s of our time, or

$N/t = 0.65 \times 10^{28}$ quantum hops/s

The displacement of the electromagnetic mass in the universe as a result of this force would be:

6×10^{74} N x $R_x = 0.28 \times 10^{84}$x 9×10^{16}

$R_x = 0.42 \times 10^{26}$m

Which means that the F_{srr1} is responsible for the trans-universe movement (displacement) of the space mass contributing to the electromagnetic force. This will also yield the same number for the cosmological clock as calculated above.

    b)   Free flowing activated open strings exerting force on the surface of the activated open string in restframe:

$F_{srr2} = 0.2 \times 10^{97}x9 \times 10^{16}x2\pi(1.61 \times 10^{-35})^2 = 2.9 \times 10^{44}$ N

This is the SRR-force applied on activated open strings in the restframe which is significantly smaller than the F_{srr1}. Therefore, the F_{srr1} force dominates the force on the structure of the extra dimension cylinder. Nevertheless, this SRR-force has the same magnitude as the gravitational force which balances out the gravitational force of the looped strings on all the structural extra dimension cylinders.

    c)   Free flowing activated looped strings exerting force on the surface of the activated looped string in restframe:

$F_{srr3} = 0.28 \times 10^{36}x9 \times 10^{16}x2\pi(2.3 \times 10^{-20})^2 = 8.3 \times 10^{13}$ N

As you can see this force is significantly smaller than the above force, by a factor of:

$1.28 \times 10^{44} / 8.3 \times 10^{13} = 1.5 \times 10^{30}$

Which means that F_{srr1} is still the dominating force.

d) Free flowing activated looped strings exerting force on the surface of the open string in restframe:

$F_{srr4} = 0.2x10^{36}x9x10^{16}x2\pi(1.61x10^{-35})^2 = 2.9 x10^{-17}$ N

This SRR-force is even smaller than F_{srr3} the force by a factor of:

$8.3x10^{13} / 2.9x10^{-17} = 2.8x10^{30}$

As you can see above, the most dominant SRR-force is the SRR1 force applied to the extra dimension cylinder of the structure of dark matter causing it to be displaced by $3x10^{-5}$m circumferentially and $0.42x10^{26}$m radially.

Recall in Chapter 3, I referred to string resistance forces that were generated on orbiting extra dimension cylinders. These resistance forces negate the collision force of activated looped and open strings which would have otherwise caused the extra dimension cylinder to collapse into a pulsating cylinder. The magnitude of the F_{srr1}, and F_{srr2} are equal to the force created by the electromagnetic and gravitational forces but in opposite directions. Hence keeping the structure of an orbiting extra dimension cylinder intact.

Weak and Strong Forces

Above, I discussed the four string resistance forces created by the activated looped and open strings flowing in infinite momentum frame against the same strings in restframe which constitute the mass of dark matter and the electromagnetic field of the universe. There is yet another SRR-force inside the spacetime cavity of fundamental particles and atoms which constitute the SRR-forces on a mass "M". Recall that inside the spacetime cavity, below a depth of about $1.4x10^{-20}$m, which happens to be the same as the radius of an activated looped string in infinite momentum frame, all strings, open and looped exist in restframe, Fig. 11-7A. This means that the flow of activated open and looped strings produce a corresponding resistance force against the cross-sectional surface of the spacetime cavity at this depth when the zone is at restframe.

The force created by the flow of activated looped strings will produce what we know in classical physics as the **Weak force,** and the force created by the activated open strings will produce the **Strong force**. As you can see from Fig. 11-7A, since these forces are resistance forces, their range is only experienced on and around the surface where the force is created. One may liken this to a pressure build up against an obstacle which dissipates as you move away from the surface of the obstacle. The pressure zone will be proportional to the magnitude of the force. As such, the zone created by the Strong force has a longer range than that of the Weak force.

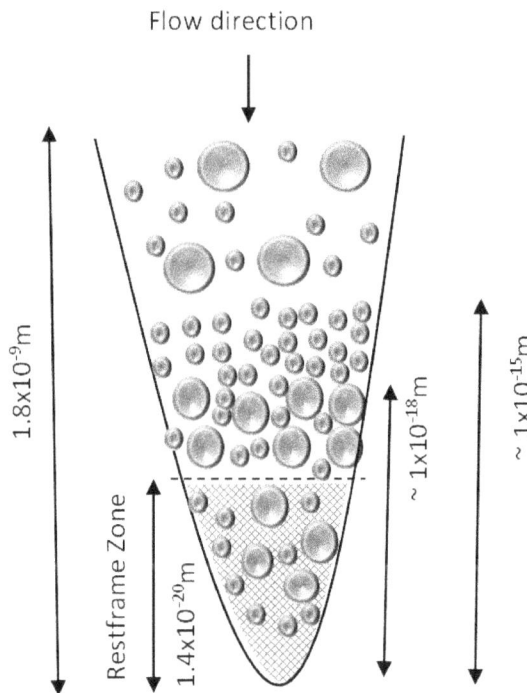

Fig. 11-7A- Conceptual representation of the flow of activated open (small spheres) and looped strings (larger spheres) inside spacetime cavity. a) resistive surface created by the strings in the restframe zone, b) creation of the strong and weak force zones.

Let us now calculate the magnitude of the two forces and their relative strength as compared to the electromagnetic and gravitational forces of the universe.

The resistance force created by the activated looped strings on the surface of the Restframe Zone inside the spacetime cavity or the Weak force is as shown in equation (11-6):

$$F_t = \rho c^2 (\pi R^2)$$

$$R = 1.4 \times 10^{-20} \times 1.6 = 2.3 \times 10^{-20} m,$$

$$\rho = 1.2 k/c^2 (1/R_a) = 0.22 \times 10^{82} \ kg/m^3$$

The magnitude of the Weak force will then be:

$$F_{ssrw} = 0.22 \times 10^{82} \ x9 \times 10^{16} x \ (\pi)x \ (2.3 \times 10^{-20})^2 = 3.35 \times 10^{59} \ N$$

The magnitude of the Strong force created by the flow of activated open strings is:

$$F_{ssrs} = 0.2 \times 10^{97} \ x9 \times 10^{16} x \ (\pi)x \ (2.3 \times 10^{-20})^2 = 3 \times 10^{74} \ N$$

Which is the same as the electromagnetic force created by the universe, $2.9 \times 10^{74} N$.

The ratio of the Strong to Weak forces would then be:

$$3 \times 10^{74} / 3.25 \times 10^{59} = 9.2 \times 10^{14}$$

This is about the same ratio measured according to the Standard Model.

The existence of the Strong force inside the spacetime cavity will impact the activated looped string attached to a Tubular Monopole. The Tubular Monopole extends deep inside the spacetime cavity, starting at a dimension of $1.1 \times 10^{-30} m$. Since activated open strings do not decay inside the Tubular Monopole, it conveys the maximum force exerted by the strong nuclear force in its tubular structure like a straw (Fig. 11-7B). As a result, the looped strings attached to the Tubular Monopole will experience the Strong force which is about 137 times higher than what they would have otherwise experienced by the electromagnetic force alone. The increased force on the activated looped strings will cause the strings to shrink in dimension which in turn will increase their mass slightly. This means that the mass of fundamental particles such as protons and neutrons, which would have been normally smaller, will increase by a factor of 137.

In essence, the Strong force, which has the same magnitude as the electromagnetic force exerted by the universe, acts to balance the force of the electromagnetic force inside the spacetime cavity and prevents the spacetime cavity from collapsing into a blackhole. In the absence of Tubular Monopoles, the 137 extra dimension cylinders with the highest vacuum energy in a spacetime cavity, each will experience the force of:

$$F_{1/137} = 3 \times 10^{74}/137 = 2.1 \times 10^{72} \text{ N}$$

As each one of the 137 extra dimension cylinders is converted into a Tubular Monopole, the electromagnetic force associated with that extra dimension cylinder is lost because activated open strings do not decay to produce this force. The Strong force makes up for this loss by increasing the forces through its capillary structure, but in the meantime, it also adds to the force that activated looped strings attached to the Tubular Monopole experience.

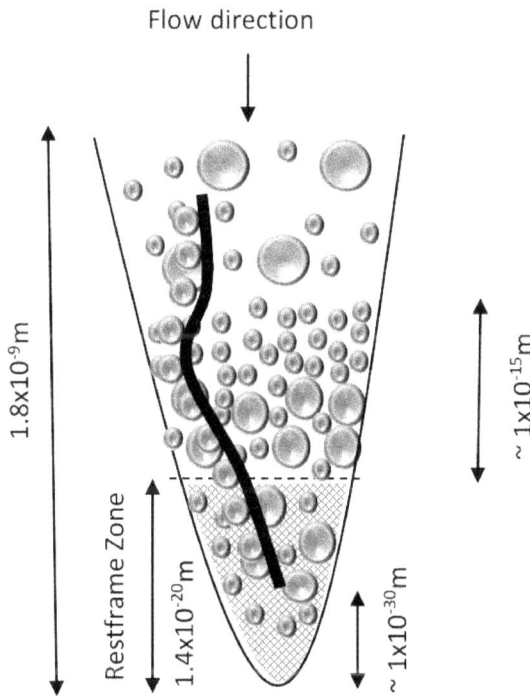

Fig. 11-7B- Conceptual representation of the strong force inside spacetime cavity and its relation to Tubular monopole. Detailed structure of the Tubular Monopole is provided in Chapter 3.

At the nucleus of the spacetime cavity, the gravitational and electromagnetic forces applied to a string attached to the extra dimension cylinders are equal and add up to:

$$\sum F = 2 \times 2.18 \times 10^{-8} \times 5.6 \times 10^{51} = 2.44 \times 10^{44}\,\text{N}$$

In which, the gravitational acceleration of activated open and looped strings inside the spacetime cavity at the nucleus is:

$$g = 9 \times 10^{16} / 1.61 \times 10^{-35} = 5.6 \times 10^{51}\,\text{m/s}^2$$

This force is balanced out by the string restframe resistance force, F_{srr2}, which was calculated to be about 2.9×10^{44} N, resulting in a net zero force on the string at the nucleus of the spacetime cavity.

Net force on the string $= 0$

Recall that at the nucleus the geometric dimension of time is also zero, $\ell = 0$. This means that there is no force to cause the displacement of the string at the precise location of the nucleus, and "time" ceases to exist.

Non-existence of "time" has a significant physical meaning. It means that the spacetime cavity of a fundamental particle becomes "anchored" to the fabric of spacetime at the nucleus because there is no net force to act on it. This is important to accumulation and growth in size of a mass. If "time" existed at the very nucleus of the spacetime cavity, the cavity would be floating in space like a photon and large masses such as atoms, stars, and planets would never form, as "matter" would drift into space and scatter. Since mass becomes anchored to the fabric of spacetime at the nucleus, it requires an enormous amount of energy to accelerate it to speeds near the speed of light. The amount of energy needed is the amount it will take to detach the string from restframe1 state back into infinite momentum frame.

The same phenomenon occurs at the singularity of a blackhole where the net force on the strings along the Schwarzschild radius will be zero, $\ell = 0$, and time ceases to exist. This allows the mass of the blackhole to be anchored across the diameter of the Schwarzschild sphere.

This brings us to a new understanding of "time". As described above, when the net force acting on a three-dimensional string is zero, time ceases to exist. Without a driving force (ΔF), strings become immobile. In our universe, in infinite momentum frame, this driving force is created by the geometric difference of activated looped and open strings which leads to a non-zero String Resistance Force. As such, three-dimensional strings remain mobile creating a non-rigid and elastic space allowing objects to move about freely.

Resistance Force on a non-structural mass "M"

Of the four F_{srr} forces described above, only F_{ssr3} is the dominant force on any mass "M". Recall that the mass of fundamental particles was described by the Tubular Monopoles inside the spacetime cavity of atoms.

Since activated open strings inside the Tubular Monopole do not decay, it creates a barrier for activated open strings in infinite momentum frame to pass around the activated looped strings in restframe and through the tube that is attached to the extra dimension cylinders. As a result, there will be no resistance force created by the activated open strings on looped strings which create the highest resistance force. This leaves us with F_{srr3}, the force created by the activated looped strings on looped strings in restframe.

However, unlike the structural mass of the dark matter described above, the dimension of activated looped strings inside the spacetime cavity of atoms changes. Therefore, the resistance force, F_{ssr3}, changes across the length of the Tubular Monopole.

In addition, an aggregate of mass "M" contains many spacetime cavities that have many different orientations as constituents of the mass of the object. For this, we must consider the quantum field of the mass described in Chapter 6. Recall that a quantum field is created by any mass "M" as a result of the addition of the forces of all spacetime cavities of individual atoms making up the mass of an object. Since the fields inside spacetime cavities are vector fields, the resultant quantum field will be the net field of the summation of all vectors pointing in different directions. The F_{srr3} force applied to any mass "M" will be applied to Tubular Monopoles that extend the entire radius of the mass R_m, starting at the center of the mass. We also had two fields, north cavity and south cavity, Fig. 8-8.

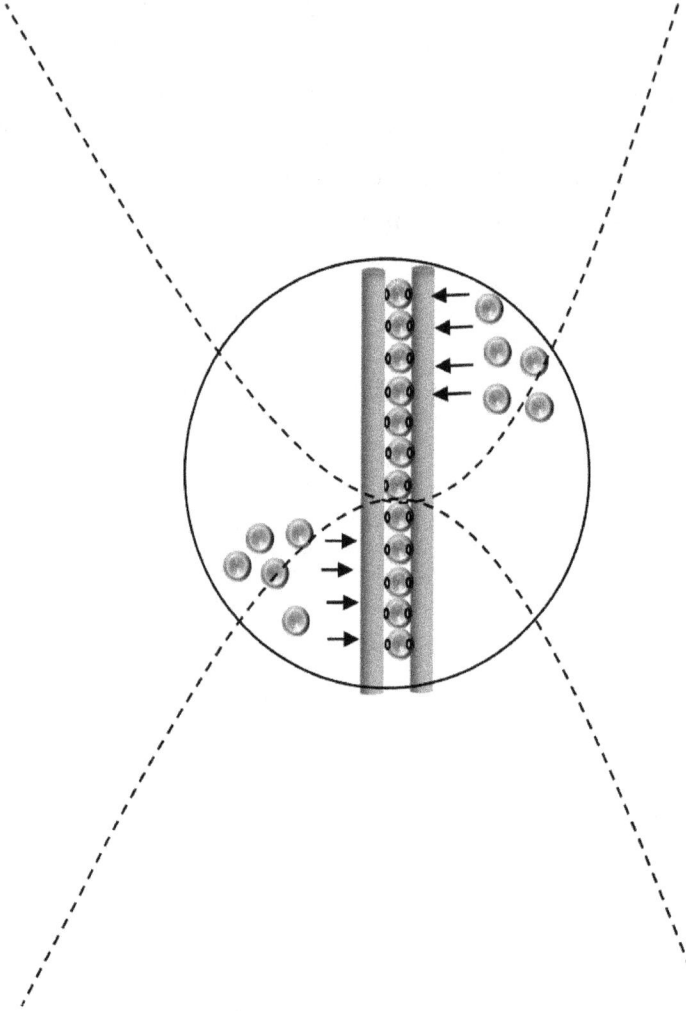

Fig. 11-8 - Conceptual representation of String Restframe Resistance Forces created on the Tubular Monopoles in a mass "M". Dotted lines represent the energy field of the mass.

Let us now calculate the String Restframe Resistance force for any mass "M" with a radius of "R_m".

$$F_{srr3} = \rho c^2 (2\pi r^2) \qquad (11\text{-}8)$$

Where ρ is the energy density of activated looped strings, and r the radius of activated looped strings. Substituting for ρ from equation (1-15b),

$$\rho = (1.2k/c^2R_a)$$

$$F_{srr3} = (1.2k/c^2R_a)c^2 (2\pi r^2) = 1.2k (2\pi r^2)/R_a$$

Substituting for r from equation, and $k = 0.24 \times 10^{79}$ N.m/m²

$$r = 0.71 \times 10^{-26} (Ra)^{1/4}$$

$$F_{srr3} = 0.91 \times 10^{27}/R_a^{1/2} \tag{11-9}$$

Recall that the Tubular Monopoles begin to form at a dimension of about 1.1×10^{-30}m inside the spacetime cavity. This will be the minimum range for the structure of the Tubular Monopoles inside the spacetime cavity of the quantum field of mass "M" which will deliver the highest force. The maximum range of R_a is the radius of the mass itself R_m, which delivers the lowest force. The median force will then fall between these two ranges. Therefore, for any mass "M", the median of the radius of the looped strings as a function of R_a will be based on the geometric mean of the two dimensions:

$$R_a = (1.1 \times 10^{-30} \times R_m)^{1/2} = 1.05 \times 10^{-15} R_m^{1/2}$$

Substituting for R_a in equation (11-9):

$$F_{srr3} = 2.82 \times 10^{34}/R_m^{1/4} \tag{11-10}$$

This is the median force applied to the extra dimension cylinders across the radius of the mass. The smallest resistance force applied to the Tubular Monopoles is on the surface of the mass "M" which is calculated from equation (11-9) by substituting R_m, for R_a:

$$F_{srr3'} = 0.91 \times 10^{27}/R_m^{1/2}$$

The maximum force across the radius of the mass will be at $R_a = 1.1 \times 10^{-30}$m which is constant for all masses:

$$F_{srr3} = 0.86 \times 10^{42} \text{ N}$$

The F_{srr3} forces across the radius of the mass "M" are responsible for inducing a motion into any mass "M". Since there will be two forces in opposite directions in the north and south spacetime cavity of the quantum field of a mass, this motion will become a spinning motion causing the mass to rotate as it moves, thus creating the coordinate of geometric time for that mass. In a way, the F_{srr3} may be called "Time-force".

Note that the F_{srr3} calculated above applies to masses that are freely suspended in space and are outside the quantum field of other masses. Masses that are in the quantum field of a larger mass become a component of the quantum field of the large mass and the $Fssr_3$ force of the larger mass will be the dominating force. As such, the orbital characteristic of the smaller mass is determined by the F_{srr3} force of the larger mass. This is because the structural Tubular Monopole extra dimension cylinders of the small mass will merge with that of the large mass in its quantum field and the higher force on the shared extra dimension cylinders is the one that dominates.

Let us now apply equation (11-8) to any mass M with a velocity of V in restframe and substitute for F_{srr3} from equation (11-6), for any field of space where r the radius of the activated looped string is constant. We can consider the space where the dimension of activated looped strings is constant as the reference frame:

$$\rho c^2 \, (2\pi r^2) \,.\, nR_a = 1/2MV^2 \qquad\qquad (11\text{-}11)$$

Where "n" is any multiple or a fraction of $R_{a'}$ and nR_a is the displacement distance. Substituting for ρ from equation (1-15b),

$$\rho = (1.2k/c^2 R_a)$$

$$(1.2k/c^2 R_a) \, c^2 \, (2\pi r^2) \,.\, nR_a = 1/2MV^2$$

$$(1.2nk \, x2\pi r^2) = 1/2M \, (d\chi/dt)^2$$

Or simply,

$$t = (M/ \, 1.2\pi kn)^{1/2} \, (\chi \, /2r) \qquad\qquad (11\text{-}12)$$

Equation (11-12) also shows that "time" for a mass M is distance χ quantized by 2r, the geometric dimension of time, which is proportional to N, the number of open string "quantum hops", as described earlier or:

$N = (\chi / 2r)$

$t = k_m N$

In which k_m is constant. Let us rewrite equation (11-11) for a mass that is moving with a velocity of V in a direction of increasing or decreasing R_a where the dimension of the activated looped string changes as r_m.

$t_m = (M_m / 1.2\pi kn)^{1/2} (\chi_m / 2r_m)$ $\qquad\qquad$ (11-13)

Equations (11-12), and (11-13) show that time is inversely proportional to r or r_m, the radius of the activated looped strings.

Meaning that as we move in a direction where r is reduced, for example moving closer to a large mass in space, "time" or the number of "quantum hops" increases. However, distance traveled is smaller since space shrinks with the dimension of activated looped strings. This will appear as if the internal clock of the moving mass has slowed down or its time has increased.

If we move away from the quantum field of a large mass, r_m increases, and "time" or the number of "quantum hops" decrease but the distance increases. This will appear as if the internal clock of the mass has sped up.

If we divide equation (11-12) by (11-13), for the same distance ($\chi = \chi_m$) traveled by mass M:

$t/t_m = r_m/r (M/M_m)^{1/2}$ $\qquad\qquad$ (11-14)

Equation (11-14) shows that if we move in the direction towards a mass where the radius of the activated looped string decreases ($r_m/r < 1$) and its mass increases, ($M/M_m < 1$):

$t/t_m < 1$

Time for the moving object increases, $t_m > t$ as compared to the mass traveling in a space where the radius of the activated looped string remains constant (reference frame).

At low velocities, we can assume the change in mass is negligible:

$$t/t_m = r_m/r \qquad\qquad (11\text{-}15)$$

Conversely, if we move away from a mass, $(r_m/r > 1)$ and $(M/M_m > 1)$, then $t_m < t$.

If we equate equations (11-12) to (11-13), we can obtain the distance traveled for the same time:

$$t = t_m = (M/1.2\pi k)^{1/2} (\chi/2r) = (M_m/1.2\pi k)^{1/2} (\chi_m/2r_m)$$

$$\chi_m/\chi = (r_m/r)(M/M_m)$$

Again, if we move in the direction towards a mass where the radius of the activated looped string decreases $(r_m/r < 1)$, its mass increases, $(M/M_m < 1)$:

$$\chi_m/\chi < 1$$

Distance traveled is less than the distance in reference frame. Conversely, if we move away from a mass, distance traveled will be more than the reference frame.

An example of a mass moving in a direction where the dimension of the activated looped string remains constant is movement on the surface of the earth, where R_a, the field radius of the mass, is constant. If we move away from the earth, the dimension of the activated looped strings will continue to increase until we reach the diameter of the quantum field of the earth which is twice the radius of the earth. The radius of the activated looped string will eventually reach its maximum, 2.3×10^{-20}m, where there is no local gravitational or quantum fields due to stars or galaxies.

Let us now obtain the relationship of time and dimension for a mass moving in the direction of reducing r as a function of its velocity V as shown in Fig. 11-7. The velocity of a moving mass creates an additional string resistance force which has the same shrinking effect on the activated looped strings as the gravitational force of a larger mass.

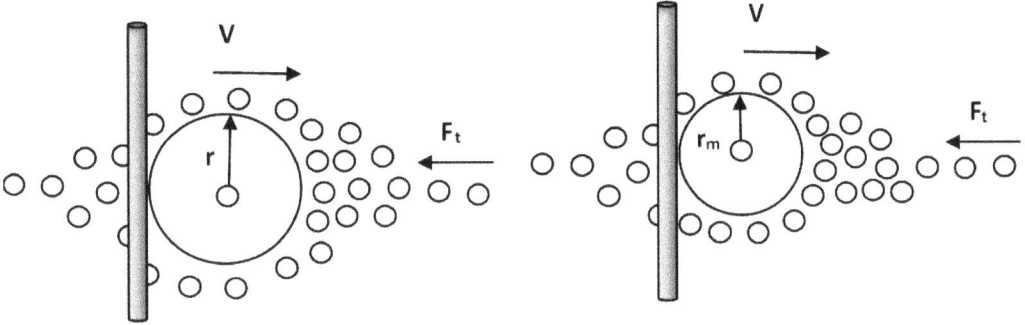

Fig. 11-9 - Conceptual representation of contraction of an activated looped string at restframe in a moving mass with a velocity of V.

Referring to Fig. 11-9, the balance of the SRR-force exerted on the activated looped string for a mass moving in the direction of reducing r, is:

$$\rho(c+V)^2 \, 2\pi r_m^2 = \rho c^2 \, 2\pi r^2 - \rho V^2 (2\pi r^2 - 2\pi r_m^2) \qquad (11\text{-}16)$$

Simplifying equation (11-16):

$$(c+V)^2 r_m^2 = c^2 r^2 - V^2(r^2 - r_m^2)$$

$$c^2 r_m^2 + V^2 r_m^2 + 2cV r_m^2 = c^2 r^2 - V^2 r^2 + V^2 r_m^2$$

$$c^2 r_m^2 - c^2 r^2 = -(V^2 r^2 + 2cV r_m^2)$$

Dividing both sides of the equation by $c^2 r^2$:

$$r_m^2/r^2 - 1 = -V^2/c^2 - 2(V/c)r_m^2/r^2$$

Since V is much smaller than c, we can ignore the term $2(V/c)r_m^2/r^2$ as compared to r_m^2/r^2:

$$r_m^2/r^2 = 1 - V^2/c^2$$

$$r_m/r = (1-V^2/c^2)^{1/2} \qquad (11\text{-}17)$$

If we substitute for r_m/r in equation (11-11) and (11-12):

$$t/t_m = (M/M_m)^{1/2} (1-V^2/c^2)^{1/2} \qquad\qquad (11\text{-}18)$$

or for no change in a mass:

$$t/t_m = (1-V^2/c^2)^{1/2} \qquad\qquad (11\text{-}19)$$

Equations (11-17), (11-18), and (11-19) are the same as those in classical special relativity for a moving object.

The above equations were developed based on the SRR-force on a single activated looped string. Since the dimensional change of the mass of an object with a radius of R is proportional to the dimensional change of the activated looped string, in equations (11-17) and (11-19) r can be replaced by R, the length of a mass, which means the entire mass of an object will shrink or expand proportionally. Since the mass of an activated looped string changes as its radius changes:

$$m = h/2\pi rc$$

The change in the mass of a moving object will be inversely proportional to the radius of the activated looped string:

$$M/M_m = (1-V^2/c^2)^{1/2} \qquad\qquad (11\text{-}20)$$

Unlike electromagnetic and gravitational forces, the SRR-forces created by the flow of activated open and looped strings are resistance forces with a short-range field. Its range is limited to the immediate vicinity of the surface that creates the resistance.

Except for Strong and Weak forces, one may consider SRR-forces an additional force of nature not identified before. Without SRR-forces, objects in the restframe have simply no inertia to move or be displaced to create the coordinate of "time".

By now, it has become clear that "time", which is "quantum hops" made by activated open strings with respect to much larger activated looped strings, is interwoven with space and a component of the geometric space. Therefore, spacetime will be simply the Hilbert space with its inherent inner product of the two vectors, which is purely geometric, and "time" having a quanta which is equal to the geometric dimension of time (or the dimension of activated looped strings in Hilbert space). The interval of this quanta is fixed anywhere in space except for the gravitational effect on the dimension of activated looped strings. In classical physics, e.g., the Minkowski space and general relativity, "time", is also a component of space but treated as the fourth dimension of a manifold. As you have seen in this theory, "time" is imbedded in the geometric dimensions of our three-dimensional universe.

Let us now follow the trail of "time" inside a blackhole and for two-dimensional strings.

Recall that inside the blackhole's Schwarzschild radius, first activated looped strings will reduce to the same dimension as the activated open string at the singularity of the blackhole. This means that the geometric dimension of time becomes zero and time will cease to exist for the three-dimensional universe.

When the strings decay into two-dimensional Planck strings, the geometric dimension of time will be based on the geometric difference between the Planck-size two-dimensional strings and the E-loop strings which propagate to create them. The force creating the geometric dimension of time is the centrifugal force propelling the E-loop strings into Planck-size disks.

$$\ell_{time} = 2(1.61 \times 10^{-35} - 2.3 \times 10^{-53})$$

Or simply,

$$\ell_{time} = 2(1.61 \times 10^{-35})$$

This means that "time" for the two-dimensional strings with Planck length outside Hilbert space exists with different metrics. As it turns out, the ratio of this geometric disparity is 7×10^{15} which is in the same order of magnitude as "time" in the three-dimensional universe, 1.4×10^{15}. This means that if we define "time" as the string "quantum hop" with a geometric and dimensional ratio of 1.4×10^{15}, the strings will maintain the same geometric proportionality in two-dimensions as in a three-dimensional state.

When two-dimensional Planck strings collapse into E-loop strings and E-portals, time for the two-dimensional strings will also cease to exist as $\ell_{time} = 0$.

Inside the singularity of the blackhole, E-loop strings (2.3×10^{-51}m) and one-dimensional strings (8.6×10^{-160}m) create a new geometric dimension of time corresponding to their geometric dimensions. The force is the torque that stretches the strings from a looped form into a one-dimensional string:

$$\ell_{time} = 2(1.61 \times 10^{-51} - 8.6 \times 10^{-160})$$

The ratio of this geometric change will be 0.18×10^{109}. This means that the internal clock of "time" at the E-loop space runs $0.18 \times 10^{109} / 1.4 \times 10^{15} = 0.13 \times 10^{94}$ times faster than the time experienced in our three-dimensional universe. In essence, time existed before the Big Bang event which marks the beginning of time for a three-dimensional universe.

In a space where all the strings are one dimensional, $\ell_{time} = 0$, it will appear that time no longer exists, and the notion of "time" will lose its meaning. This will be the universe prior to the formation of E-loop strings before the Big Bang.

String resistance force and wave properties of fundamental particles

In Chapter 1, I demonstrated the curvature of spacetime of the universe by plotting the field of activated three-dimensional strings for positive and negative values of Ra with the radius of the field of restframe as the reference. In Chapter 2, the energy field and curvature of spacetime of a mass was obtained by plotting the positive and negative values of R, the field of restframe with respect to Ra as the reference axis. The universe creates equal amounts of right and left handed spinning three-dimensional open and looped strings establishing an inherent symmetry with respect to the fields of activated strings in restframe (mass) and infinite momentum frame. To describe the system as a whole we must account for the energy of these fields for both right and left handed spinning activated strings.

In this chapter we learned that string resistance forces are responsible for creating the motion of the strings and masses they create. If we now plot the energy field of activated strings of the opposite rotation with respect to both reference axis, restframe and infinite momentum frame, we will obtain a curvature of spacetime that describes the motion of the fields restframe and infinite momentum frame with respect to its own structures but with opposing spins, Fig. 11-10A. This means that the energy fields, restframe and infinite momentum frame can both be in positive and negative territory simultaneously with respect to itself which describes the motion of a wave. This motion is created by the opposing forces of "string resistance force". Fig. 11-10A indicates that the motion of all objects in the universe at quantum and string scale is a wave.

Since the graph is "naturalized" with respect to dimensions, it can be scaled accordingly to represent masses and objects at quantum scale or cosmic scale. For example, at a cosmic scale, the wave corresponding to the activated looped strings will represent the background gravitational wave of the universe, and the wave corresponding to the activated open string represents the background cosmic magnetic wave of the universe.

When rescaled accordingly, it will represent the wave of a charge, photon, dark matter, and other fundamental particles such as neutrinos.

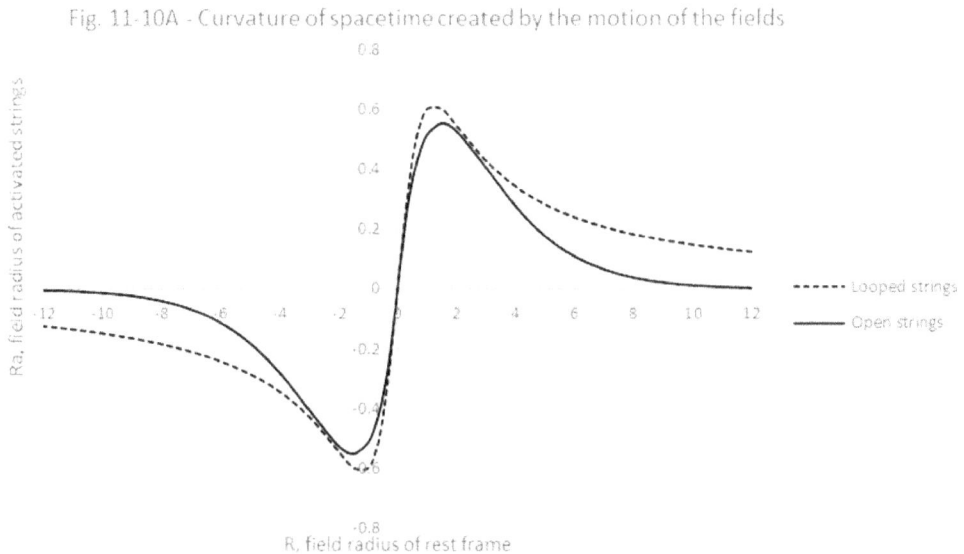

Fig. 11-10A - Curvature of spacetime created by the motion of the fields

In the absence of an external force, string resistance force F_{srr3}, acting on the Tubular Monopole of a fundamental particle, is responsible for inducing a motion and establishment of the geometric dimension of time. In the previous sections, I described this effect on a macro-mass "M". Tubular Monopoles are elongated structures with a length of about 13×10^{-12}m orbiting the center axis of the spacetime cavity at the speed of light and a high frequency.

Recall from Chapters 3 and 4, when two activated looped strings attached to the Tubular Monopole decay, two entangled two-dimensional strings are created. The energy of one string is conserved and forms the high-vac extra dimension cylinder. The energy of the second string forms the low-vac extra dimension cylinders which then propagate as two-dimensional strings. However, as seen before, the energy of the low-vac string. which can be obtained from the Planck relation and the orbital radius of the string, is far below the total energy of the actual string when it decays.

The energy of the second decayed string is vastly used for the displacement of the Tubular Monopole with the string resistance force F_{srr3}, created by the activated looped strings behind it as its driving force. The displacement of the Tubular Monopole as a result of the string resistance force, creates a wave. As such, one may treat the motion of a Tubular Monopole as a wave, Fig. 11-10B.

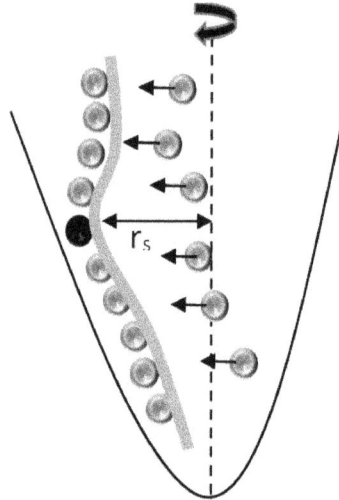

Fig. 11-10B – Conceptual representation of String Resistance Force and the wave behavior of Tubular Monopoles inside the spacetime cavity of fundamental particles. Coordinate "r_s" represents the center of mass.

As seen from Fig. 11-10B, the orbital radius of the Tubular Monopole as a wave changes across the length of the spacetime cavity. Which means that the wave will have different frequencies across the length of the wave. The radius of the orbital rotation can be calculated from equation (3-19):

$$r_s = 4.26 \times 10^{-18}(R_a)^{1/2}$$

The frequency of the wave will be:

$$f = c/2\pi r_s$$

The energy of the orbital rotation of the strings at any orbit from the Planck-Einstein relation is:

$$E = hc/2\pi r_s$$

$$E = hc/8.52\pi \times 10^{-18}(R_a)^{1/2} \qquad\qquad (11\text{-}21)$$

This energy is far below the total energy of the string before decay:

$$E = hc/2\pi r_a$$

In which, r_a is the activated looped string radius:

$$r_a = 0.71 \times 10^{-26}(R_a)^{1/4}$$

$$E = hc/1.42\pi \times 0.71 \times 10^{-26}(R_a)^{1/4} \qquad\qquad (11\text{-}22)$$

Let us now calculate the total energy of the wave, the string resistance force, and the displacement value of the wave (amplitude).

The total energy of the wave, which in classical physics is referred to as the Hamiltonian of the wave, is the sum of the energy of all activated looped strings attached to the Tubular Monopole across its 13×10^{-12}m length. In classical physics, the energy of each string is referred to as the energy eigenstate of the wave.

Since the diameter of the activated looped string varies inside the spacetime cavity, we will take an average dimension based on the mid-point of the wave:

$r_a = 0.71x10^{-26}(6.5x10^{-12})^{1/4} = 1.13x10^{-29}m$

The average energy of the activated looped string will then be:

$E = hc/2\pi r_a = 2.8x10^3 J$

The total number of strings attached to the Tubular Monopole is about $4x10^{23}$ as calculated in Chapter 3:

$N = 13x10^{-12}/2x1.31x10^{-29} = 0.575x10^{18}$

The total energy or the Hamiltonian of the wave will then be:

$\sum E = 2.8x10^3 \ x0.575x10^{18} = 1.6x10^{21} J$

Let us now calculate the string resistance force:

$F = \rho c^2 (2\pi r^2)$

The energy density of the activated looped strings will be:

$\rho = 1.2k/c^2(1/R_a)$

$\rho = 1.2x0.24x10^{79}/9x10^{16}x6.5x10^{-12} = 0.5x10^{73} \ kg/m^3$

$F = 0.5x10^{73} \ x9x10^{16} \ (2\pi)x(1.13x10^{-29})^2 = 3.6x10^{32}N$

We can now find the displacement value of the wave:

$F . R = \sum E$ (11-23)

$3.6x10^{32} \ x \ R = 1.6x10^{21}$

$R = 4.4x10^{-12}m$

Since this is the total displacement for the SRR-forces applied from two sides of the wave, the amplitude of the wave will be half of the above, or 2.2x10^{-12}m, Fig.11-11.

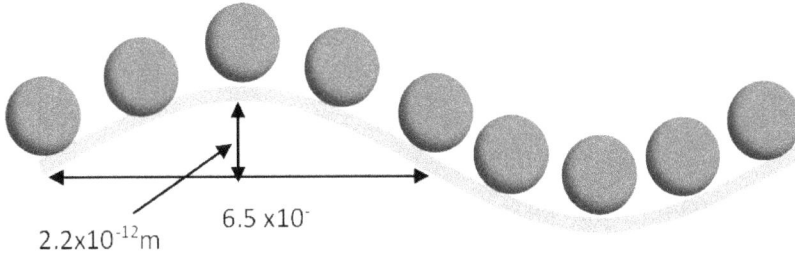

2.2x10^{-12}m   6.5 x10$^-$

Fig. 11-11 – Conceptual representation of wave property of a Tubular Monopole.

Attachment of two-dimensional open strings, which are connected to and driven by the pulsating strings which then drive the orbital rotation of the wave, is random along the length of the wave except in one location, the center of mass, as shown in Fig. 11-10. As described in Chapter 4, this is the location where the mass of the particles manifests itself to our universe and becomes measurable.

In classical physics, the energy eigenstate of this position is described as "stationary state" and the wave as a "standing wave". This is because the attachment of the open strings to the activated looped strings at this point is permanent.

The energy eigenstate of the standing wave can be calculated from equation (11-21).

If we replace the SRR force applied to the wave by the momentum of the wave:

$$F = dp/dt \qquad\qquad (11\text{-}24)$$

And substitute for the momentum:

$$p = h/\lambda$$

Substituting in (11-24),

$F = dp/dt = d(h/\lambda)/dt = h \, d(1/\lambda)/dt = (h/2\pi) \, d(2\pi/\lambda)/dt$

Since $2\pi/\lambda = k =$ *wave number*

$F = \hbar dk/dt$

Substituting in equation (11-23),

$F \cdot R = \hbar(R dk)/dt = \sum E$ (11-25)

The term *(Rdk)* is the amplitude of the wave multiplied by its wave number. This defines the state of a wave and can be replaced by its wave function ψ (t).

If we replace the total energy of the wave $\sum E$, by the Hamiltonian of the wave function in equation (11-25):

$\hbar \, \|\psi(t)/ \, dt = H \, \|\psi(t)$ (11-26)

Equation 11-26 is the famous Schrödinger equation.

Photon

Let us apply the above calculations to a photon which has a length of about 16×10^{-12}m (see Chapter 3):

$N = 16 \times 10^{-12}/2 \times 1.61 \times 10^{-35} = 5 \times 10^{23}$ activated open strings

The total energy of the photon as a wave will be:

$\sum E = 5 \times 10^{23} \times 1.96 \times 10^{9} = 9.8 \times 10^{32}$ J

The SRR-force calculated earlier for activated open strings exerting a resistance force on activated open strings is 2.9×10^{44}N.

$$F . R = \sum E$$

$$2.9x10^{44} \ x \ R = 9.8x10^{32}$$

$$R = 3.38x10^{-12}m$$

The displacement distance which is the amplitude of the photon as a wave will then be half the above or about 1.7 x10^{-12}m.

String Resistance Force and Blackholes

Let us now look at various forces acting on the singularity of a blackhole. The maximum gravitational acceleration was calculated to be:

$$g = 1.45x10^{52}m/s^2$$

The force on the strings attached to the singularity is:

$$F = 2.8x10^{-8}x1.45x10^{52} = 4x10^{44}N$$

Since all strings at the singularity are at Planck length, the string resistance force is the same as $F_{srr2} = 2.9x10^{44}N$ calculated earlier. This means that the two forces which are equal in magnitude and opposite direction cancel each other out.

The electromagnetic force of the universe ($4x10^{74}N$) is balanced out by the force of the mass of the E-portals inside the singularity which was calculated to be about $6.5x10^{22}kg$ (Chapter 10):

$$F = 1.45x10^{52}x6.5x10^{22} = 9.4x10^{74}N$$

The net force acting on the strings attached to the extra dimension cylinder (singularity) will be zero which means no string displacement, and $\ell = 2(r - 1.61x10^{-35}) = 0$, t = 0 or time will cease to exist.

In summary, the string restframe resistance (SRR) forces not only work at the subatomic level to create Weak and Strong nuclear forces, but also work on the cosmic level as the driving force behind the movement of strings, mass and expansion of the universe.

As mentioned earlier, unlike electromagnetic and gravitational forces, the SRR-forces do not create a vector field at the macro-level and its range similar to Strong and Weak nuclear forces is small. The strong and weak nuclear forces are simply one type of SRR-force that can be detected at a subatomic level. The four SRR-forces acting on the restframe structure of the universe and on the strings in infinite momentum frame all have very small ranges to be detectable as compared to the other two major forces in the universe.

Strings and Expansion of Universe

I have postponed the discussion of the expansion of the universe until the regeneration mechanism of the activated strings and the various forces of the universe have been discussed which will help the understanding of the underlying mechanism behind the expansion of universe from the moment of the Big Bang.

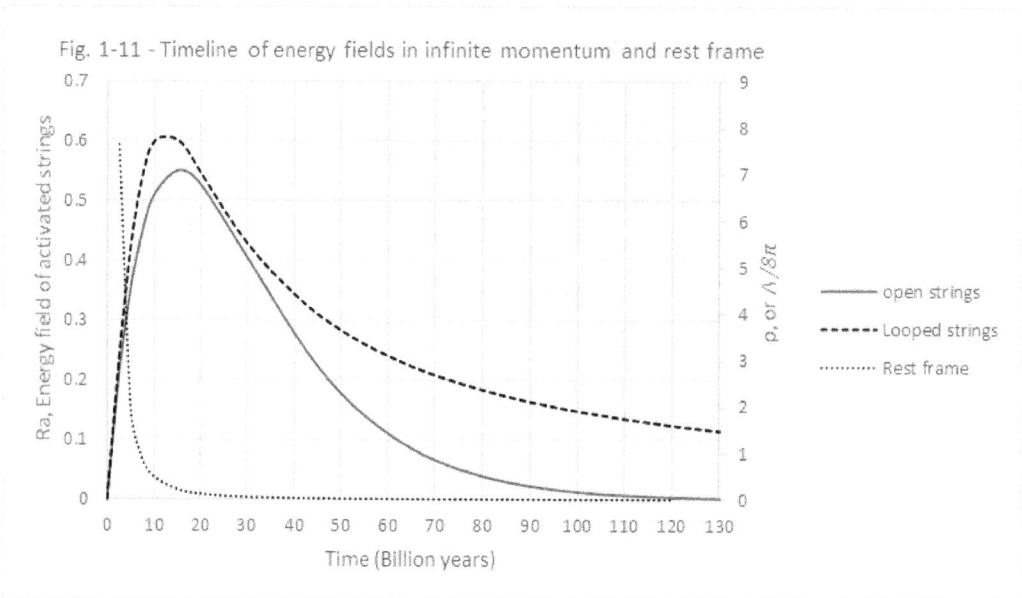

Fig. 1-11 - Timeline of energy fields in infinite momentum and rest frame

- open strings
- Looped strings
- Rest frame

Referring to Fig. 1-11, Chapter 1, one can see that the universe went through a rapid inflation immediately after the Big Bang. The energy density of the universe in rest-frame drops rapidly but coincides with creation of activated strings discussed in the previous sections which cause the field of looped and open strings to expand in infinite momentum frame. During this initial expansion, "matter" is being formed and activated strings are being consumed as they decay sustaining the state of "matter". While the rate of generation of activated strings is faster than the rate of consumption, the field of infinite momentum frame, which is our observable universe, will continue to grow and expand. At around 12-14B years, which interestingly coincides with our era, the field of activated looped and open strings peak and begin to fall. This is when the rate of generation of strings begins to lag their consumption and decay, likely because of the formation of new matter and black holes.

As I mentioned in Chapter 1, and the criteria I used to set up the fundamental equations, two-dimensional looped strings that form as the activated strings decay, exit our universe (Hilbert space), and form an outer shell layer around the periphery of the universe. One may view this as a layer that prevents the activated looped and open strings from dispersing into the space rapidly, confining the energy as it is being formed. As such, our universe will remain as a bubble from the time of the Big Bang until all the activated strings inside the sphere of the bubble are consumed.

The decline of the fields of the activated strings after peaking means that the fields of infinite momentum frame (R_a), which is our observable universe, will begin to shrink, but the field of two-dimensional strings in restframe, designated as R will continue to expand. In fact, after peaking, the rate of generation of two-dimensional strings in restframe accelerates as more activated strings are being converted to two-dimensional strings than can be generated. This coincides with an accelerated expansion of the universe in the restframe and accelerated shrinkage of radius of the observable universe. "Matter" wherever it is in the universe acts as a "sink" to draw activated strings toward it as activated strings making up the matter decay.

Recall when two activated strings (open or looped) decay, two entangled extra dimension objects are created having an opposite spin rotation. I termed the corresponding extra dimensions as high and low-vac extra dimensions. The high-vac extra dimensions become components of fundamental particles with a spin that creates a downward torque, and the low-vac extra dimensions propagate instantaneously to become a component of the fabric of spacetime with a spin rotation that creates an upward force. The two entangled forces work to create a balance between the universe collapsing on itself or expanding at an exceedingly high expansion rate (dispersion). I also discussed the forces in the universe that drive the expansion. These included the string resistance forces which act to move the masses in the universe.

As the field of activated strings shrink, future generations will see less stars in the sky. It is likely that stars and galaxies that reside outside the fields of infinite momentum frame will move faster than the speed of light and decay quickly without three-dimensional strings to sustain them.

It is a remarkable coincidence that the peak of the dark energy field (both looped and open strings), and the second accelerated expansion of the universe occur at about 13B years which is concurrent with our time.

Number of dimensions in the universe

As you have seen throughout this book, the dimensions of the vacuum species that play a significant role in creation of our universe are far too small to be measured experimentally, at least given the current state of technology. These species range from a Z-string with a dimension of about 8.9×10^{-160}m to 2.3×10^{-20}m, the dimension of activated looped strings.

We may count the total number of dimensions in our universe by lumping them as major groupings of spacetime dimensions, a) the three-dimensional space making up the universe we can see and measure including the largest three-dimensional looped string making up the macro-space and spacetime cavity, b) the Planck space, all strings with Planck dimensions, and c) E-loop space, all strings with E-loop dimensions including E-portals.

Dimensional Space	No. of dimensions (as major categories)
Three-dimensional macro-universe- Subcategory: (Activated looped string 1.61×10^{-20} m, Spacetime cavity 1.8×10^{-9}m)	3
Planck Space – 1.61×10^{-35} m Subcategory: (3D looped, 3D open, 2D looped, 2D open)	3
E-loop Space – 2.3×10^{-51}m Sub-category: (2D E-loop string)	3
Z-string - 8.9×10^{-160} m Subcategory: (One dimensional string - 1.3×10^{58} m)	1
Time	1
Total	11

Depending on how one counts the number of dimensions, one may find an equivalent definition in classical string theory. For example, the total number of dimensions can be counted as 10 without the Z-string as in classical theories, or as 11 including the Z-string which is consistent with the conjecture made by Professor Edward Witten in "M–theory".

If we include the total number of dimensions in each sub-category given in the parentheses in the above table, the total number of dimensions will equal 26 which is consistent with the classical bosonic string theory.

CHAPTER 12
The Big Bang

A Brief overview of the Big Bang

LET US NOW CONSTRUCT THE universe before the Big Bang from what we have learned so far by reversing the process that occurs inside a blackhole.

The activated three-dimensional looped and open strings are converted to two-dimensional Planck strings and eventually become E-loop strings inside the singularity of a blackhole.

In the last stage of a blackhole, the E-loop strings are stretched to one-dimensional strings with a thickness of 8.9×10^{-160}m and a length of about 1.3×10^{58}m. If we assume this is the end point, one-dimensional strings become the fundamental structure of the background vacuum in our universe. This one-dimensional string is also the vacuum energy I have referred to as Bulk vacuum in our universe. It is the vacuum at the smallest dimension the universe will allow.

How are the one-dimensional strings, as the background vacuum of the universe, produced before the Big Bang?

To answer this question, we will revert to the two fundamental equations that were derived in Chapter 1 and the resultant energy fields which we rescaled four times to arrive at 8.9×10^{-160}m:

1) $R_a = 3R/(2R^2 +3)$ Looped strings
2) $R_a = Re^{-(2/3)R}$ Open strings

Of the two equations, equation (2) describes the universe in its entirety as a system. Recall that the string for string energy balance of the E-loop strings in the creation of the volume of activated strings is similar to that of activated open strings in which the momentum energy of two strings is used to create new space volume but the energy of one is returned back to the system. In Chapter 1, if we substitute the field of activated open strings with a field of E-loop strings, and the volume of the extra dimension objects created by the two-dimensional open strings with the total volume of space created, we will obtain the same energy field equation as equation (2) above describing the universe, from E- loop string to activated strings.

In Chapter 5, we saw that the area encompassing R =0 to R= 12, by the curve representing equation (2), is proportional to the total vacuum energy of the energy field scaled to correspond to a mass "m":

$$\text{Area} = \int_0^{12} Re^{-2/3R}\, dR = -3/2\, Re^{-2/3R} - 9/4e^{-2/3R}+C = 22.5 \text{ (units of } R_a^2) \qquad (12\text{-}1)$$

The above equation indicates that a small but constant amount of energy is produced regardless of how small the dimension of space is. In order words, we can never reach a space where R_a is truly zero. R_a equals zero, is a place in space where nothing exists, where the vacuum is absolute nothingness. Equation (12-1) tells us this place in space does not exist, or nothingness is not allowed.

So, what happens below the dimension of about 8.9×10^{-160}m? If we were able to rescale the dimensions of space beyond this point mathematically, we would be able to do so infinitely. If we were able to continue rescaling the dimensions of space, the species that form are likely unstable on their own and collectively coagulate and collapse creating a small amount of energy with a dimension of about 12.3×10^{-160}m and thickness of about 8.9×10^{-160}m or a zero-dimensional string we can call a Z-string.

In fact, the universe uses the space collapsing mechanism several other times to generate energy in our current three-dimensional universe. Recall when the two-dimensional Planck-size looped strings collapse into E-loop strings to create the rotation energy of the pulsating strings. This collapse and implosion of the expanded activated looped strings results in the generation of many new activated looped strings.

As such, at every point in the universe, empty space produces an energy field proportional to its dimension which leads to an energized vacuum specie with a dimension of about 12.3×10^{-160} to 8.9×10^{-160}m at its nucleus rotating left- or right-handed.

I termed this string, the Z-string. A conceptual representation of this process in the universe is shown in Fig 12-1. The Z-string, similar to the E-loop and two-dimensional Planck string, is still a three-dimensional object. However, its dimension as compared to a two-dimensional E-loop string is so much smaller, earning the name of a zero-dimensional string. Accumulation of the energized species of Z-strings in a continuum then leads to the formation of a one-dimensional vacuum specie with an infinitesimally small internal rotational energy. Formation of Z-strings accompanies an inherent spin similar to all other strings that are created in the same manner. However, the linear speed of its spin is unclear.

Equation (12-1) demonstrates that a constant energy is created at the smallest dimension with a width of about 12.3×10^{-160}m. If we substitute for R_a as the dimension of the Z-string, the elemental surface area of the tiniest vacuum specie is:

$$\pi \, (6.1 \times 10^{-160})^2 = 1.16 \times 10^{-318} \text{m}^2$$

This will be approximately equal to the mass of the Z-string if they were to gain the speed of light converting its energy to mass according to the process described earlier.

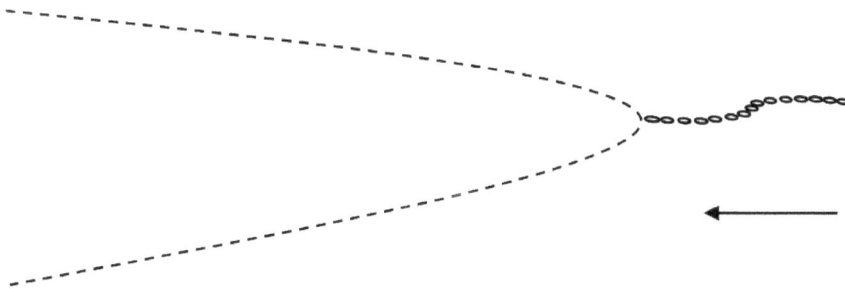

Fig. 12-1- Conceptual representation of generation of one-dimensional space vacuum at a dimension of 8.9×10^{-160}m.

This means that empty space without its smallest vacuum specie, the Z-string or the one-dimension string comprised of the Z-strings, does not exist. If we were able to stretch the one-dimensional string, it would create Z-strings indefinitely. It will follow the $\rho =KR$ rule with no limitations on the value of R. This is the universe at its smallest dimension.

Recall from Chapter 8, the length of the radius of the universe is limited to about 1.15×10^{26}m for three-dimensional strings, about 2.5×10^{26}m for two-dimensional Planck strings, and about 1.3×10^{58}m for E-loops strings. Z-strings as a one-dimensional object can stretch indefinitely without limitation.

The one-dimensional strings form the background vacuum of the universe at a dimension of about 8.9×10^{-160}m. Its structure is a random web without a pattern. It is vacuum at its lowest energy level extending infinitely in all directions. The Z-strings forming a one-dimensional string will likely all have the same spin or at least in certain segments of the long strand of strings. However, since the universe does not have a preference, production of one-dimensional strings with an equal amount of left and right-handed spinning Z-strings are the most likely scenario.

In absolute quiescence and away from disturbances of an existing universe, a small segment of these one-dimensional strings, an area with the dimension of an E-loop string or 1.6×10^{-101}m^2 packed with one-dimensional strings, will stretch a distance of about 1.3×10^{58}m as a packed column. The Z-string of the one-dimensional strings in each strand will have the same spin rotation (left- or right-handed). One may envision that creation of a long column of one-dimensional strings with the like Z-string spin in such a small area of space is a random event that is inevitable in a space that is infinitely long.

The torque created by the Z-strings of like spin in the long strand of the one-dimensional strings will continue to add up and eventually causes it to curl up and break off at about 1.3×10^{58} m. The rotation speed and the torque value of the Z-strings is unclear at this point. However, this information is not necessary to construct our current universe.

The tension energy that is built into the one-dimensional strings will be released and the one-dimensional strings of 1.3×10^{58}m with a thickness of 8.9×10^{-160}m will collapse into an E-loop string with a radius of 2.3×10^{-51}m, creating the surface area or vacuum space as an E- loop string. The area of the E-loop string will be equal to the sum of the area of all Z-strings comprising the one-dimensional string of 1.3×10^{58}m. The tension energy of the string will then be converted to a rotational energy of the E-loop string creating its spin which can be left- or right-handed depending on the natural rotation of the parent one-dimensional string, Fig. 12-2.

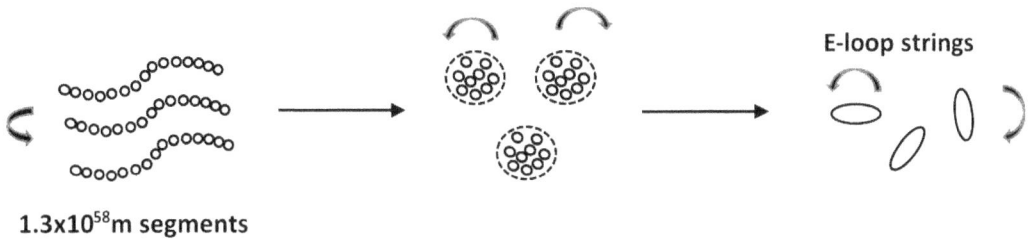

1.3×10^{58}m segments

Fig. 12-2- Transformation of fragments of one dimensional strings into two-dimensional E-loop strings.

If we designate the mass of the one-dimensional string to be "m", the tension constant of the one-dimensional string, K_1 will be:

$$K_1 = m/1.3 \times 10^{58}m$$

And the tension constant of the E-loop string will be:

$$K_2 = m/2 \times 2.3 \times 10^{-51}$$

If we further assume the elastic velocity of the one-dimensional string collapsing into an E- loop string is the same as its inflation speed of 0.7×10^{63}m which is the maximum speed an E-portal can expand, we can calculate the spinning velocity of the E-loop string generated:

$$E = \tfrac{1}{2}(m/1.3 \times 10^{58}) \times (0.7 \times 10^{63})^2 = (m/4.6 \times 10^{-51})\times(V)^2$$

$$V = 2.944 \times 10^{8}m/s$$

Which is the speed of light c. In essence, the breakage of a single strand of a one-dimensional string creates an E-loop string with a spin of about 3×10^{8}m/s. Therefore, each E-loop string will have an inherent spin of a linear speed equal to that of the speed of light.

The number of one-dimensional strings that comprise the packed column with a surface area of $1.6 \times 10^{-101} m^2$ is approximately:

$$1.6 \times 10^{-101} / \pi (6.1 \times 10^{-160})^2 = 0.13 \times 10^{218}$$

The break-off of one-dimensional strings of the packed column all at once will create 0.13×10^{218} two-dimensional E-loop strings, spinning left- and right-handed due to the collapse of the one-dimensional strings to set off the process that leads to what we know as the Big Bang. Note that we did not know the rotation speed of the Z-strings before the collapse of the one-dimensional string, but we know the kinetic energy of the collapse induces a rotational speed of $2.94 \times 10^8 m/s$ into the Z-strings which then become a component of an E-loop string. A key parameter of this process is that the Z-strings of each strand are predominately rotating in the same direction, creating the necessary torque to initiate the break-off process of the long chain of the strings making up the one-dimensional string. In a space where the Z-strings are created with random spins, the odds of creating a chain of Z-strings $1.3 \times 10^{58} m$ long with predominately the same rotation is extremely low but not impossible. Given that the Z-strings can extend infinitely, one may envision this process happening more than once.

The process preceding the Big Bang begins by attachment of two spinning E-loop strings (EL_0 and EL_1) as described in the previous chapters, which then allows transformation of potential energy of one E-loop string into kinetic energy of the other.

In this process, as the released E-loop strings pair up, the layers of E-loop strings that are the anchoring E-loops (EL_0) and those that are the rotating E-loops (EL_1) will continue to increase until their length reaches $1.3 \times 10^{58} m$ to fill the void created. The E- loop strings form two columns, one that contains the anchoring E-loops and one that contains the rotating E-loops as shown in Fig. 12-3. Since equal numbers of left- and right-handed spinning E-loop strings are expected, each produces its own spacetime cavity to attract the E-loop string of the same rotation.

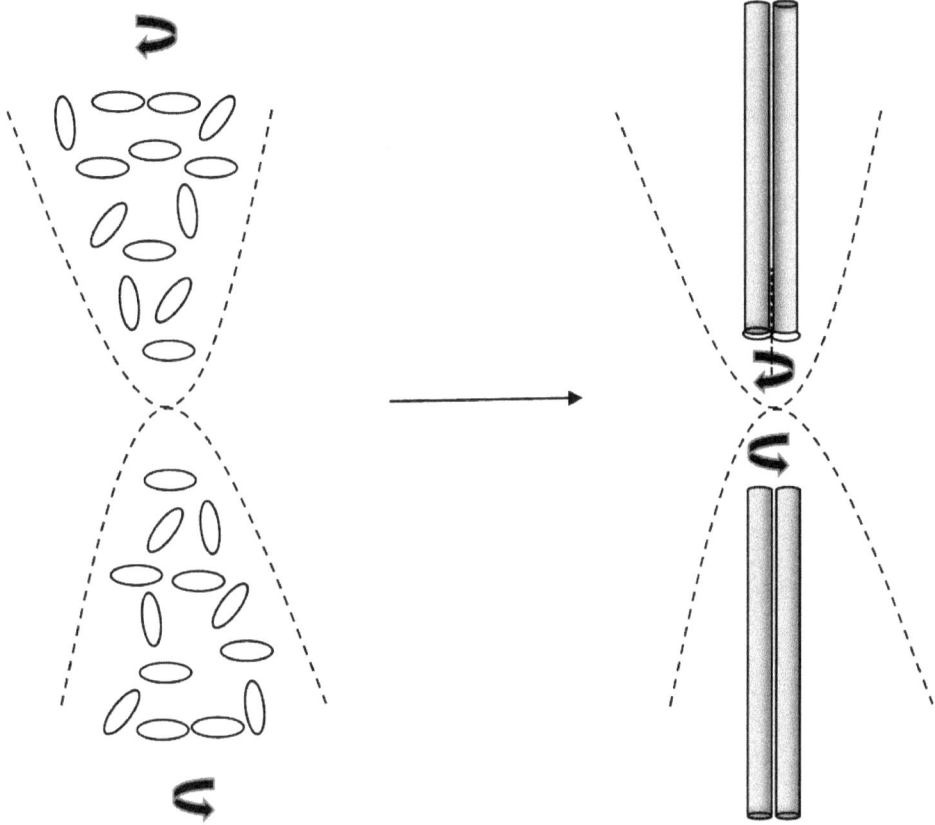

Fig. 12-3- Conceptual representation of random E-loop strings forming aggregate of strings with the same spin rotation, leading to creation of maximally compacted E-portals prior to the Big Bang.

We can calculate the total number of E-loop strings required to fill the length of each column of E-portals:

$$1.3\text{x}10^{58}/8.9\text{x}10^{-160} = 0.14\text{x}10^{218}$$

In which $8.9\text{x}10^{-160}$m is the thickness of the E-loop string. Therefore, just prior to the Big Bang, there will be two columns of E-portals with a length of $1.3\text{x}10^{58}$m, each containing $0.14\text{x}10^{218}$ E-loop strings. Two spacetime cavities are required, north and south, that accommodate the left- and right-hand spinning columns of E-loops since we have equal numbers of right- and left-handed spinning E-portals.

Once again, the torque created by the rotating E-loops will cause the two adjacent E-portal columns to collapse and defragment into many pairs of layered E-loops of the same length, but with a smaller number of individual E-loop strings, as shown in Fig. 12-4. The total volume of the E-loop strings compacted in the two columns in each spacetime cavity will be $4.1 \times 10^{-43} \text{m}^3$ ($2 \times 1.3 \times 10^{58} \times 1.6 \times 10^{-101}$) if it were to collapse into a single point prior to defragmentation.

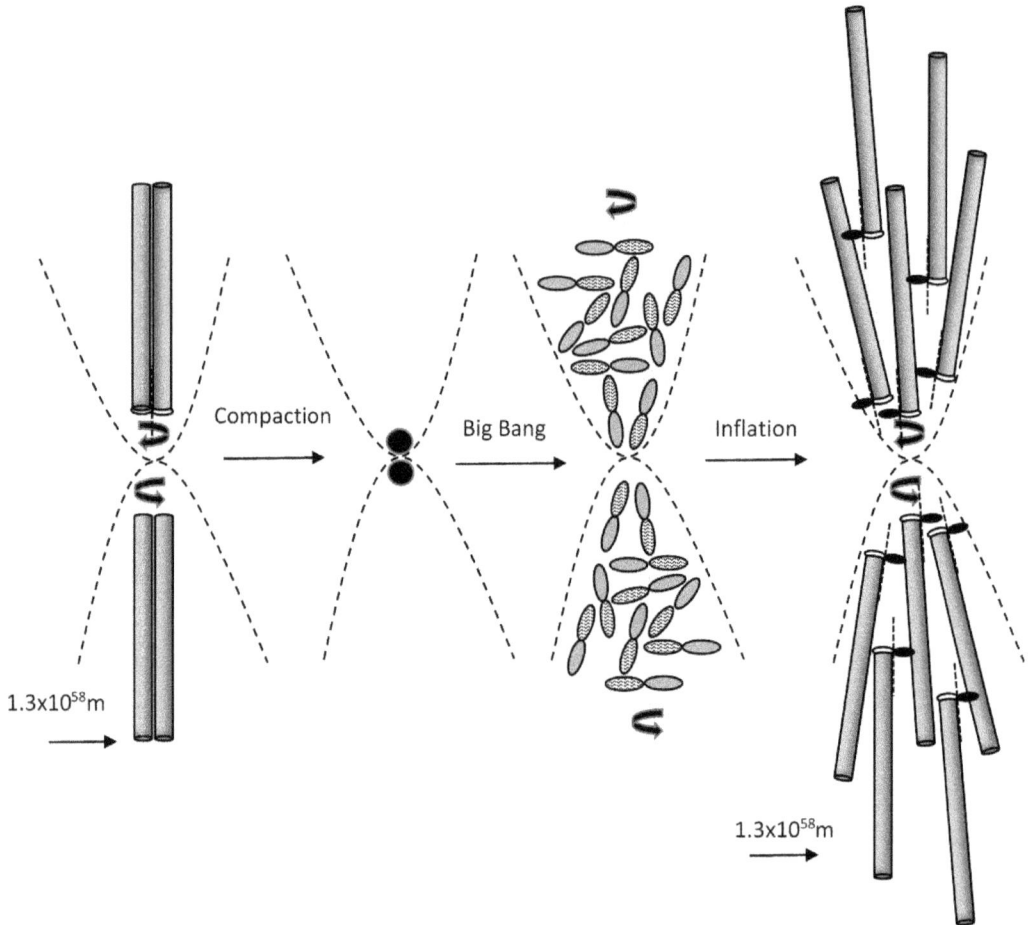

$1.3 \times 10^{58} \text{m}$

Compaction

Big Bang

Inflation

$1.3 \times 10^{58} \text{m}$

Fig. 12-4- Conceptual representation of transformation of maximally compacted E-portals into fragmented pair of energized E-loop strings initiating the process of the big bang, and inflation of the energized E-loops into E-portals.

The break-up and fragmentation of the two columns in each spacetime cavity which contained 0.14×10^{218} E-loops creates an enormous amount of momentum from defragmented layers of E-loop strings:

$E \approx M^2$

$(0.14 \text{x} 10^{218})^{1/2} = 0.37 \text{x} 10^{109}$

Meaning that two fully compacted E-portals, each containing $0.14 \text{x} 10^{218}$ E-loops, breaks-up into $0.37 \text{x} 10^{109}$ pairs of layered E-loops, each containing $0.37 \text{x} 10^{109}$ individual E-loops. The energy of one pair becomes the mass of the E-portal adjacent to the pair.

Recall that this was the number of E-loop strings that create a three-dimensional Planck string with a mass of $2.18 \text{x} 10^{-8}$ kg. In other words, upon the breakup of each column, each fragment has $0.37 \text{x} 10^{109}$ E-loop strings that will create a three-dimensional Planck string according to the process described in Chapter 8.

As stated in the previous chapters, the natural uncompact state of the E- loop strings of the E-portal is when each E-loop is separated by its diameter. Therefore, each pair of layered E-loops containing $0.37 \text{x} 10^{109}$ individual E-loops will immediately expand to its maximum length of about $1.3 \text{x} 10^{58}$ m. This is inflation of the fragments of the layered E-loops which will expand to its maximum length in the presence of bulk or background vacuum of space.

$2 \text{x} 2.3 \text{x} 10^{-51} \text{ x } 0.37 \text{x} 10^{109} = 1.7 \text{x} 10^{58}$ m

Therefore, the process of the Big Bang begins by disintegration or fragmentation of fully compacted E-portals with a length of about $1.3 \text{x} 10^{58}$ m into $0.37 \text{x} 10^{109}$ E-portals each containing about $0.37 \text{x} 10^{109}$ E-loop strings which is equivalent to a Planck mass in our universe.

Since we have two E-portals with E-loops of opposite rotation, we will have $2 \text{x} 0.37 \text{x} 10^{109}$ E-portals.

The total mass of the E-loops contained in the E-portals is:

$2.18 \text{x} 10^{-8} \text{x} 2 \text{x} 0.37 \text{x} 10^{109} = 1.6 \text{x} 10^{101}$ kg (in Planck space)

This will be the mass of the universe at the time of the Big Bang.

The volume of the two E-portals containing this mass will be:

$$2 \times 1.3 \times 10^{58} \times 1.6 \times 10^{-101} = 4.1 \times 10^{-43} m^3$$

Which is about the same volume as two Planck-size orbiting extra dimension cylinders extending the length of the universe.

The energy density of the E-portal prior to the Big Bang is:

$$1.61 \times 10^{101} / 4.1 \times 10^{-43} = 0.39 \times 10^{144} kg/m^3$$

Which is the same as the energy density of a singularity.

To initiate the process of the Big Bang, we need $4 \times 0.13 \times 10^{218}$ one-dimensional strings which have a volume of only $2 \times 4.1 \times 10^{-43} m^3$ to break off and deliver the energy needed to create our universe. After the creation of the universe, this energy increases by the process of looped and open string generation discussed in the previous chapter.

The inflated E-portals with stored E-loop strings then produce the three-dimensional looped and open strings as described in the previous chapters.

In essence, the E-loop string, the two and three-dimensional open and looped Planck-size strings, and the one-dimensional E-loop string created by Z-strings, are all vacuum species with different dimensions and energy levels. This means that the universe is made up of nothing but vacuum and its associated energy, and that these vacuum species arise from a process that does not allow "absolute" vacuum to exist.

The evolution and the corresponding space transformation for each species of vacuum is demonstrated in Fig. 12-5.

E-loop strings

2D Planck string   **3D Planck string**

8.9×10^{-160}m $\longrightarrow$ 2.3×10^{-51}m $\longrightarrow$ 1.6×10^{-35} m $\longrightarrow$

$\longrightarrow$ 1.8×10^{-9} m $\longrightarrow$ 1.1×10^{26} m

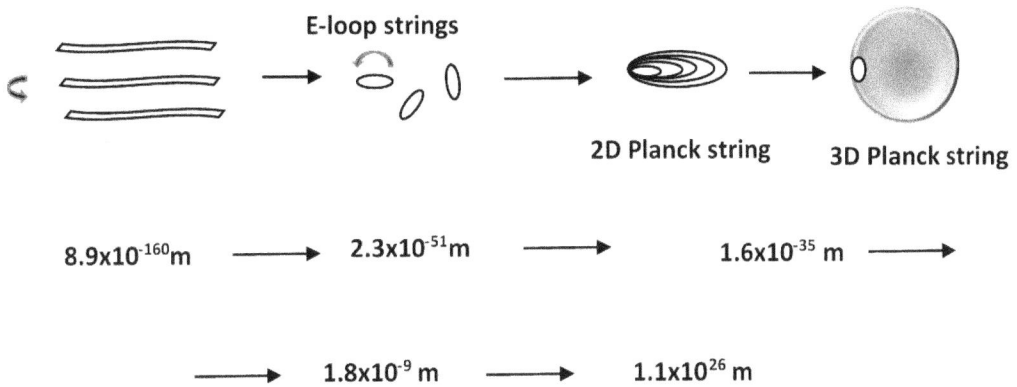

Fig. 12-5- Sequential stages of string formation and corresponding geometric space transformation.

If we consider the Big Bang event to be the point when the compacted E-portals defragment into many pairs of multi-layered E-loop strings, the process leading to this event would have started long before the actual event. The Big Bang event would appear as arising from nothing because the process preceding it is a random detachment of one-dimensional strings forming the background vacuum of the universe.

As mentioned earlier, the universe has no preference in creating left- or right-handed spinning Z-strings which will translate into left- or right-handed spinning three-dimensional activated looped and open strings. Therefore, one can expect creation of equal amounts of left- and right-handed spinning strings in the universe at every dimension. As you may have observed throughout this book, this is critical to maintaining the balance of forces in the universe.

Temperature at the time of the Big Bang

As we saw earlier, thermal energy is the energy that is lost to creation of space for the formation of three-dimensional Planck strings. Since one E-loop string is lost to this process for every two E-loop strings used, the change in the energy of the system contributing to the thermal energy is:

$$\Delta E = E_s = 1.61 \times 10^{101} \times 9 \times 10^{16} = 14.5 \times 10^{117} J$$

Number of degrees of freedom = total number of strings in each energy field **x** two degrees of freedom = $2 \times 0.37 \times 10^{109}$

$$S = 2 \times 0.37 \times 10^{109} k_b$$

$$T = \Delta E / S$$

$$k_b = 1.38 \times 10^{-23} J/K$$

$$T = 14.5 \times 10^{117} / 2 \times 0.37 \times 10^{109} k_b$$

$$T = 1.41 \times 10^{32} K$$

Closing Remarks

AS YOU HAVE SEEN THROUGHOUT this book, I have described a theory in which our universe consists of two important three-dimensional, spherical, vacuum objects that I have termed as open and looped strings. These three-dimensional objects contain a substantial amount of bulk vacuum which is yet another vacuum specie of significantly lower energy content. I termed these as one-dimensional vacuum objects, which in turn consists of zero-dimensional vacuum strings of about 8.9×10^{-160}m. In essence, we live on the surface of these three-dimensional objects which have surface gauges of the above dimension. As such, they behave as species without a surface and are impossible to observe by any means, at least for now.

When the three-dimensional spherical objects decay, they release or expose their entrapped vacuum containing the two-dimensional string and its mass in their core. The decay of these three-dimensional strings is most predominant within the structure of what we know as baryonic mass. The exposed bulk vacuum with the conserved energy and momentum of the two-dimensional strings creates a network of extra-dimension objects in the universe. These extra dimension objects, some of which extend the entire length of the universe, coexist in the same space as the three-dimensional open and looped strings and are indeed three-dimensional objects themselves with different properties. I referred to the space they create as the "extra dimension space" because they create a vector field that is unlike that of the three-dimensional looped and open strings. Interaction of the extra dimension objects with the three-dimensional looped and open strings creates many new species such as baryonic mass of fundamental particles, photons, charges, magnetic monopoles, mass of Dark Matter, etc. These extra dimension objects are also invisible to our universe because the two-dimensional strings that make them also have a thickness of about 8.9×10^{-160}m, although its length or diameter is as large as Planck length (1.61×10^{-35}m).

With coexistence of the three-dimensional open and looped strings and the extra dimension objects in the same space, our universe becomes a network of three-dimensional strings roaming freely at the speed of light with a framework of extra dimension objects comprising its backbone structure. I designated these as "infinite momentum frame" and restframe" respectively.

The decay of three-dimensional looped and open strings, which is predominant in matter with baryonic mass, creates a momentum change which creates the force of gravity and electromagnetic fields respectively. I used this concept to set up and derive the fundamental equations in Chapter 1.

The force of the gravity created by the momentum change of the three-dimensional looped strings impacts the curvature of spacetime which is created by the extra dimension objects forming the backbone structure of the universe. From the same equations, quantum gravity emerges taking into account dimensional transformation of space from our macro-universe to subatomic scale. As you saw in Chapter 6, three-dimensional looped strings seamlessly integrate quantum gravity with what we have come to know as "general relativity" in one equation.

Three-dimensional open strings on the other hand, while their decay creates the electromagnetic force, have no impact on the curvature of spacetime.

As you saw, one key feature of a three-dimensional looped string is that its dimension changes from 1.61×10^{-35}m at the nucleus of fundamental particles and atoms to about 2.3×10^{-20}m in open spaces outside the gravitational and quantum field of a large mass. This dimensional change brings about many interesting physical phenomena.

One is the shrinkage of the space itself around large masses. This means that the way we measure "time" as a coordinate for moving objects changes depending on the location of space itself. We see this as the time dilation effect around large masses and blackholes.

The second important feature of an expanding three-dimensional looped string is the thermal effect we observe as heat. As you saw in Chapter 9, expansion of space at a local level means generation of thermal energy. For this reason, the surface of the stars will be cooler than its atmosphere because it takes time for the emitted activated looped string to expand which in turn creates the heat effect we observe and measure.

I also demonstrated how the generation of this heat due to the creation of space is related to the entropy of the strings and how we can then calculate the background temperature of the universe, entropy, and temperature of a black hole.

The third important feature of an expanding three-dimensional looped string is the change it creates in the force of a magnetic monopole (Chapter 4). This change results in electromotive force of a charge and the flow of electric current in a magnetic field.

In Chapter 11, I described a new and important force in the universe, the "String Resistance Force". This force is created by the flow of smaller open strings against the much larger looped strings, or open and looped strings in free space against those at restframe. String Resistance Force is responsible for the movement and displacement of masses in the universe, such as the rotation of planets, stars, and galaxies. This force necessitates the geometric dimension of time which we calibrate to measure the relative position of moving objects as a coordinate we know as "Time". I described one sub-category of the String Resistance Force as Strong and Weak forces inside atoms and fundamental particles.

I also described this geometric dimension of time as the dimension of an activated looped string which is the distance a string will travel in space as a "quantum hop". We then calibrate the number of quantum hops as the rotation cycle of objects in space as "Time". As such, time becomes a coordinate imbedded in the geometric dimension of spacetime.

As mentioned above, what seems a vast open and empty space in our universe is, in reality, a space filled with three-dimensional, spherical vacuum species with a radius of 1.61×10^{-35}m (open strings) and 2.3×10^{-20}m (looped strings) with extra dimension objects filling the interstitial spaces between them also at Planck dimension. In Classical physics, this unknown energy is referred to as Dark Energy.

The universe goes through four dimensional transformations beginning with vacuum species with a dimension of about 8.9×10^{-160}m prior to the Big Bang event. In Chapter 12, I described a model of how these tiny strings become an aggregate of multilayered and significantly larger E-loop strings of radius 2.3×10^{-51}m by reversing the process inside a blackhole's singularity. E-loop strings are the fundamental building block of our existence.

In Chapter 8, I described a model of how the E-loop strings are transformed to two, and then three-dimensional vacuum species of looped and open strings of Planck length and how the looped strings can be stretched to the much larger dimension of 2.3×10^{-20}m.

Since the Big Bang event, the universe has managed to produce a vast quantity of the three-dimensional vacuum species according to the process I described in Chapter 8. The generation of additional three-dimensional vacuum species, I termed "activated strings", means expansion of our three-dimensional universe to its current radius of about 1.15×10^{26}m or what is our observable universe. However, as more matter has been formed, the rate of decay of the three-dimensional strings is beginning to exceed the rate of generation. This means that as the universe with the extra dimension objects will continue to expand, the field radius of what creates our three-dimensional universe is beginning to shrink. In Chapter 1, I estimated based on the conversion rate we have seen so far, this will take about 120B years before all three-dimensional vacuum species are consumed.

Remarkably, after only about 14 billion years from the Big Bang event, we currently live in an era where the radius of the field of both three-dimensional open and looped strings is at its maximum and in a few billion years it will begin its accelerated decline. This will coincide with an accelerated rate of increase in the radius of the field of extra dimension objects which appears to have already started.

Three-dimensional vacuum species that make up our universe store their energy and mass in an extra dimension object I termed an E-portal, which resides inside the sphere of the vacuum specie. The change in this energy is too small to detect and as such we do not observe the "space mass" which is significantly larger than baryonic mass. The change in energy of baryonic mass is significantly higher and thus becomes observable. I defined "space mass" as a mass in which the change of energy remains in our universe. Recall, this was the change in energy that converted an E-loop string to a Planck string and created the entropy of the string. I also defined baryonic mass as a change of energy exiting our three-dimensional universe, otherwise known as Hilbert space. All strings, three and two-dimensional after decay revert to E-loop strings which then form the outer circumference of the universe containing the extra dimension objects. Inside a blackhole's singularity, a portion of the E-portal energy is used to stretch the E-loop strings to their most fundamental form, a one-dimensional string, which I described as the constituent of bulk vacuum in the universe.

Lastly in Chapter 4, I introduced a new model forming the structure of fundamental particles and atoms with significantly more clarity and refinement of its details than the century old Standard Model will allow. In Chapter 10, I described the energy source and underlying mechanisms that make the extra dimension objects forming fundamental particles behave like a wave and the fundamentals behind Schrodinger's famous equation.

As I mentioned in the preface, my goal with this book is to make its concepts and mathematics understandable to a wide audience, including those in the fields of engineering and science outside the physics community, in hopes that the scientific community as a whole can adapt its teachings and apply it to their research further corroborating the fundamentals of this theory and work.

Professor Ray R Eshraghi

Selected References

1. Georges Obied, Hirosi Ooguri, Lev Spodyneiko, Cumrun Vafa, "De Sitter Space and the Swampland", CALT-TH 2018-020, IPMU18-0100

2. Souvik Banerjee, Ulf Danielsson, Giuseppe Dibitetto, Suvendu Giri, and Marjorie Schillo, "Emergent de Sitter Cosmology from Decaying Anti–de Sitter Space" Physical Review Letters 121, 261301 (2018)

3. Nikolay Bobev, Pieter Bomans and Friðrik Freyr Gautason, "Spherical Branes", (2018) arXiv:1805.05338.

4. Bachas, C. P. "Lectures on D-branes" (1998). arXiv:hep-th/9806199.

5. Sean M. Carroll, Grant N. Remmen, "A Nonlocal Approach to the Cosmological Constant Problem", Phys. Rev. D 95, 123504 – Published 7 June 2017.

6. Sean M. Carroll, "DARK ENERGY AND THE PREPOSTEROUS UNIVERSE", Enrico Fermi Institute and Department of Physics, University of Chicago.

7. *Goldhabar, Gerson, "The Acceleration of the Expansion of the Universe: A Brief Early History of the Supernova Cosmology Project (SCP)". Proceedings of the 8th UCLA Dark Matter Symposium.* **1166***: 53–72 (2009).*

8. Moshe Carmeli and Tanya Kuzmenko, "Value of the Cosmological Constant: Theory versus Experiment", Feb., 4, 2001. https://arxiv.org/abs/astro-ph/0102033v2

9. S. Weinberg, "The Cosmological Constant Problem," Rev. Mod. Phys. 61 (1989) 1.

10. S. W. Hawking, "The Cosmological Constant Is Probably Zero," Phys. Lett. B134 (1984) 403.

11. Paul J. Steinhardt, "A Quintessential Introduction to Dark Energy", Published online 17 September 2003.

12. Adam G. Riess, et.al, "a 2.4% Determination of the Local Value of the Hubble Constant", The Astrophysical Journal, 826:56 (31pp), 2016 July 20.

13. A. G. Riess et al., Supernova Search Team Collaboration, "Observational evidence from supernovae for an accelerating universe and a cosmological constant," Astron.J. 116 (1998) 1009–1038, arXiv:astro-ph/9805201].

14. S. Perlmutter et al.,Supernova Cosmology Project Collaboration, "Measurements of Omega and Lambda from 42 high redshift supernovae," Astrophys.J. 517 (1999) 565–586, arXiv:astro-ph/9812133 [astro-ph].

15. Behar, S., and Carmeli, M., *Intern. J. Theor. Phys.* 39, 1375 (2000). (astro-ph/0008352)

16. Helen Quinn, et.al, "Teachers' Resource Book on Fundamental Particles and Interactions", SLAC-PUB-4879, LBL-26669, January, 1989.

17. Robert Byron Bird, Warren E. Stewart, Edwin N. Lightfoot, "Transport Phenomena", 1960, John Wiley and Sons.

18. Chas A. Egan, and Charles H. Lineweaver, "A Larger Estimate of the Entropy of the Universe" , the Astrophysical Journal, 710:1825–1834, 2010 February 20.

19. Leonard Susskind, John Uglum, (Nov 30, 1995) arXiv:hep-th/9511227v1

20. Leonard Susskind, Publications, lectures, books, and media.

www.ingramcontent.com/pod-product-compliance
Lightning Source LLC
Chambersburg PA
CBHW051749200326
41597CB00025B/4495